Green Grafting: Innovations in Polymer Functionalization for Sustainable Solutions in Pharmaceutical and Healthcare Industry
(Part 1)

Edited by

Kuldeep Vinchurkar
Department of Pharmaceutics
Sandip Institute of Pharmaceutical Sciences (SIPS)
Affiliated To Savitribai Phule Pune University (SPPU, Pune)
Nashik, Maharashtra 422213, India

Satish Polshettiwar
Department of Pharmaceutical Sciences
School of Health Sciences and Technology
Dr. Vishwanath Karad MIT-World Peace University
Pune, Maharashtra 411038, India

Nilesh Mahajan
Department of Pharmaceutics
Dadasaheb Balpande College of Pharmacy
Besa, Nagpur, Maharashtra 440037
India

&

Yogeshwar Bachhav
Adex Pharmaceutical Consultancy Services
Mumbai, Maharashtra 400089, India

Green Grafting: Innovations in Polymer Functionalization for Sustainable Solutions in Pharmaceutical and Healthcare Industry - *(Part 1)*

Editors: Kuldeep Vinchurkar, Satish Polshettiwar, Nilesh Mahajan & Yogeshwar Bachhav

ISBN (Online): 979-8-89881-168-6

ISBN (Print): 979-8-89881-169-3

ISBN (Paperback): 979-8-89881-170-9

© 2026, Bentham Books imprint.

Published by Bentham Science Publishers Pte. Ltd. Singapore, in collaboration with Eureka Conferences, USA. All Rights Reserved.

First published in 2026.

BENTHAM SCIENCE PUBLISHERS LTD.
End User License Agreement (for non-institutional, personal use)

This is an agreement between you and Bentham Science Publishers Ltd. Please read this License Agreement carefully before using the ebook/echapter/ejournal (**"Work"**). Your use of the Work constitutes your agreement to the terms and conditions set forth in this License Agreement. If you do not agree to these terms and conditions then you should not use the Work.

Bentham Science Publishers agrees to grant you a non-exclusive, non-transferable limited license to use the Work subject to and in accordance with the following terms and conditions. This License Agreement is for non-library, personal use only. For a library / institutional / multi user license in respect of the Work, please contact: permission@benthamscience.org.

Usage Rules:

1. All rights reserved: The Work is the subject of copyright and Bentham Science Publishers either owns the Work (and the copyright in it) or is licensed to distribute the Work. You shall not copy, reproduce, modify, remove, delete, augment, add to, publish, transmit, sell, resell, create derivative works from, or in any way exploit the Work or make the Work available for others to do any of the same, in any form or by any means, in whole or in part, in each case without the prior written permission of Bentham Science Publishers, unless stated otherwise in this License Agreement.
2. You may download a copy of the Work on one occasion to one personal computer (including tablet, laptop, desktop, or other such devices). You may make one back-up copy of the Work to avoid losing it.
3. The unauthorised use or distribution of copyrighted or other proprietary content is illegal and could subject you to liability for substantial money damages. You will be liable for any damage resulting from your misuse of the Work or any violation of this License Agreement, including any infringement by you of copyrights or proprietary rights.

Disclaimer:

Bentham Science Publishers does not guarantee that the information in the Work is error-free, or warrant that it will meet your requirements or that access to the Work will be uninterrupted or error-free. The Work is provided "as is" without warranty of any kind, either express or implied or statutory, including, without limitation, implied warranties of merchantability and fitness for a particular purpose. The entire risk as to the results and performance of the Work is assumed by you. No responsibility is assumed by Bentham Science Publishers, its staff, editors and/or authors for any injury and/or damage to persons or property as a matter of products liability, negligence or otherwise, or from any use or operation of any methods, products instruction, advertisements or ideas contained in the Work.

Limitation of Liability:

In no event will Bentham Science Publishers, its staff, editors and/or authors, be liable for any damages, including, without limitation, special, incidental and/or consequential damages and/or damages for lost data and/or profits arising out of (whether directly or indirectly) the use or inability to use the Work. The entire liability of Bentham Science Publishers shall be limited to the amount actually paid by you for the Work.

General:

1. Any dispute or claim arising out of or in connection with this License Agreement or the Work (including non-contractual disputes or claims) will be governed by and construed in accordance with the laws of Singapore. Each party agrees that the courts of the state of Singapore shall have exclusive jurisdiction to settle any dispute or claim arising out of or in connection with this License Agreement or the Work (including non-contractual disputes or claims).
2. Your rights under this License Agreement will automatically terminate without notice and without the

need for a court order if at any point you breach any terms of this License Agreement. In no event will any delay or failure by Bentham Science Publishers in enforcing your compliance with this License Agreement constitute a waiver of any of its rights.

3. You acknowledge that you have read this License Agreement, and agree to be bound by its terms and conditions. To the extent that any other terms and conditions presented on any website of Bentham Science Publishers conflict with, or are inconsistent with, the terms and conditions set out in this License Agreement, you acknowledge that the terms and conditions set out in this License Agreement shall prevail.

Bentham Science Publishers Pte. Ltd.
No. 9 Raffles Place
Office No. 26-01
Singapore 048619
Singapore
Email: subscriptions@benthamscience.net

CONTENTS

PREFACE	i
LIST OF CONTRIBUTORS	iii
CHAPTER 1 INTRODUCTION TO GREEN GRAFTING	1

Taufik Mulla, Dipti Patel, Drashti Dave, Kuldeep Vinchurkar and *Satish Polshettiwar*

INTRODUCTION	2
Overview of Green Grafting	2
Importance of Green Grafting in the Pharmaceutical and Healthcare Industries	3
Sustainable Drug Delivery Systems	3
Green Grafting in Medical Devices	3
Environmental Impact and Waste Reduction	4
Principles of Green Chemistry in Grafting	4
METHODS OF GREEN GRAFTING	6
Use of Green Solvents	6
Biocatalysts and Organocatalysts	8
Renewable Resources	9
APPLICATIONS OF GREEN GRAFTING IN THE PHARMACEUTICAL AND HEALTHCARE INDUSTRIES	10
Drug Delivery Systems	10
Medical Devices	11
Tissue Engineering	13
CASE STUDIES	14
Biodegradable Polymers for Drug Delivery	14
Antimicrobial Surfaces for Medical Devices	15
Functionalized Scaffolds for Tissue Engineering	15
FUTURE DIRECTIONS AND CHALLENGES	16
CONCLUSION	17
ACKNOWLEDGEMENTS	18
REFERENCES	18
CHAPTER 2 FUNDAMENTALS OF POLYMER GRAFTING: BASICS AND TECHNIQUES	22

Akanksha Phondke, Meghana Jagtap, Priya Patne, Popat Mohite, Premlata Ambre, Sudarshan Singh and *Kuldeep Vinchurkar*

INTRODUCTION	23
POLYMER GRAFTING APPROACHES	23
Grafting onto (Grafting to)	24
Grafting from (Surface-initiated Polymerization)	25
Grafting Through (Macromonomer Method)	26
TECHNIQUES OF POLYMER GRAFTING	26
Chemical Grafting	27
Free Radical Polymerization (FRP)	28
Ionic Grafting	31
Physical Grafting	34
Photochemical Grafting	35
Plasma Induced Polymerization Technique	37
Radiation Grafting	39
Biological Grafting	40
Enzymatic Grafting	40
FACTORS CONTRIBUTING TO GRAFT POLYMERIZATION	42

Nature of Backbone	42
Effect of Monomer	42
Effect of Solvent	43
Effect of Initiator	44
Effect of Temperature	44
RECENT APPLICATIONS IN POLYMER GRAFTING TECHNIQUES	**45**
Surface Grafting Polymer	45
Conducting Polymer	46
Conductive Polymer Covalent Grafting	49
Conducting Polymer Noncovalent Grafting	49
Graft Polymerization via Click Chemistry	50
FUTURE PERSPECTIVES AND CHALLENGES	**51**
CONCLUSION	**52**
REFERENCES	**52**

CHAPTER 3 PRINCIPLES OF GREEN CHEMISTRY IN GRAFTING OF POLYMERS **60**
A.R. Chabukswar, S.C. Jagdale, Yash D. Kale and S.A. Polshettiwar

INTRODUCTION	**61**
Overview of Green Chemistry	61
Definition and Goals	61
Historical Context and Development	61
Introduction to Polymer Grafting	62
Definition and Significance	62
Types of Grafting Techniques	64
Importance of Green Chemistry in Polymer Grafting	65
Environmental Impact	66
Economic and Safety Benefits	68
PREVENTION OF WASTE IN POLYMER GRAFTING	**68**
Efficient Grafting Techniques	68
Grafting Through Chemical	68
Free-radical Grafting	69
Atom Transfer Radical Polymerization (ATRP)	69
Reversible Addition-Fragmentation Chain Transfer (RAFT) Polymerization	71
Applications	72
Challenges and Solutions	74
ATOM ECONOMY IN GRAFTING REACTIONS	**75**
Principles of Atom Economy	75
Strategies for Achieving a High Atom Economy in Grafting	76
Selective Grafting Techniques	76
Controlled Polymerization Methods	80
USE OF RENEWABLE FEEDSTOCKS	**81**
Renewable and Natural Polymers	81
Cellulose, Starch, and Chitosan as Grafting Substrates	81
Cellulose	81
Starch	82
Chitosan	82
Benefits and Challenges of Using Renewable Feedstocks	83
Overcoming Supply Chain Issues	84
Addressing Technological Limitations	84
Case Studies and Industrial Applications	84
Applications of Biopolymers	85

Biopolymers for Medical Applications	85
Biopolymers for Industrial Applications	85
SAFER SOLVENTS AND REACTION CONDITIONS	86
Environmentally Benign Solvents	86
Water	86
Supercritical Fluids	87
Examples of Greener Solvents in Polymer Grafting	88
Comparative Analysis with Traditional Solvents	89
ENERGY EFFICIENCY IN GRAFTING PROCESSES	90
Importance of Reducing Energy Consumption	90
Microwave-Assisted Grafting	91
Ultrasound-Assisted Grafting	92
Ultrasound-assisted Liquid-liquid Microextraction	92
Case Studies and Practical Applications	93
REDUCTION OF DERIVATIVES IN GRAFTING	93
Direct Grafting Methods	94
Benefits of Reducing Derivatization Steps	95
Examples and Case Studies	95
DESIGN FOR DEGRADATION	97
Principles of Degradable Polymers	97
Grafting Techniques for Biodegradable Polymers	99
Environmental Impact and Applications	101
Applications of Biodegradable Polymer	101
Environmental Fate And Assessment	101
Case Studies of Biodegradable Grafted Polymers	101
Case Study I: Resorbable Sutures	101
CATALYSIS IN GRAFTING REACTIONS	102
Role of Catalysis in Green Chemistry	102
Enzyme-Catalyzed Grafting	103
Metal-Free Catalytic Systems	103
Case Studies and Examples	104
Importance of Real-time Monitoring	105
Techniques for Monitoring Grafting Reactions	106
Examples of Implementation and Benefits	107
INHERENTLY SAFER CHEMISTRY FOR ACCIDENT PREVENTION	108
Designing Safer Grafting Processes	108
Less hazardous chemical syntheses	109
Reduce derivatives	109
Real-time pollution prevention	109
Use of renewable feedstocks	109
Design for degradation	109
Design for energy efficiency	109
Case Studies of Safety in Polymer Grafting	109
FUTURE TRENDS AND PERSPECTIVES	110
Emerging Techniques in Green Grafting	110
Potential Applications and Innovations	111
REACTIVE EXTRUSION	112
Challenges and Opportunities	112
Challenges	113
Opportunities	113
CONCLUSION	113

Summary of Key Points	113
The Role of Green Chemistry in the Future of Polymer Grafting	113
Final Thoughts on Sustainability and Innovation	114
Social Challenges	116
Energy Consumption	116
Impact on the Environment	116
REFERENCES	116

CHAPTER 4 RENEWABLE RESOURCES FOR GREEN GRAFTING: TYPES, BENEFITS, AND CHALLENGES ... 124
Deepali D. Bhandari, Dattatraya M. Shinkar, Ramanlalal N. Kachave, Unmesh G. Bhamare and *Sunil V. Amrutkar*

INTRODUCTION	125
GRAFTING	127
Grafting by Chemical Process	128
Grafting with Free Radicals	128
Ionic Grafting	129
Living Polymerization-induced Grafting	129
Photochemical Grafting	130
Radiation-based Grafting Approach	131
Enzymatic Grafting	131
Plasma- Radiation-Induced Grafting	131
Radiation Grafting	131
FACTORS AFFECTING GRAFTING	133
The Polymer Backbone's Nature	133
The Impact of the Monomer	133
Solvent's Effect	133
The Initiator's Impact	133
Temperature's Effect	133
Additives' Impact on Grafting	134
ASSESSMENT OF GRAFTED POLYMERS USING ANALYTICAL TECHNIQUES	135
THE PHARMACEUTICAL ADVANTAGES OF GRAFTED POLYMERS	135
To Modify a Drug's Biological Carrying Capacity	135
To Use Polymer Grafting to Obtain Customized Physicochemical Attributes	136
To Attain the Intended Features of the Dosage Form	136
To Accomplish Site-specific Distribution by Polymer Grafting	137
APPLICATIONS OF NATURAL POLYMERS	137
Catalytic Applications	137
Lignocellulose	138
Hemicellulose	139
Nanocellulose	140
Corrosion Inhibitor's Application	140
Green Obstructers (inhibitors)	140
Experimental and Computational Analysis	141
Natural Polymer Applications in the Food Industry	142
Biomedical and Biotechnological Applications	144
Regeneration and Tissue Engineering	144
Modified Drug Delivery	145
Environmental Remediation	148
Water Purification	148
Air Filters	148

Pesticides and Fertilizers with Controlled Release	148
Applications in Food Packaging	149
Cell Encapsulation	149
Prodrug Activation	150
Electrospinning Fiber Diameter	150
Environmental Impact and Sustainability	150
NEWLY SUBMITTED PATENTS CONCERNING POLYMER GRAFTING	151
FUTURE PERSPECTIVES AND CHALLENGES	152
CONCLUSION	154
REFERENCES	155

CHAPTER 5 TYPES OF GREEN SOLVENTS IN GRAFTING PROCESSES 169
Shashikant V. Bhandari, Shambhavi S. Sing, Neha A. Raut, Sagar R. Ghanwa, Abhishek V. Shitol and *Sharayu P. Ninaew*

TYPES OF GREEN SOLVENTS IN GRAFTING PROCESSES	170
GRAFTING TECHNIQUES	173
Grafting Through Chemical	173
Free-Radical Grafting	174
Click Chemistry in Polymerization and Polymer Modification	174
Other Methods	175
Esterification	176
Chain Transfer	176
COMMON GREEN SOLVENTS	177
Significance and Key Principles	177
Ethanol and Other Biobased Alcohols	177
Principles	177
Ionic Liquids	178
Principles	178
Deep Eutectic Solvents (DES)	178
Principles	178
Key Principles of Green Chemistry in Solvent Selection	178
Applications and Benefits	178
APPLICATIONS OF THE GRAFTING PROCESS IN POLYMER SCIENCE	179
Surface Modification	179
Material Strengthening	179
Functionalization	179
Adhesion Improvement	180
Nanotechnology	180
CHALLENGES IN THE GRAFTING PROCESS	180
Control of Grafting Density	180
Reproducibility	180
Complexity of Synthesis	180
Surface Characterization	180
Environmental and Health Concerns	181
Scalability	181
Cost	181
Degradation and Stability	181
Compatibility with Base Polymer	181
CASE STUDY: TYPES OF GREEN SOLVENTS IN THE GRAFTING PROCESS	181
Introduction	181
Case Study: Grafting of Polypropylene Using Green Solvents	182

DISCUSSION	183
CONCLUSION	184
RECOMMENDATIONS	184
REFERENCES	184

CHAPTER 6 RADIATION-INDUCED GREEN GRAFTING: MECHANISM, APPLICATIONS AND BENEFITS ... 187
Shweta Bhandari, Sumeet Dwivedi, Vishal Garg and *Idress Hamad Attitalla*

INTRODUCTION	188
IMPORTANCE OF RADIATION-INDUCED GREEN GRAFTING IN PHARMACEUTICALS	188
Sustainability of the Environment	188
Accuracy and Regulation	189
Improved Properties of the Material	189
Flexibility and Wide Range of Use	189
Better Systems for Delivering Drugs	189
Less Immunogenicity and Biocompatibility	189
Antimicrobial Characteristics	190
TYPES OF RADIATION SOURCES	190
Gamma	190
Electron Beam Radiation	190
Plasma Irradiation	191
UV	191
METHODS:	191
Simultaneous/Mutual Procedure	192
Method of Pre-Irradiation	193
TYPE OF POLYMER	195
Types of Polymers Suitable for Grafting	195
Fluoropolymers	195
Polyolefins	195
Acrylic Polymers	195
Cellulosic Polymers	196
Polyamide and Polyester	196
Characterization	196
Advantage Over Conventional Methods	196
APPLICATIONS IN BIOMEDICINE	197
Hydrogels in Medical Applications	198
Immobilization Techniques for Hydrogels	198
Hydrogels in Wound Dressing	199
Therapeutic Application of Hydrogel	199
Applications of Crosslinked Hydrogels	199
Thermoresponsive Gels in Medical Devices	199
Polymeric Drug Delivery Systems	199
Cancer	200
Immobilization of Enzymes	200
Sterilization	201
CHALLENGES OF RADIATION-INDUCED GREEN GRAFTING	202
FUTURE PERSPECTIVES OF RADIATION-INDUCED GREEN GRAFTING	204
CONCLUSION	205
REFERENCES	206

CHAPTER 7 PLASMA-ASSISTED GREEN GRAFTING: PRINCIPLES, ENVIRONMENTAL BENEFITS, AND USES 209
Pratiksha Bramhe, Prafulla Sabale, Suchita Waghmare, Pramod Khedekar, Vidya Sabale, Nilesh Rarokar and Mahavir Chougule
- **INTRODUCTION** 210
- **PRINCIPLES OF PLASMA-ASSISTED GREEN GRAFTING:** 215
  - Minimization of Environmental Impact 215
  - Energy Efficiency 215
  - Use of Environmentally Benign Plasma Gases 216
  - Selective Surface Modification 216
  - Reduction of Toxic Catalysts and Solvents 216
  - Sustainable Feedstocks and Monomers 216
  - Scalability and Industrial Relevance 216
- **PLASMA TECHNOLOGY AND SURFACE ACTIVATION/ MODIFICATION** 217
  - Modification of Polymer Surfaces for Water Purification 217
  - Grafting of Biocompatible Coatings on Medical Devices 218
  - Functionalization of Cellulose for Sustainable Packaging 219
  - Grafting of Antifouling Layers on Marine Equipment 219
  - Enhancement of Catalytic Surfaces for Green Chemistry 220
- **MECHANISM OF PLASMA ASSISTED GRAFTING** 220
  - Plasma Generation 221
  - Surface Activation 222
  - Grafting of Functional Groups or Polymers 222
  - Control of Grafting Density and Thickness 223
  - Post-Grafting Treatment 223
- **COMPARATIVE ANALYSIS OF PLASMA-ASSISTED SYNTHESIS AND TRADITIONAL SYNTHESIS METHODS IN THE FRAMEWORK OF GREEN CHEMISTRY** 223
  - Energy Efficiency 224
  - Use of Hazardous Chemicals 224
  - Reaction Selectivity and Yield 224
  - Scalability and Practical Application 225
  - Environmental Impact 225
- **ENVIRONMENTAL BENEFITS OF PLASMA-ASSISTED GREEN GRAFTING** 225
  - Elimination of Hazardous Chemicals 226
  - Energy Efficiency and Process Optimization 226
  - Minimal Environmental Impact and Waste Reduction 226
  - Enhanced Material Properties and Recyclability 226
  - Water Conservation and Process Sustainability 227
  - Reduced Environmental Footprint in Surface Engineering 227
- **APPLICATION OF PAGG IN THE PHARMACEUTICAL INDUSTRY** 227
  - Enhance Biocompatibility of Medical Implants 227
  - Development of Antimicrobial Surfaces 228
  - Improved Drug Delivery Systems 228
- **CONCLUSION** 229
- **ABBREVIATIONS** 229
- **ACKNOWLEDGEMENTS** 230
- **REFERENCES** 230

CHAPTER 8 ENZYMATIC GREEN GRAFTING: ROLE OF ENZYMES AND THEIR SUSTAINABLE ADVANTAGES 238

Kishor Danao, Deweshri Nandurkar, Vijayshri Rokde, Atul Bendale, Nilesh Karande, Akhil Nagar and *Viren Soni*

INTRODUCTION	239
METHODS OF GRAFTING [46]	242
MECHANISMS OF ENZYME-INITIATED GREEN GRAFTING	242
Sustainable Wood Preserving Method	242
Radical Polymerization Method [50 - 54]	244
OXIDATIVE POLYMERISATION [55 - 60]	246
ROLE OF ENZYMATIC GREEN GRAFTING	246
Green Approaches for Starch Modification	246
Polymer-grafting from a Metallo-Centered Enzyme	246
ENZYMATIC GRAFTING ON EUCALYPTUS GLOBULUS WOOD BY LACCASE	248
RAFT POLYMERIZATION FOR COMPLICATED POLYMER ARCHITECTURE	248
ENZYMATIC GRAFTING OF NATURAL PHENOLS TO FLAX FIBRES	249
NATURAL GAS CONVERSION	250
ADVANTAGES OF ENZYMATIC GREEN GRAFTING	250
Future Scope of Enzymatic Green Grafting	257
Applications of Enzymatic Green Grafting	258
Challenges and Considerations of Enzymatic Green Grafting	259
Scalability: Developing Cost-Effective Methods to Scale Up Enzymatic Grafting Processes	260
Enzyme Stability: Ensuring Enzymes Remain Active and Stable During the Grafting Process	260
Regulatory Approvals: Meeting Regulatory Standards for New Materials and Applications	261
CONCLUSION	262
REFERENCES	262
CHAPTER 9 MECHANOCHEMICAL GREEN GRAFTING: METHODS, ENVIRONMENTAL BENEFITS, AND USES	271
Vijayshri Rokde, Kishor Danao, Deweshri Nandurkar, Shweta Saboo and *Ujwala Mahajan*	
INTRODUCTION	272
METHODS OF MECHANOCHEMICAL GREEN GRAFTING	273
ROLE OF MECHANOCHEMICAL GREEN GRAFTING	278
Ball-milled Seashells as A Nano-Biocomposite	278
Mechanochemical Transformations of Biomass into Functional Materials	279
Solvent-free Synthesis of Amorphous Salts of Folic Acid	279
Mechanochemical Design of Nanomaterials	281
ADVANTAGES OF ENZYMATIC GREEN GRAFTING	281
FUTURE SCOPE	281
CONCLUSION	282
ABBREVIATIONS	283
ACKNOWLEDGEMENTS	284
REFERENCES	284
SUBJECT INDEX	289

PREFACE

The first part of *Green Grafting: Innovations in Polymer Functionalization for Sustainable Solutions in the Pharmaceutical and Healthcare Industry* marks a significant step forward in promoting sustainability within polymer science, drug delivery, and healthcare materials. This part serves as a comprehensive introduction to the fundamental principles of green grafting, integrating concepts of polymer chemistry, eco-friendly synthesis, and material functionalization with the overarching goals of environmental stewardship and technological advancement.

The chapters in this volume establish the core theoretical foundation for understanding how grafting techniques can be adapted to meet the tenets of green chemistry, focusing on minimizing hazardous substances, utilizing renewable feedstocks, and enhancing biodegradability. The content encompasses the basic mechanisms, methods, and characterization techniques of polymer grafting while illustrating their relevance to pharmaceutical and healthcare innovations.

In addition, Part **1** explores the relationship between chemistry and sustainability, showcasing how the fusion of polymer science and green chemistry principles enables the creation of safer, more efficient, and environmentally responsible materials. The authors highlight pioneering work in renewable resources, eco-friendly solvents, and sustainable grafting methods that not only improve product performance but also reduce environmental impact.

This first part serves as the foundation for advanced exploration in Part **2**, where the focus shifts toward specialized applications, industrial scalability, and future prospects. Together, both parts form a unified and forward-looking resource that embodies innovation, responsibility, and interdisciplinary collaboration in sustainable polymer research.

We extend our sincere gratitude to all the contributors, reviewers, and the Bentham Books editorial team for their invaluable contributions and support throughout the preparation of this volume. Their collective efforts have helped shape this book into a meaningful scientific reference for academicians, researchers, and industry professionals striving toward a more sustainable and innovative future.

Kuldeep Vinchurkar
Department of Pharmaceutics
Sandip Institute of Pharmaceutical Sciences (SIPS)
Affiliated To Savitribai Phule Pune University (SPPU, Pune)
Nashik, Maharashtra 422213, India

Satish Polshettiwar
Department of Pharmaceutical Sciences
School of Health Sciences and Technology,
Dr. Vishwanath Karad MIT-World Peace University,
Pune, Maharashtra 411038, India

Nilesh Mahajan
Department of Pharmaceutics
Dadasaheb Balpande College of Pharmacy
Besa, Nagpur, Maharashtra 440037
India

&

Yogeshwar Bachhav
Adex Pharmaceutical Consultancy Services,
Mumbai, Maharashtra 400089,
India

List of Contributors

Akanksha Phondke	Department of Pharmaceutical Chemistry, AETs St. John Institute of Pharmacy and Research, Palghar, Maharashtra 401404, India
A.R. Chabukswar	Department of Pharmaceutical Sciences, School of Health Sciences and Technology, Dr. Vishwanath Karad MIT World Peace University, Pune, Maharashtra 411038, India
Abhishek V. Shitol	Department of Pharmaceutical Chemistry, AISSMS College of Pharmacy, Pune, Maharashtra 411001, India
Atul Bendale	Department of Pharmaceutical Chemistry, Mahavir Institute of Pharmacy, Nashik, Maharashtra 422202, India
Akhil Nagar	Department of Pharmaceutical Chemistry, R.C. Patel Institute of Pharmaceutical Science and Education, Shirpur, Maharashtra 425405, India
Dipti Patel	Department of Pharmaceutics, Faculty of Pharmacy, Institute of Pharmaceutical Sciences, Parul University, Vadodara, Gujarat 391760, India
Drashti Dave	Department of Pharmaceutics, Sigma Institute of Pharmacy, Sigma University, Vadodara, Gujarat 391760, India
Deepali D. Bhandari	Department of Pharmaceutical Chemistry, GES's Sir Dr. M. S. Gosavi College of Pharmaceutical Education and Research, Nashik, Maharashtra 422001, India
Dattatraya M. Shinkar	Department of Pharmaceutics, GES's Sir Dr. M. S. Gosavi College of Pharmaceutical Education and Research, Nashik, Maharashtra 422001, India
Deweshri Nandurkar	Department of Pharmaceutical Chemistry, Dadasaheb Balpande College of Pharmacy, Nagpur, Maharashtra 440037, India
Idress Hamad Attitalla	Department of Microbiology, Faculty of Science, Omar Al-Mukhtar University, Al-Bayda, Libya
Kuldeep Vinchurkar	Department of Pharmaceutics, Sandip Institute of Pharmaceutical Sciences (SIPS), Affiliated To Savitribai Phule Pune University (SPPU, Pune), Nashik, Maharashtra 422213, India
Kishor Danao	Department of Pharmaceutical Chemistry, Dadasaheb Balpande College of Pharmacy, Nagpur, Maharashtra 440037, India
Meghana Jagtap	Department of Pharmaceutical Chemistry, Bombay College of Pharmacy, Kalina, Santacruz (E), Mumbai, Maharashtra 400098, India
Mahavir Chougule	Department of Pharmaceutical Sciences, Mercer University College of Pharmacy, Atlanta, Georgia 30341, USA
Neha A. Raut	Department of Pharmaceutical Chemistry, AISSMS College of Pharmacy, Pune, Maharashtra 411001, India
Nilesh Rarokar	Department of Pharmaceutics, G. H. Raisoni Institute of Life Sciences, Nagpur, Maharashtra 440016, India
Nilesh Karande	Department of Pharmaceutical Chemistry, Institute of Pharmaceutical Science and Education, Wardha, Maharashtra 442001, India
Priya Patne	Department of Pharmaceutical Chemistry, Bombay College of Pharmacy, Kalina, Santacruz (E), Mumbai, Maharashtra 400098, India

Popat Mohite	Department of Pharmaceutical Chemistry, AETs St. John Institute of Pharmacy and Research, Palghar, Maharashtra 401404, India
Premlata Ambre	Department of Pharmaceutical Chemistry, Bombay College of Pharmacy, Kalina, Santacruz (E), Mumbai, Maharashtra 400098, India
Pratiksha Bramhe	Department of Pharmaceutical Sciences, Rashtrasant Tukadoji Maharaj Nagpur University, Nagpur, Maharashtra 440033, India
Prafulla Sabale	Department of Pharmaceutical Sciences, Rashtrasant Tukadoji Maharaj Nagpur University, Nagpur, Maharashtra 440033, India
Pramod Khedekar	Department of Pharmaceutical Sciences, Rashtrasant Tukadoji Maharaj Nagpur University, Nagpur, Maharashtra 440033, India
Ramanlalal N. Kachave	Department of Pharmaceutical Chemistry, GES's Sir Dr. M. S. Gosavi College of Pharmaceutical Education and Research, Nashik, Maharashtra 422001, India
Satish Polshettiwar	Department of Pharmaceutical Sciences, School of Health Sciences and Technology, Dr. Vishwanath Karad MIT World Peace University, Pune, Maharashtra 411038, India
Sudarshan Singh	Faculty of Pharmacy, Chiang Mai University, Chiang Mai, 50200, Thailand
S.C. Jagdale	Department of Pharmaceutical Sciences, School of Health Sciences and Technology, Dr. Vishwanath Karad MIT World Peace University, Pune, Maharashtra 411038, India
S.A. Polshettiwar	Department of Pharmaceutical Sciences, School of Health Sciences and Technology, Dr. Vishwanath Karad MIT World Peace University, Pune, Maharashtra 411038, India
Sunil V. Amrutkar	Department of Pharmaceutical Chemistry, GES's Sir Dr. M. S. Gosavi College of Pharmaceutical Education and Research, Nashik, Maharashtra 422001, India
Shashikant V. Bhandari	Department of Pharmaceutical Chemistry, AISSMS College of Pharmacy, Pune, Maharashtra 411001, India
Shambhavi S. Sing	Department of Pharmaceutical Chemistry, AISSMS College of Pharmacy, Pune, Maharashtra 411001, India
Sagar R. Ghanwa	Department of Pharmaceutical Chemistry, AISSMS College of Pharmacy, Pune, Maharashtra 411001, India
Sharayu P. Ninaew	Department of Pharmaceutical Chemistry, AISSMS College of Pharmacy, Pune, Maharashtra 411001, India
Shweta Bhandari	Department of Pharmacy, Sumandeep Vidyapeeth (Deemed to be University), Vadodara, Gujarat 391760, India
Sumeet Dwivedi	Department of Pharmacognosy, Acropolis Institute of Pharmaceutical Education & Research, Indore, Madhya Pradesh 453771, India
Suchita Waghmare	Department of Pharmaceutical Sciences, Rashtrasant Tukadoji Maharaj Nagpur University, Nagpur, Maharashtra 440033, India
Shweta Saboo	Department of Pharmacognosy, Government College of Pharmacy, Chhatrapati Sambhajinagar, Maharashtra 431005, India
Taufik Mulla	Department of Pharmaceutics, Faculty of Pharmacy, Institute of Pharmaceutical Sciences, Parul University, Vadodara, Gujarat 391760, India

Unmesh G. Bhamare	Department of Pharmaceutics, GES's Sir Dr. M. S. Gosavi College of Pharmaceutical Education and Research, Nashik, Maharashtra 422001, India
Ujwala Mahajan	Department of Pharmacognosy, Dadasaheb Balpande College of Pharmacy, Nagpur, Maharashtra 440037, India
Vishal Garg	Department of Pharmaceutics, Jaipur School of Pharmacy, Maharaj Vinayak Global University, Jaipur, Rajasthan 302028, India
Vidya Sabale	Department of Pharmaceutics, Dadasaheb Balpande College of Pharmacy, Besa, Nagpur, Maharashtra 440037, India
Vijayshri Rokde	Department of Pharmaceutical Chemistry, Dadasaheb Balpande College of Pharmacy, Nagpur, Maharashtra 440037, India
Viren Soni	Department of Pharmacology and Experimental Therapeutics, Thomas Jefferson University, Philadelphia, PA 19107, USA
Vijayshri Rokde	Department of Pharmaceutical Chemistry, Dadasaheb Balpande College of Pharmacy, Nagpur, Maharashtra 440037, India
Yash D. Kale	Department of Pharmaceutical Sciences, School of Health Sciences and Technology, Dr. Vishwanath Karad MIT World Peace University, Pune, Maharashtra 411038, India

CHAPTER 1

Introduction to Green Grafting

Taufik Mulla[1]**, Dipti Patel**[1,*]**, Drashti Dave**[2]**, Kuldeep Vinchurkar**[3] **and Satish Polshettiwar**[4]

[1] *Department of Pharmaceutics, Faculty of Pharmacy, Institute of Pharmaceutical Sciences, Parul University, Vadodara, Gujarat 391760, India*

[2] *Department of Pharmaceutics, Sigma Institute of Pharmacy, Sigma University, Vadodara 391760, Gujarat, India*

[3] *Department of Pharmaceutics, Sandip Institute of Pharmaceutical Sciences (SIPS), Affiliated To Savitribai Phule Pune University (SPPU, Pune), Nashik, Maharashtra 422213, India*

[4] *Department of Pharmaceutical Sciences, School of Health Sciences and Technology, Dr. Vishwanath Karad MIT World Peace University, Pune, Maharashtra 411038, India*

Abstract: Green grafting is an innovative approach in the realm of polymer functionalization, focusing on the integration of eco-friendly processes and sustainable materials to advance the pharmaceutical and healthcare industries. This introductory chapter provides a comprehensive overview of green grafting, outlining its principles, methods, and applications. Emphasis is placed on the environmental and economic benefits of adopting green grafting techniques, such as reduced carbon footprint, enhanced material efficiency, and minimized use of hazardous chemicals. The chapter delves into the chemistry of green grafting, exploring various green solvents, catalysts, and renewable resources employed in the functionalization of polymers. Key advancements in the field, including the development of biodegradable polymers and biobased grafting techniques, are thoroughly discussed. Furthermore, the chapter highlights the significant role of green grafting in enhancing drug delivery systems, medical devices, and tissue engineering. Through a series of case studies and real-world applications, the transformative impact of green grafting on sustainability in the pharmaceutical and healthcare sectors is demonstrated. The chapter concludes with a discussion on future directions and potential challenges, emphasizing the need for continued innovation and interdisciplinary collaboration to fully realize the benefits of green grafting in creating sustainable solutions.

Keywords: Biodegradable polymers, Drug delivery systems, Green chemistry, Polymer functionalization, Sustainable solutions.

---

[*] **Corresponding author Dipti Patel:** Department of Pharmaceutics, Sigma Institute of Pharmacy, Sigma University, Vadodara 391760, Gujarat, India; E-mail: dipti.patel@paruluniversity.ac.in

Kuldeep Vinchurkar, Satish Polshettiwar, Nilesh Mahajan & Yogeshwar Bachhav (Eds.)
All rights reserved-© 2026 Bentham Science Publishers

INTRODUCTION

Overview of Green Grafting

Green grafting refers to the integration of green chemistry principles into the process of grafting, which involves the attachment of one molecular fragment (graft) onto a main polymer chain (backbone) or substrate. The concept of grafting has been extensively used in material sciences and biotechnology to alter or enhance the properties of substrates, such as improving their chemical resistance, mechanical properties, or biocompatibility [1].

Traditional grafting methods often involve the use of toxic chemicals, solvents, and energy-intensive processes that pose significant environmental risks. These concerns have driven research towards more sustainable practices, giving rise to green grafting methods. Green grafting adopts the principles of green chemistry to reduce the environmental impact of the grafting process. It focuses on using environmentally benign chemicals, reducing hazardous waste, and minimizing energy consumption during synthesis. Green grafting offers a sustainable alternative to traditional methods by reducing environmental risks and energy consumption while using safer chemicals. Table **1** provides a comparative overview of traditional and green grafting approaches, detailing the types of chemicals used, health risks, energy demands, and techniques applied. As shown, green grafting utilizes eco-friendly solvents and catalysts, significantly lowering the environmental impact and health hazards associated with traditional grafting methods.

Table 1. Traditional *vs.* green grafting [3, 4].

Aspects	Traditional Grafting	Green Grafting
Chemicals Used	Hazardous chemicals like benzene, toluene, and heavy metals	Safer alternatives like water-based solvents or bio-based catalysts
Environmental/Health Risks	Significant risks due to toxic chemicals and heavy metals	Reduced risks with eco-friendly chemicals
Energy Consumption	High temperatures and prolonged reaction times	Reduced, using milder conditions
Techniques Used	Free radical polymerization, atom transfer radical polymerization (ATRP)	Optimized methods using microwaves, enzymes, or catalytic systems
Reaction Conditions	Requires high temperatures, leading to greater energy demand	Milder conditions with reduced energy demands
Catalysts/Solvents	Solvents like benzene and toluene, contribute to environmental hazards	Bio-based catalysts and water-based solvents

The shift to green grafting aligns with global efforts to promote sustainability and reduce industrial pollution. It has also opened new avenues for the development of biocompatible materials in industries such as pharmaceuticals, biotechnology, and healthcare, where there is a growing demand for materials that are not only functional but also environmentally friendly [2].

Importance of Green Grafting in the Pharmaceutical and Healthcare Industries

Green grafting plays a crucial role in the pharmaceutical and healthcare industries, where the need for sustainable practices is growing in response to stricter regulations and increasing public awareness about environmental issues. The adoption of green grafting techniques provides a means to develop more sustainable materials for drug delivery systems, medical devices, and tissue engineering applications [3].

Sustainable Drug Delivery Systems

One of the significant applications of green grafting in the pharmaceutical industry is the development of advanced drug delivery systems. Polymers and other biomaterials used in drug delivery must meet stringent safety and biocompatibility standards. Green grafting allows for the synthesis of materials that degrade safely within the body, reducing toxic byproducts and lowering the overall environmental footprint. Grafted polymers can improve the solubility, stability, and targeted release of drugs, enhancing their therapeutic efficacy. By incorporating green chemistry principles, pharmaceutical companies can manufacture drug delivery systems that are not only effective but also aligned with sustainable production goals [4].

For instance, hydrogels and nanoparticles have been explored as carriers for controlled drug release. Through green grafting, researchers can design hydrogels that are synthesized using non-toxic, bio-based solvents and biodegradable polymers. This ensures that the materials break down into harmless byproducts after fulfilling their function in the body.

Green Grafting in Medical Devices

Green grafting is also pivotal in the development of medical devices such as catheters, implants, and wound dressings. These devices often require surface modifications to enhance their biocompatibility, reduce infection risks, or promote tissue integration. Traditional surface grafting techniques used in medical devices can leave residual toxic substances that are detrimental to patient health [5].

By employing green grafting methods, manufacturers can create safer, biocompatible surfaces that reduce adverse reactions and improve patient outcomes. For example, silver nanoparticles grafted onto polymers are widely used in wound dressings due to their antimicrobial properties. Using green grafting techniques, these nanoparticles can be synthesized in a way that minimizes the use of harmful chemicals while maintaining their efficacy [6].

Environmental Impact and Waste Reduction

Another key benefit of green grafting is the reduction of waste and pollution in pharmaceutical manufacturing. Traditional grafting processes often generate significant amounts of hazardous waste, which must be carefully managed to avoid environmental contamination. Green grafting, by contrast, reduces or eliminates the use of toxic solvents and reagents, leading to a cleaner and more sustainable production process [7].

Furthermore, the ability to design biodegradable grafted materials through green chemistry reduces the environmental burden of medical waste. Single-use medical devices, packaging materials, and drug delivery systems that incorporate green-grafted polymers can be designed to degrade naturally over time, reducing the volume of waste sent to landfills or incineration [8].

Principles of Green Chemistry in Grafting

Green grafting is underpinned by the different principles of green chemistry, which were developed to guide the design of chemical processes and products that reduce environmental harm. The application of these principles in grafting has led to significant improvements in sustainability across various industries, particularly in the pharmaceutical and healthcare sectors [9].

Prevention: The principle of prevention emphasizes minimizing waste during chemical processes. Green grafting achieves this by using catalysts that enhance reaction efficiency and reduce the amount of unreacted starting materials. Additionally, by designing materials that are biodegradable or easily recyclable, green grafting minimizes the long-term environmental impact of waste materials. For example, in the synthesis of biodegradable drug delivery systems, green grafting methods can produce polymers that break down into non-toxic components, reducing the need for post-use treatment and disposal [10].

Atom Economy: Atom economy refers to the efficient use of atoms in a chemical reaction to ensure that the final product contains the maximum number of atoms from the starting materials. In green grafting, atom economy is achieved by selecting reaction pathways that minimize byproducts and waste. This not only improves the efficiency of the grafting process but also reduces the cost and environmental impact of raw material consumption.

Less Hazardous Chemical Syntheses: Green grafting prioritizes the use of non-toxic chemicals and reagents in synthesis. For example, water-based solvents or bio-based solvents are preferred over traditional organic solvents like benzene or toluene, which are hazardous to both human health and the environment. Enzyme-catalyzed grafting is one example where less hazardous chemicals are used. Enzymes act as biological catalysts that can operate under mild conditions, eliminating the need for toxic initiators or extreme reaction conditions [11].

Designing Safer Chemicals: The design of safer chemicals is crucial in the development of pharmaceuticals and medical devices. By incorporating green chemistry principles, chemists can design grafting agents and polymers that are not only effective but also less harmful to human health. This is especially important for materials used in medical applications, where biocompatibility and safety are paramount.

Safer Solvents and Auxiliaries: Traditional grafting techniques often require large amounts of organic solvents, many of which are harmful to both the environment and human health. Green grafting aims to reduce or eliminate the use of these solvents, opting instead for safer alternatives such as water, ethanol, or ionic liquids. Water, in particular, is an excellent solvent for many green grafting processes due to its abundance, low cost, and non-toxic nature [12].

Design for Energy Efficiency: The energy consumption of chemical processes is a significant contributor to their environmental impact. Green grafting methods are designed to minimize energy use by employing reaction conditions that require lower temperatures, shorter reaction times, or less intensive mixing. For instance, microwave-assisted grafting is an energy-efficient alternative to traditional heating methods. It allows for rapid, uniform heating of the reaction mixture, reducing both energy consumption and reaction time.

Use of Renewable Feedstocks: Green grafting emphasizes the use of renewable feedstocks—materials that can be replenished naturally over time. In the context of pharmaceuticals and medical devices, renewable polymers derived from plant-based sources such as cellulose, chitosan, and starch are increasingly being used in green grafting processes. These bio-based polymers are biodegradable, reducing the long-term environmental impact of medical waste.

Reduce Derivatives: The reduction of unnecessary derivatization—such as the use of protecting groups or complex intermediate steps—helps streamline chemical processes and reduces waste. Green grafting techniques are designed to avoid these additional steps whenever possible, improving both the efficiency and sustainability of the grafting process.

Catalysis: Catalysis is a cornerstone of green chemistry. In green grafting, catalysts are used to accelerate reactions while minimizing the need for excess reagents and harsh reaction conditions. Enzyme-catalyzed grafting is a prime example, where biological catalysts enable reactions to occur under mild, environmentally friendly conditions [13].

Design for Degradation: In line with the principle of designing materials for degradation, green grafting focuses on creating polymers and materials that can safely degrade into non-toxic byproducts. This is particularly important in healthcare, where medical devices, drug delivery systems, and packaging materials should break down naturally without harming the environment.

Real-time Analysis for Pollution Prevention: Real-time monitoring of grafting processes allows for the immediate detection and correction of inefficiencies or hazardous byproducts. Techniques such as *in situ* spectroscopy or chromatography enable researchers to optimize green grafting processes, ensuring that they adhere to environmental safety standards throughout production.

Inherently Safer Chemistry for Accident Prevention: Green grafting reduces the risk of accidents by minimizing the use of hazardous substances and extreme reaction conditions. For example, enzyme-catalyzed grafting or the use of water-based solvents significantly lowers the risk of explosions, fires, or toxic spills compared to traditional grafting methods that rely on volatile organic solvents [14, 15].

METHODS OF GREEN GRAFTING

The development of green grafting methods is critical to achieving more sustainable practices in industries such as pharmaceuticals, healthcare, and materials science. By incorporating green chemistry principles, these methods reduce environmental harm, minimize toxic waste, and increase the efficiency of chemical reactions. Green grafting methods can be broadly classified into three main categories: the use of green solvents, the application of biocatalysts and organic catalysts, and the utilization of renewable resources [16].

Use of Green Solvents

Green solvents are one of the most significant developments in sustainable chemistry. Solvents traditionally used in grafting reactions, such as benzene, toluene, and chloroform, are highly toxic and pose risks to both human health and the environment. In response, green grafting focuses on using safer alternatives that are non-toxic, biodegradable, and derived from renewable sources.

Water as a Green Solvent: Water is often regarded as the ideal green solvent due to its non-toxic nature, availability, and cost-effectiveness. It serves as an excellent medium for grafting reactions, especially in processes that involve hydrophilic polymers. Water-based grafting is extensively used in synthesizing hydrogels, where polymers like polyvinyl alcohol (PVA) or polyethylene glycol (PEG) are cross-linked in aqueous environments to form biocompatible and biodegradable materials. The use of water not only reduces the need for hazardous organic solvents but also aligns with the principle of waste minimization in green chemistry.

In grafting techniques like radical polymerization, water can also serve as a dispersion medium, especially for grafting hydrophobic monomers onto water-soluble polymers. This results in less volatile organic compound (VOC) emissions and significantly reduces the environmental impact [17].

Ionic Liquids: Ionic liquids are another category of green solvents used in grafting processes. These solvents, composed of ions, have low vapor pressures and are considered environmentally friendly due to their non-volatile nature. Ionic liquids have tunable properties, meaning their chemical and physical characteristics can be adjusted to suit the specific requirements of a grafting process. This adaptability makes them suitable for complex grafting reactions, especially in synthesizing advanced polymeric materials. For instance, ionic liquids can be used to graft chitosan or cellulose with functional groups, creating biodegradable and sustainable materials. Due to their low toxicity and recyclability, ionic liquids contribute to a more sustainable grafting process by reducing the use of volatile organic solvents and minimizing waste generation [18].

Supercritical CO_2: Supercritical carbon dioxide (CO_2) is another solvent used in green grafting. Supercritical CO_2 is a state of carbon dioxide that exhibits both liquid and gas-like properties, making it an efficient solvent for various grafting reactions. One of the major advantages of using supercritical CO_2 is that it can be easily recovered and reused, reducing waste and solvent consumption [19].

Supercritical CO_2 is particularly useful in the synthesis of drug delivery systems, where it can be used to graft polymers onto drug molecules without the need for harmful organic solvents. Its ability to dissolve a wide range of materials, combined with its non-toxic and non-flammable nature, makes it an attractive option for green grafting.

Biocatalysts and Organocatalysts

Catalysis is a crucial aspect of green grafting, as it enables reactions to proceed more efficiently under milder conditions, thereby reducing energy consumption and waste. Traditional grafting methods often rely on metal-based catalysts, which can be toxic and difficult to remove from the final product. Green grafting, however, employs biocatalysts and organo-catalysts, which are environmentally benign and offer several advantages over conventional catalysts.

Biocatalysts: Biocatalysts, such as enzymes, are increasingly being used in green grafting due to their high specificity, efficiency, and ability to function under mild conditions. Enzymes are natural catalysts that accelerate chemical reactions without the need for extreme temperatures or pressures, making them ideal for green chemistry applications. Enzymatic grafting is particularly advantageous in the pharmaceutical and biomedical fields, where biocompatibility is a critical concern [20].

For instance, lipase enzymes can be used to graft polyesters onto natural polymers, such as starch or cellulose, creating biodegradable materials suitable for use in drug delivery systems. Enzymatic grafting reactions are often carried out in water, further reducing the need for harmful organic solvents. In addition, biocatalysts are renewable and biodegradable, aligning with the principles of sustainability. Unlike metal-based catalysts, which can leave behind toxic residues, enzymes are easily separated from the reaction mixture and do not pose environmental hazards [21].

Organocatalysts: Organocatalysts are small organic molecules that catalyze chemical reactions without the need for metal catalysts. They offer several advantages in green grafting, such as high selectivity, ease of handling, and the ability to work under mild reaction conditions. Organocatalysis eliminates the need for transition metals, which are often toxic and difficult to remove from grafted materials. In polymer grafting, organocatalysts are used to promote polymerization reactions without the need for metal-based catalysts. For example, the use of organic catalysts in atom transfer radical polymerization (ATRP) has been explored to create polymer-grafted surfaces with enhanced properties. By eliminating metals, organocatalytic processes reduce the environmental impact of the grafting process and make the materials more suitable for medical and pharmaceutical applications [22]. The versatility of organocatalysts also extends to renewable polymer grafting, where natural compounds such as amino acids or alkaloids are used as catalysts to promote environmentally friendly grafting reactions. These organocatalysts can efficiently graft renewable polymers such as cellulose or lignin, creating bio-based materials with a wide range of applications.

Renewable Resources

The use of renewable resources in grafting is a key component of green chemistry. Renewable resources, such as bio-based polymers and natural monomers, provide an environmentally friendly alternative to petroleum-based materials traditionally used in grafting processes. By incorporating renewable resources, green grafting contributes to reducing the reliance on finite resources and decreasing the carbon footprint of the production process [23].

Bio-based Polymers: Bio-based polymers, such as cellulose, chitosan, starch, and lignin, are derived from natural sources and are widely used in green grafting. These polymers are biodegradable, non-toxic, and renewable, making them ideal candidates for sustainable grafting applications. Cellulose, the most abundant natural polymer, has been extensively grafted with functional groups to improve its properties for applications in pharmaceuticals, textiles, and packaging. For example, cellulose can be grafted with hydrophobic monomers to create water-resistant coatings or with bioactive molecules to enhance its antimicrobial properties for medical use. Chitosan, derived from chitin, is another renewable polymer used in green grafting. Chitosan's biodegradability and biocompatibility make it particularly suitable for medical and pharmaceutical applications, such as drug delivery systems and wound healing materials. Grafting chitosan with functional groups can improve its solubility and mechanical properties, extending its potential applications [24].

Renewable Monomers: In addition to bio-based polymers, renewable monomers derived from natural sources are used in green grafting processes. These monomers include lactic acid, itaconic acid, and furan derivatives, which can be polymerized and grafted onto natural or synthetic polymers to create sustainable materials. Lactic acid, for example, is a renewable monomer that can be polymerized to form polylactic acid (PLA), a biodegradable polymer widely used in packaging and biomedical applications. Grafting lactic acid onto cellulose or other natural polymers can enhance the material's mechanical properties, biodegradability, and thermal stability [25].

Itaconic acid, derived from the fermentation of sugars, is another renewable monomer used in green grafting. Itaconic acid can be grafted onto natural polymers to create bio-based materials with enhanced thermal and chemical resistance. These materials have potential applications in coatings, adhesives, and medical devices [26].

Lignin-based Grafting: Lignin, a byproduct of the paper and biofuel industries, is an abundant and renewable resource that can be utilized in green grafting. Lignin's complex structure and abundance of functional groups make it an attractive candidate for grafting with synthetic or natural polymers. Lignin can be grafted with synthetic polymers like polyethylene or polystyrene to improve their mechanical properties and reduce their environmental impact. The resulting materials are biodegradable and possess enhanced properties suitable for packaging and construction materials [27, 28].

By utilizing renewable resources such as bio-based polymers and natural monomers, green grafting contributes to the development of sustainable materials that reduce the environmental impact of industrial processes. This aligns with the broader goals of green chemistry, which aims to create products and processes that are safer, more efficient, and less harmful to the environment [29, 30].

APPLICATIONS OF GREEN GRAFTING IN THE PHARMACEUTICAL AND HEALTHCARE INDUSTRIES

Green grafting has emerged as a pivotal method in developing sustainable, efficient, and biocompatible materials for the pharmaceutical and healthcare industries. The incorporation of green chemistry principles into grafting processes allows for the creation of advanced materials that align with both ecological and medical goals. These materials are designed to reduce environmental impact while enhancing patient safety and efficacy. Key applications of green grafting in the pharmaceutical and healthcare sectors include drug delivery systems, medical devices, and tissue engineering [31].

Drug Delivery Systems

Drug delivery systems (DDS) are one of the most prominent applications of green grafting in healthcare. The primary goal of DDS is to deliver therapeutic agents to specific sites in the body in a controlled and sustained manner, improving the efficacy of drugs while minimizing side effects. Grafting techniques allow for the modification of polymers and biomaterials, improving their properties for effective drug delivery [32].

Biodegradable Polymers for Drug Delivery: Green grafting enables the development of biodegradable and biocompatible polymers that can be used as carriers in drug delivery systems. By grafting functional groups onto these polymers, their properties can be tailored to control drug release, target specific tissues, and reduce toxicity. For example, polymers like polylactic acid (PLA), polycaprolactone (PCL), and chitosan are often used in drug delivery applications. Grafting these polymers with hydrophilic or bioactive molecules enhances their solubility, biocompatibility, and drug-carrying capacity. In particular, green grafting has been used to improve the solubility and bioavailability of poorly water-soluble drugs. Hydrophilic polymers grafted with bioactive agents can encapsulate hydrophobic drugs, ensuring controlled release at the target site. This reduces the need for frequent dosing and minimizes the risk of drug accumulation, thus enhancing patient compliance and reducing side effects [33].

Targeted Drug Delivery: One of the most innovative applications of green grafting in drug delivery systems is the creation of targeted DDS. Targeted drug delivery involves directing the therapeutic agent to a specific site in the body, minimizing its interaction with healthy tissues, and reducing systemic toxicity. Green grafting techniques enable the attachment of targeting ligands, such as peptides, antibodies, or aptamers, to polymeric drug carriers. These ligands can recognize specific receptors overexpressed in diseased tissues, such as tumors, ensuring that the drug is delivered precisely where it is needed [35]. For instance, grafting polyethylene glycol (PEG) onto drug-loaded nanoparticles (a process known as PEGylation) is a well-established method for enhancing the circulation time of drugs in the bloodstream. PEGylation helps to "shield" the drug from the immune system, preventing its premature clearance and allowing it to reach the target site. Additionally, green grafting methods have been used to attach targeting ligands like folic acid or monoclonal antibodies to PEGylated nanoparticles, improving the specificity of the drug delivery system for cancer cells [36].

Stimuli-Responsive Drug Delivery: Green grafting is also employed to create stimuli-responsive DDS, where drug release is triggered by external factors such as temperature, pH, or enzymes. For example, polymers grafted with pH-sensitive groups can release drugs in response to the acidic microenvironment of a tumor or an inflamed tissue. This ensures that the drug is released only at the desired site, reducing systemic side effects. Similarly, temperature-sensitive polymers, such as those based on poly(N-isopropyl acrylamide) (PNIPAM), have been grafted with functional groups to create thermoresponsive drug delivery systems. These systems can release their payload when exposed to body heat or an external heat source, making them ideal for localized drug delivery in hyperthermia treatments for cancer [37].

Medical Devices

Green grafting plays a significant role in the development of eco-friendly and biocompatible materials for medical devices. Medical devices, such as implants, catheters, and wound dressings, require materials that are not only durable but also biocompatible and non-toxic. Traditional materials often involve synthetic polymers and metals, which can cause adverse reactions or lead to the accumulation of toxic byproducts in the body. Green grafting techniques help to address these challenges by incorporating biocompatible and biodegradable materials into medical device design.

Antimicrobial Coatings: One of the key applications of green grafting in medical devices is the creation of antimicrobial coatings. Hospital-acquired infections (HAIs) are a major concern in healthcare settings, often resulting from bacterial colonization on medical devices such as catheters, surgical implants, and wound dressings. Green grafting techniques have been employed to graft antimicrobial agents onto the surface of medical devices, creating a barrier that prevents bacterial adhesion and biofilm formation. For instance, chitosan, a natural biopolymer with inherent antimicrobial properties, can be grafted onto the surface of medical devices to inhibit bacterial growth. Grafting chitosan with bioactive molecules like silver nanoparticles or antimicrobial peptides further enhances its effectiveness, providing a long-lasting antimicrobial effect. These coatings are not only effective against a wide range of pathogens but are also biodegradable and environmentally friendly, aligning with the principles of green chemistry [38].

Biocompatible Implants: Biocompatibility is a critical factor in the design of implants, as the material must integrate with the surrounding tissue without triggering an immune response or rejection. Green grafting allows for the modification of implant surfaces to improve biocompatibility and reduce the risk of complications. For example, polymers like polyvinyl alcohol (PVA) and poly(lactic-co-glycolic acid) (PLGA) are commonly used in grafting processes to create biodegradable and bioresorbable implants. Grafting bioactive molecules such as growth factors or peptides onto the surface of implants promotes tissue regeneration and accelerates the healing process. This is particularly important in orthopedic implants, where grafted materials can stimulate bone growth and improve the integration of the implant with the surrounding bone tissue. By using biodegradable and biocompatible materials, green grafting reduces the need for secondary surgeries to remove the implant, minimizing the risk of complications and improving patient outcomes [39].

Wound Healing Materials: Green grafting has also revolutionized the development of wound-healing materials, such as dressings and scaffolds. These materials need to be biocompatible, promote healing, and prevent infection. Grafting natural polymers like collagen, alginate, and chitosan with bioactive molecules has led to the creation of advanced wound healing materials that promote cell proliferation, angiogenesis, and tissue regeneration. For example, alginate dressings grafted with growth factors can enhance the healing of chronic wounds, such as diabetic ulcers, by promoting cell migration and tissue repair. Additionally, the antimicrobial properties of grafted chitosan help prevent infection, creating an ideal environment for wound healing. These materials are biodegradable and environmentally friendly, reducing the environmental impact associated with traditional wound dressings made from synthetic polymers.

Tissue Engineering

Tissue engineering is a rapidly growing field that seeks to create functional tissues to replace damaged or diseased organs. Green grafting plays a crucial role in developing biomaterials for tissue engineering, as it allows for the creation of scaffolds that support cell growth, differentiation, and tissue regeneration. These scaffolds must be biocompatible, biodegradable, and capable of mimicking the natural extracellular matrix (ECM) to provide an optimal environment for cell attachment and proliferation.

Biodegradable Scaffolds: Green grafting techniques are widely used to create biodegradable scaffolds for tissue engineering. These scaffolds are designed to degrade over time as new tissue forms, eliminating the need for surgical removal. Polymers such as polylactic acid (PLA), polycaprolactone (PCL), and collagen are commonly used in grafting processes to create scaffolds with tailored properties, such as porosity, mechanical strength, and degradation rate. By grafting bioactive molecules such as growth factors, peptides, or proteins onto these scaffolds, the materials can promote cell adhesion, proliferation, and differentiation. For example, scaffolds grafted with vascular endothelial growth factor (VEGF) can stimulate the formation of blood vessels (angiogenesis), which is critical for tissue regeneration in large defects or injuries.

3D Printing and Green Grafting: The combination of 3D printing and green grafting has opened new possibilities in tissue engineering. 3D printing allows for the precise fabrication of scaffolds with complex geometries that mimic the architecture of natural tissues. Green grafting techniques can be applied to functionalize these scaffolds, enhancing their biocompatibility and promoting tissue regeneration. For example, 3D-printed scaffolds made from biodegradable polymers like PLA or PCL can be grafted with bioactive molecules to create a supportive environment for cell growth. These scaffolds can be designed to degrade at a controlled rate, allowing the newly formed tissue to replace the scaffold over time. This approach has been used in the engineering of various tissues, including bone, cartilage, and skin.

Regenerative Medicine: In regenerative medicine, green grafting is used to develop biomaterials that support the repair and regeneration of damaged tissues. By grafting natural polymers like collagen or hyaluronic acid with growth factors or stem cell attractants, researchers have been able to create biomaterials that promote the healing of tissues such as skin, cartilage, and nerves. For instance, collagen scaffolds grafted with nerve growth factor (NGF) have been used to support nerve regeneration in patients with peripheral nerve injuries. These grafted materials provide a biocompatible scaffold that guides the growth of new nerve fibers, restoring function to damaged tissues. Similarly, green grafting has been used to create cartilage scaffolds that promote the regeneration of damaged cartilage in patients with osteoarthritis or traumatic injuries [40].

CASE STUDIES

In this section, we explore three real-world applications of green grafting, focusing on biodegradable polymers for drug delivery, antimicrobial surfaces for medical devices, and functionalized scaffolds for tissue engineering. These case studies illustrate how green grafting techniques have been applied in the pharmaceutical and healthcare industries, demonstrating the potential of environmentally friendly methods to improve medical outcomes.

Biodegradable Polymers for Drug Delivery

Case Study: Poly(lactic-co-glycolic acid) (PLGA) Grafted with Polyethylene Glycol (PEG) for Cancer Therapy [41]

PLGA is one of the most widely used biodegradable polymers in drug delivery due to its biocompatibility, controlled degradation, and FDA approval. However, the hydrophobicity of PLGA limits its effectiveness in delivering hydrophilic drugs. In response, researchers have explored green grafting methods to improve its properties. In a notable study, PLGA was grafted with PEG to enhance its hydrophilicity and circulation time in the body. PEGylation of PLGA nanoparticles allowed for the sustained release of anticancer drugs, improving drug solubility and targeting tumor cells more effectively. Moreover, this approach reduced the use of organic solvents typically required in conventional drug formulation methods, aligning with green chemistry principles.

Impact: The green grafting of PLGA with PEG has led to more efficient drug delivery systems, reducing the required dosage and frequency of administration while minimizing side effects. This method has shown great promise in the treatment of cancers, particularly for drugs with low water solubility, such as paclitaxel.

Challenges: Despite its success, the scalability of PEGylated nanoparticles remains a challenge due to the complexity of grafting techniques. Future research is needed to streamline manufacturing processes and enhance the bioavailability of drugs encapsulated in these grafted polymers.

Antimicrobial Surfaces for Medical Devices

Case Study: Grafting Chitosan with Silver Nanoparticles for Catheters [42]

Hospital-acquired infections (HAIs) often result from bacterial colonization on medical devices, leading to severe complications and increased healthcare costs. To combat this, researchers have focused on developing antimicrobial coatings for medical devices using green grafting techniques. Chitosan, a natural biopolymer derived from chitin, is biodegradable and possesses inherent antimicrobial properties. In a breakthrough study, chitosan was grafted with silver nanoparticles to create antimicrobial coatings for catheters. This combination provided long-lasting antimicrobial effects without the use of toxic chemicals or non-biodegradable materials. The green synthesis of silver nanoparticles ensured minimal environmental impact, making the entire process eco-friendly.

Impact: The chitosan-silver nanoparticle grafted coating significantly reduced bacterial adhesion and biofilm formation on catheters, reducing the incidence of catheter-related infections. This green grafting approach offers a sustainable alternative to conventional antimicrobial coatings, which often rely on harmful chemicals.

Challenges: One challenge is the potential cytotoxicity of silver nanoparticles at high concentrations. Although the grafting process reduces this risk, further research is needed to optimize the balance between antimicrobial efficacy and biocompatibility.

Functionalized Scaffolds for Tissue Engineering

Case Study: Collagen Grafted with Growth Factors for Skin Tissue Engineering [43]

Tissue engineering aims to develop scaffolds that support the regeneration of damaged tissues, and green grafting techniques have played a crucial role in advancing this field. Collagen, a natural polymer abundant in the extracellular matrix (ECM), is frequently used in tissue engineering due to its biocompatibility and biodegradability. In a recent study, collagen scaffolds were grafted with epidermal growth factor (EGF) to enhance skin regeneration. The green grafting process involved the use of water as a solvent and enzyme-catalyzed reactions, minimizing the need for toxic chemicals. The functionalized scaffolds promoted cell proliferation, angiogenesis, and wound healing *in vitro* and *in vivo*.

Impact: The collagen-EGF grafted scaffolds demonstrated enhanced healing properties, making them ideal for treating chronic wounds, such as diabetic ulcers.

The green grafting process also reduced the environmental impact of scaffold fabrication, contributing to the sustainability of tissue engineering practices.

Challenges: Despite these promising results, large-scale production of functionalized scaffolds remains difficult due to the high cost of growth factors and the complexity of grafting procedures. Additionally, the long-term stability of growth factor-functionalized scaffolds needs further investigation.

FUTURE DIRECTIONS AND CHALLENGES

While green grafting has made significant strides in the pharmaceutical and healthcare industries, the field is still in its early stages. To fully realize its potential, several key areas need to be addressed, and various challenges must be overcome [44].

Scaling Up Green Grafting Techniques: One of the primary challenges in green grafting is scaling up laboratory-based techniques for industrial applications. Many grafting processes, such as the use of biocatalysts or renewable resources, are not yet optimized for large-scale production. The difficulty lies in ensuring that the eco-friendly nature of the process is maintained while producing materials in sufficient quantities to meet industry demands. Future research should focus on developing scalable green grafting methods that maintain the efficiency and sustainability of laboratory techniques. Automation, high-throughput screening, and advances in biocatalysis may help overcome this hurdle.

Reducing Costs and Enhancing Affordability: While green grafting offers numerous advantages, including sustainability and reduced environmental impact, the cost of these processes is often higher than traditional methods. This is particularly true when using biocatalysts, renewable resources, or functionalizing materials with bioactive molecules like growth factors. The high cost of raw materials, coupled with the complexity of grafting techniques, can limit the widespread adoption of green grafting in industries like pharmaceuticals and healthcare. To address this, future research should explore ways to reduce the cost of green grafting processes. This could involve developing more cost-effective biocatalysts or finding cheaper alternatives to expensive bioactive molecules. Additionally, improving the efficiency of green grafting methods through process optimization could reduce the overall cost of production.

Enhancing Material Performance: While green grafting has already led to the development of advanced materials for drug delivery, medical devices, and tissue engineering, there is still room for improvement in terms of material performance. For example, biodegradable polymers used in drug delivery systems may degrade too quickly or too slowly, affecting the release profile of the drug. Similarly, antimicrobial coatings for medical devices may lose their effectiveness over time, leading to the recurrence of infections. To enhance material performance, future research should focus on fine-tuning the properties of grafted materials. This could involve modifying the chemical structure of polymers, optimizing the density of grafted bioactive molecules, or developing stimuli-responsive materials that can adapt to changes in the body [45].

Addressing Regulatory and Safety Concerns: As green grafting techniques become more prevalent in the pharmaceutical and healthcare industries, regulatory and safety concerns must be addressed. While green grafting typically reduces the use of toxic chemicals and non-biodegradable materials, there are still potential risks associated with some of the materials used. For example, the long-term safety of grafted nanoparticles or bioactive molecules remains a concern, particularly when used in medical devices or tissue engineering scaffolds.

Regulatory agencies, such as the FDA and EMA, will need to develop guidelines for the approval of green-grafted materials, ensuring that they meet safety and efficacy standards. Additionally, researchers must continue to investigate the long-term biocompatibility and biodegradability of grafted materials to ensure they do not cause harm to patients or the environment [46].

Expanding Applications in Healthcare: Green grafting has already demonstrated significant potential in drug delivery, medical devices, and tissue engineering, but its applications could extend even further. For example, green grafting could be used to develop new biomaterials for regenerative medicine, such as scaffolds for organ regeneration or materials for the controlled release of stem cells. Additionally, the use of green grafting in developing biosensors and diagnostic tools could lead to innovations in healthcare.

Future research should explore these emerging applications, focusing on how green grafting can be used to create sustainable, biocompatible materials that improve patient outcomes and reduce environmental impact [47].

CONCLUSION

Green grafting represents a transformative approach in the field of polymer functionalization, prioritizing environmental sustainability and efficiency. By integrating principles of green chemistry, green grafting methods reduce the reliance on hazardous chemicals, lower energy consumption, and minimize waste, thereby contributing to the development of safer and more sustainable materials. The applications of green grafting are broad, impacting pharmaceutical drug delivery systems, medical devices, and tissue engineering. These advancements not only enhance the performance and safety of medical products but also align with global efforts to reduce industrial pollution and environmental degradation. A notable achievement of green grafting is its ability to produce biodegradable polymers and biocompatible materials that offer significant advantages in drug delivery, such as improved solubility, targeted drug release, and minimized toxicity. The adoption of eco-friendly solvents, bio-based catalysts, and renewable resources in grafting processes highlights the growing importance of sustainable practices in modern pharmaceutical and healthcare industries. While green grafting has made remarkable progress, challenges remain, particularly in scaling up laboratory methods for industrial production and reducing costs. Continued research and collaboration across scientific disciplines will be essential in addressing these challenges and advancing the field of green grafting. Green grafting provides a promising pathway towards a more sustainable and environmentally friendly future in materials science, especially in the pharmaceutical and healthcare sectors.

ACKNOWLEDGEMENTS

I would like to express my deepest gratitude to the Institute of Pharmaceutical Sciences, Faculty of Pharmacy, Parul University, Vadodara, Gujarat, India, for their continuous support and encouragement throughout this research. My sincere thanks go to my co-authors, Mr. Taufik Mulla, Ms. Drashti Dave, and Dr. Kuldeep Vinchurkar, whose expertise and guidance have been invaluable to the successful completion of this work. Finally, I would like to thank my family and friends for their unwavering support throughout this research.

REFERENCES

[1]   Matyjaszewski K, Spanswick J. Controlled/living radical polymerization. Mater Today 2005 Mar; 8(3): 26–33.
[http://dx.doi.org/10.1016/S1369-7021(05)00745-5]

[2]   Michalchuk AAL, Boldyreva EV, Belenguer AM, Emmerling F, Boldyrev VV. Tribochemistry, mechanical alloying, mechanochemistry: what is in a name? Front Chem 2021; 9: 685789.
[http://dx.doi.org/10.3389/fchem.2021.685789]

[3]   Kajdas C. General approach to mechanochemistry and its relation to tribochemistry. In: Pihtili H, ed. Tribology in Engineering. Rijeka (Croatia): InTech; 2013. p. 209–40.
[http://dx.doi.org/10.5772/50507]

[4]   Rijpma D, Pijpe A, Claes K, *et al*. Outcomes of Meek micrografting versus mesh grafting: study protocol. PLOS One. 2023;18(2):e0281347.
[http://dx.doi.org/10.1371/journal.pone.0281347]

[5]   Do JL, Friščić T. Mechanochemistry: a force of synthesis. ACS Cent Sci 2017; 3(1): 13–9.
[http://dx.doi.org/10.1021/acscentsci.6b00277]

[6]   Kerton F, Marriott R. Alternative Solvents for Green Chemistry. London: The Royal Society of Chemistry; 2013.
[http://dx.doi.org/10.1039/9781849736824]

[7]   Lancaster M. Green Chemistry: An Introductory Text. 3rd ed. Cambridge: The Royal Society of

[8] Teli MD. Applications of Green Grafting in Textile and Medical Sciences. J Ind Text 2020.
[http://dx.doi.org/10.1177/1528083720913456]

[9] Pizzi D, Humphries J, Morrow JP, *et al.* Poly(2-oxazoline) macromonomers as building blocks for functional and biocompatible polymer architectures. Eur Polym J 2019 Dec; 121: 109258.
[http://dx.doi.org/10.1016/j.eurpolymj.2019.109258]

[10] Loupit G, Cookson SJ. Identifying molecular markers of successful graft union formation and compatibility. Frontiers Plant Sci 2020;11:610352.
[http://dx.doi.org/10.3389/fpls.2020.610352]

[11] Kumar P, Lucini L, Rouphael Y, Cardarelli M, Kalunke RM, Colla G. Vegetable grafting as a tool to improve drought tolerance: physiological and molecular insights. Frontiers in Plant Science. 2017;8:1130.
[http://dx.doi.org/10.3389/fpls.2017.01130]

[12] McKeen LW. Plastics in Medical Devices: Properties, Requirements, and Applications. William Andrew 2017.

[13] Sano S, Kawabata S, Horiuchi Y, *et al.* Graft-induced phenotypic changes via mobile small RNAs: experimental evidence and implications. Plant Journal. 2014;80(5):629–641.
[http://dx.doi.org/10.1111/tpj.12637]

[14] Parlett CMA, *et al.* Catalysis in Green Chemistry: A Review. Chem Soc Rev 2011.
[http://dx.doi.org/10.1039/c0cs00195c]

[15] Sutter EJ, Desveaux D, Foster KR. Long-distance movement of signaling peptides and their roles in graft compatibility. Plant Signaling & Behavior. 2016;11(9):e1234567.
[http://dx.doi.org/10.1080/15592324.2016.1234567]

[16] Anastas, P.T., Warner, J.C. Green Chemistry: Theory and Practice. Oxford University Press, 1998.

[17] Rouphael Y, Kyriacou MC, Colla G. Vegetable grafting: a toolbox for securing yield stability under multiple stress conditions. Frontiers in Plant Science. 2018;8:2255.
[http://dx.doi.org/10.3389/fpls.2017.02255]

[18] Sheldon RA. Green Chemistry and Catalysis. J Chem Soc 2005.
[http://dx.doi.org/10.1039/b418069k]

[19] Wang MR, Bettoni JC, Zhang AL, Lu X, Zhang D, Wang QC. In vitro micrografting of horticultural plants: method development and the use for micropropagation. Horticulturae. 2022;8(7):576.
[http://dx.doi.org/10.3390/horticulturae8070576]

[20] Assunção M, *et al.* Understanding the molecular mechanisms underlying grafting: perspectives and challenges. BMC Plant Biol. 2019;19: (article).
[http://dx.doi.org/10.1186/s12870-019-1967-8]

[21] Isikgor FH, Becker CR. Green Polymers and Biopolymers: Current Research Trends. Mater Chem Phys 2015.
[http://dx.doi.org/10.1016/j.matchemphys.2015.05.001]

[22] Allen MJ, *et al.* Sustainable Biomaterials for Medical Devices. Nat Biomed Eng 2018.
[http://dx.doi.org/10.1038/s41551-018-0232-3]

[23] Rahimi A, *et al.* Recycling and Upcycling of Polymers through Green Chemistry. ACS Sustain Chem& Eng 2019.
[http://dx.doi.org/10.1021/acssuschemeng.9b01246]

[24] Luo Y, Wu Y, Huang C, Menon C, Feng SP, Chu PK. Plasma modified and tailored defective electrocatalysts for water electrolysis and hydrogen fuel cells. Eco Mat 2022 Mar 07; 4(4): e12197.
[http://dx.doi.org/10.1002/eom2.12197]

[25] Geyer R, Jambeck JR, Law KL. Production, use, and fate of all plastics ever made. Sci Adv 2017; 3(7): e1700782.
[http://dx.doi.org/10.1126/sciadv.1700782] [PMID: 28776036]

[26] Sun X, Bao J, Li K, *et al.* Advance in using plasma technology for modification or fabrication of carbon-based materials and their applications in environmental, material, and energy fields. Adv Funct Mater 2021 Feb 10; 31(7): 2006287.
[http://dx.doi.org/10.1002/adfm.202006287]

[27] Notaguchi M, Kurotani KI, Sato Y *et al.* T. Cell-cell adhesion in plant grafting is facilitated by β-1,--glucanases. Science 2020; 369 (6504): 698-702.
[http://dx.doi.org/10.1126/science.abc3710]

[28] Primc G. Recent advances in surface activation of polytetrafluoroethylene (PTFE) by gaseous plasma treatments. Polymers (Basel) 2020 Oct 7; 12(10): 2295.
[http://dx.doi.org/10.3390/polym12102295]

[29] Hati S, Patel M, Yadav D. Food bioprocessing by non-thermal plasma technology. Curr Opin Food Sci 2018; 19: 85–91.
[http://dx.doi.org/10.1016/j.cofs.2018.03.011]

[30] Di L, Zhang J, Zhang X, *et al.* Cold plasma treatment of catalytic materials: a review. J Phys D Appl Phys 2021; 54(33): 333001.
[http://dx.doi.org/10.1088/1361-6463/ac0269]

[31] Zaborniak I, Chmielarz P. Polymer-modified regenerated cellulose membranes: following the atom transfer radical polymerization concepts consistent with the principles of green chemistry. Cellulose 2023; 30: 1–38.
[http://dx.doi.org/10.1007/s10570-022-04880-4]

[32] Park SJ, Lee SY, Jin FL. Surface modification of carbon nanotubes for high-performance polymer composites. In: Kar K, Pandey J, Rana S, eds. Handbook of Polymer Nanocomposites. Processing, performance and application; 2015. p. 13–59.
[http://dx.doi.org/10.1007/978-3-642-45229-1_34]

[33] Liu K, Zhang JJ, Cheng FF, Zheng TT, Wang C, Zhu JJ. Green and facile synthesis of highly biocompatible graphene nanosheets and its application for cellular imaging and drug delivery. J Mater Chem. 2011; 21(32): 12034–12040.
[http://dx.doi.org/10.1039/C1JM10749F]

[34] Koo JH. Polymer nanocomposites. New York: McGraw-Hill; 2006.

[35] Dang M, Deng Q, Fang G, Zhang D, Liu J, Wang S. Preparation of novel anionic polymeric ionic liquid materials and their potential application to protein adsorption. J Mater Chem B. 2017; 5(31): 6339–6347.
[http://dx.doi.org/10.1039/C7TB01234A]

[36] Weerasinghe UA, Wu T, Chee PL, *et al.* Deep eutectic solvents towards green polymeric materials. Green Chem 2024; 26: 8497–527.
[http://dx.doi.org/10.1039/D4GC00532E]

[37] Wang J, Wang D, Du W, *et al.* Synthesis of functional polypropylene via solid-phase grafting soft vinyl monomer and its mechanism. J Appl Polym Sci 2009 Aug 05; 113(3): 1803–10.
[http://dx.doi.org/10.1002/app.30179]

[38] Baláž M. Environmental mechanochemistry: recycling waste into materials using high-energy ball milling. Cham (Switzerland): Springer International Publishing; 2021.
[http://dx.doi.org/10.1007/978-3-030-75224-8]

[39] El-Eskandarany MS. Mechanical alloying. 2nd ed. Amsterdam: Elsevier; 2015.

[40] Tavares, L.M. A Review of Advanced Ball Mill Modelling. KONA Powder Part J 2017; 34, 106-124.

[http://dx.doi.org/10.14356/kona.2017015]

[41] Mülhaupt R. Green polymer chemistry and bio-based plastics: opportunities and challenges. Macromol Chem Phys 2013; 214(2): 159–74.
[http://dx.doi.org/10.1002/macp.201200439]

[42] Melnyk CW. Plant grafting: insights into tissue regeneration. Regeneration. 2016;4(1):3–13.
[http://dx.doi.org/10.1002/reg2.71]

[43] Liland KH, Øvrebø B, Munz M. An integrated approach to predicting graft compatibility using anatomical, biochemical and transcriptomic markers. Front Plant Sci 2021;12:667482.
[http://dx.doi.org/10.3389/fpls.2021.667482]

[44] Kundu A, Venkatesh S, Sharma M, *et al*. Compatible graft establishment in fruit trees and its potential markers. Agronomy. 2022;12(8):1981.
[http://dx.doi.org/10.3390/agronomy12081981]

[45] Colla G, Cardarelli M, Lucini L. Use of grafting to improve ion uptake and salinity tolerance in horticultural crops. Front Plant Sci 2016;7:1434.
[http://dx.doi.org/10.3389/fpls.2016.01434]

[46] He W, Wang Y, Chen Q, Wang X. Dissection of the mechanism for compatible and incompatible graft combinations of *Citrus grandis (L.)* ('Hongmian Miyou'). Int J Mol Sci. 2018;19(2):505.
[http://dx.doi.org/10.3390/ijms19020505]

[47] Liu S, Zhou Y, Wang J. Eco-friendly strategies for polymer surface modification: a review. Prog Org Coat 2021; 156: 106265.

CHAPTER 2

Fundamentals of Polymer Grafting: Basics and Techniques

Akanksha Phondke[1], Meghana Jagtap[2], Priya Patne[2], Popat Mohite[1,*], Premlata Ambre[2], Sudarshan Singh[3] and Kuldeep Vinchurkar[4]

[1] *Department of Pharmaceutical Chemistry, AETs St. John Institute of Pharmacy and Research, Palghar, Maharashtra 401404, India*

[2] *Department of Pharmaceutical Chemistry, Bombay College of Pharmacy, Kalina, Santacruz (E), Mumbai, Maharashtra 400098, India*

[3] *Faculty of Pharmacy, Chiang Mai University, Chiang Mai 50200, Thailand*

[4] *Department of Pharmaceutics, Sandip Institute of Pharmaceutical Sciences (SIPS), Affiliated to Savitribai Phule Pune University (SPPU, Pune), Nashik, Maharashtra 422213, India*

Abstract: Polymers are gaining more attention due to a diverse range of applications. Crosslinking, co-polymerization, curing, mixing, and blending are various methods of polymer modification. However, the most effective and appealing method for enhancing the morphology, chemical composition, and physical properties of polymers is polymer grafting. Grafting permits modification of the parent structure, which enhances physicochemical properties, compatibility, thermal stability, multiphase response, flexibility, and reactivity of the parent polymer. Grafting includes the insertion of a functional group into the polymer backbone, grafting a sidechain to the polymer backbone, and combining one or more monomers on the polymer with or without an initiator by employing a variety of chemical, biological, and physical activators. The different technique employed in polymer grafting involves ionic grafting, radical generation using free radical polymerization (FRP), reversible addition-fragmentation chain transfer (RAFT), and nitroxide-mediated polymerization (NMP), as well as physical methods like plasma irradiation, UV radiation, photochemical grafting, and biological methods like enzymatic grafting. Polymers functionalized using the grafting technique can be employed in various fields like pharmaceuticals, drug delivery applications, wastewater treatment, tissue engineering, diagnosis, the textile industry, biotechnology, and other fields.

Keywords: Enzymatic polymerization, Free radical polymerization, Grafting polymer, Initiator, Plasma irradiation.

---

* **Corresponding author Popat Mohite:** Department of Pharmaceutical Chemistry, AETs St. John Institute of Pharmacy and Research, Palghar, Maharashtra 401404, India; E-mail: mohitepb@gmail.com

Kuldeep Vinchurkar, Satish Polshettiwar, Nilesh Mahajan &Yogeshwar Bachhav (Eds.)
All rights reserved-© 2026 Bentham Science Publishers

INTRODUCTION

Polymers are made up of chemically identical units called monomers. Modified polymers have been gaining greater attention in pharmaceuticals, material sciences, textiles, and electronic fields due to their unique applications. Often, polymers are modified by altering their surfaces or by introducing end-terminal functional groups. There are various methods for altering polymer properties, including copolymer synthesis, crosslinking, polymer blending, polymer interpenetrating networks, and grafting. Among these methods, the grafting technique is widely used to alter the physicochemical properties of a polymer. When two or more different monomers are combined to form a polymer, it is called a copolymer, whereas a graft copolymer is created by the process of connecting one or more homopolymer blocks as branches to a primary polymer chain. As a consequence, a branched configuration is formed, where the main backbone is connected to separate homopolymers that serve as side chains [1]. When a graft copolymer has only one branch, it is referred to as a *miktoarm* star copolymer. These graft copolymers might be heteropolymers, which have distinct chemical structures, or homopolymers, which have the same chemical composition as the main polymer chains [2, 3]. Whereas when it is composed of two or more branches, it is referred to as a multi-arm star-shaped copolymer [4]. Grafting is a desirable method for adding different functional groups to a polymer [5]. Therefore, numerous techniques have been demonstrated to enhance the fundamental properties of the polymer backbone under specific conditions [6]. Graft copolymers are primarily utilized for altering polymer characteristics due to their distinct mechanical and thermal properties [2, 3]. Graft polymerization modifies the parent polymer, enabling the adjustment of specific properties without altering its core structure. This technique is commonly applied to both natural and synthetic polymers. Grafted polymers provide a simple method for introducing new properties while maintaining the original qualities of the base polymer. As a result, these polymers play a vital role in modifying physicochemical properties. This chapter discusses various approaches to graft polymerization, key parameters for controlling graft reactions, and recent advancements in the process, with relevant examples provided in each section.

POLYMER GRAFTING APPROACHES

Grafting of polymers is carried out using three main approaches: "grafting through," "grafting on," and "grafting from". The following section covers the details of grafting approaches using the diagrammatic representation illustrated in Fig. (**1**).

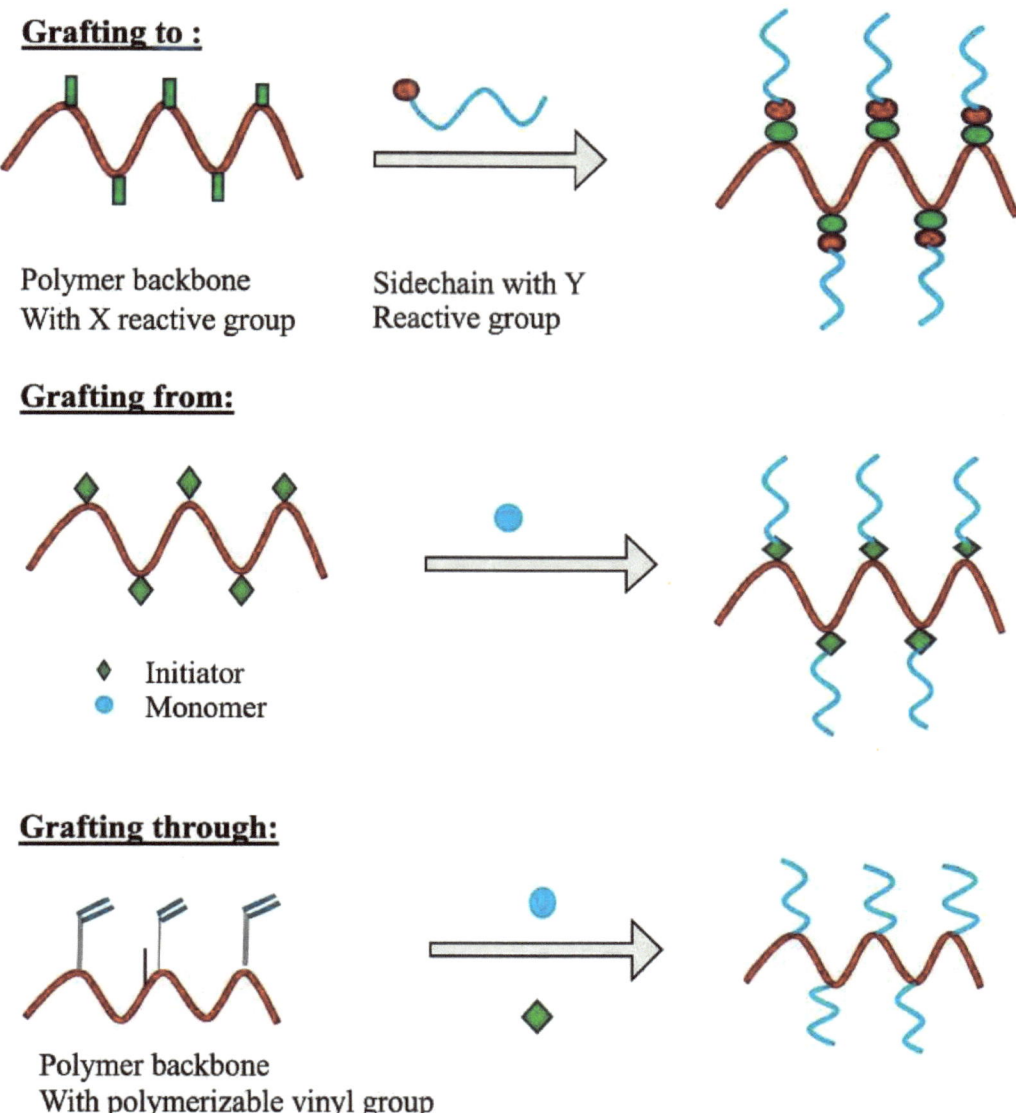

Fig. (1). Illustration of polymer grafting approaches.

Grafting onto (Grafting to)

In this approach, the polymer backbone is reacted with a single monomer. A monomer with a functional group is grafted to the single unit of polymer backbone that has a complementary functional group, which is suitable for attachment of the monomer [7]. A covalent bond is formed between the polymer backbone and monomer functional groups *via* chemical reaction. This approach is primarily employed for the synthesis of reactive polymers. These polymers are typically synthesized *via* radical polymerization, cationic/anionic polymerization, living polymerization, and other methods [8]. Simple examples include the chloro or bromo methylation reaction in polystyrene (PS). The chloromethylation reaction of PS is carried out with Tin (IV) chloride ($SnCl_4$) as a catalyst, carbon tetrachloride (CCl_4) is used as a solvent, and chloromethyl methyl ether as a reactive side chain. Fig. **(2)** represents the chloromethyl reaction of PS. This technique has been utilized for the synthesis of graft polymer PS-g-poly(ethylene oxide). The degree of chloromethylation of polystyrene was controlled to be less than 10% before grafting another polymer, polyethylene oxide (PEO).

Fig. (2). Chloromethylation reaction of polystyrene [7].

Grafting from (Surface-initiated Polymerization)

This type of grafting allows us to graft more than one monomer onto the polymer backbone in a single step. It requires three components: a polymer backbone, which contains an initiator that acts as an anchor site for the attachment of monomer. Once the monomer is attached to the active site, it allows the growth of the graft chain on the polymer backbone. Sequential grafting and simultaneous grafting approaches are used to obtain "grafting from" polymers. Fig. **(1)** depicts the "grafting from" technique. Highly dense Polymer grafting can be achieved by using appropriate polymer brushes and controlling the number of active sites on the polymer backbone.

Atom transfer radical polymerization (ATRP) is the best example of this grafting from approach. Lacerda *et al.* in 2020 synthesized Three-Dimensional (3D) printing material by grafting poly(Methyl Methacrylate) (PMMA) from the Cork powder surface by ATRP [8]. They have modified cork residue by grafting a technique to enhance thermal and mechanical properties. In this, cork is reacted with atom radical initiator 2-bromoisobutyryl (BiB) moieties. The monomer of methyl methacrylate (MMA) is grafted onto cork by replacing isobutyryl moieties. Cork-attached BiB is used as a macroinitiator, and radical polymerization is performed under an inert atmosphere using solvent water in DMF (80:20) at 50°C. Cork-g-poly(Methyl Methacrylate) polymeric material has a smooth surface and can be utilized to produce high-quality 3D printed products.

Grafting Through (Macromonomer Method)

Macromonomers are low molecular weight (LMW) polymer chains or oligomers that are used to carry out polymerization because they contain functional groups. Vinyl groups are more preferred in the "grafting through" polymerization method. The addition of a co-monomer, either with or without an initiator, results in the formation of various complex structures of graft polymers. LMW monomer is radically copolymerized with methacrylate-type functionalized macromonomer in a controlled manner. Macromonomers used for incorporating into polystyrene or polymethacrylate backbones include PEO, polycaprolactone (PCL), polylactic acid (PLA), polyethylene, and polysiloxane, among others. These polymers are polymerized under controlled conditions. Pizzi *et al.* in 2019 synthesized a graft polymer poly(methyl methacrylate-g-α-methyl-β-alanine) [PMMA-g-mBA] using a grafting-through approach [9]. The poly(α-methyl-β-alanine) (PmBA) polymer was synthesized using sodium tertiary butoxide (anionic initiator) and methacrylamide monomer. The polymer PmBA with a smaller molecular weight (oligomer) possesses olefinic functional groups at the end terminal, which act as a macromonomer. In the second step, poly(MMA-g-mBA) was synthesized using a grafting-through approach by incorporating MMA onto macromonomer PmBA. PmBA and poly(MMA-g-mBA) synthesis are discussed in the Fig. (3). A specified amount of PmBA macromonomer, MMA monomer, and azobisisobutyronitrile (AIBN) initiator is reacted for 24 hours at 70°C. This allows for the simultaneous polymerization of MMA and its attachment to PmBA, resulting in the synthesis of poly(MMA-g-mBA) graft polymer.

TECHNIQUES OF POLYMER GRAFTING

The process of combining monomers and polymers is elucidated by methods such as "Physisorption," "grafting," and "crosslinking". The term physisorption refers to a process that is driven by physical attractive forces. Physisorption is a surface phenomenon that involves functionalizing the surface of a polymeric backbone. Because this is solely caused by deposition on polymeric material, which causes weak interaction with the surface, these reactions are reversible. The process of grafting entails forming covalent bonds between functionalized monomers, which serve as side chains, and the polymeric backbone. Grafting is considered irreversible because it results in the alteration of different physical, chemical, and biological characteristics of the polymer. The bonding of polymers through a chemical bond is referred to as "crosslinking." Typically, crosslinking is an irreversible process that can occur within the same or between different polymer chains. The following sections describe the chemical, physical, and biological graft techniques used during polymerization.

Step 1- Synthesis of poly(α- methyl- β- alanine) [PmBA] macromonomer with olefinic end group

Step 2 - Graft copolymerization of methyl methacrylate and poly(α- methyl- β- alanine)

Fig. (3). Synthesis of Poly(methyl methacrylate-g-α-methyl-β-alanine) [PMMA-g-PmBA)] graft copolymer [10].

Chemical Grafting

Chemical grafting is a process that involves modifying the surface of a polymer or its backbone by adding a specific functional group or molecule. This process imparts properties to the polymer such as biocompatibility, solubility, and chemical resistance. Chemical grafting reactions involve either free radical generation or ion generation, and are achieved through the living polymerization method [11].

Free Radical Polymerization (FRP)

In this type of polymerization, free radicals are formed using an initiator, and it is transferred to the substrate, where they react with the monomer to produce a graft copolymer. FRP techniques are subclassified as a) Conventional Free radical polymerization (CFRP), b) Reversible deactivation radical polymerization (RDRP), 3) Reactive extrusion polymerization (REP) methods. CFRP works *via* the initiation step, propagation step, and termination step. Ammonium persulfate (APS), AIBN, Ceric ammonium nitrate (CAN), Potassium persulfate (KPS) are the frequently used initiators for FRP. Preferably, a free radical initiator decomposes during the polymerization process while it remains stable at room temperature. Because variations in the solvent/monomer mixture can significantly affect the rate of decomposition of initiators, the choice of an appropriate initiator is dependent on the solvent/monomer system being employed. CFRP of vinyl chloride using AIBN has been discussed in Fig. (**4i**). Betraoui *et al.* in 2023 have [12] reported the synthesis of FRP grafts using acrylic acid -g- agar (AA-g-Agar) and acrylamide-g-agar (AAm-g-agar). Ammonium peroxodisulfate was used as an initiator. The graft hydrogels were analyzed for their competence to adsorb methylene blue dye, and the results indicated superior adsorption properties for AA-g-Agar than AAm-g-Agar. The flexibility and ease of fabrication that polymers provide through the FRP method make it most significant on an industrial scale. However, FRP is constrained by the generation of grafted polymers that exhibit a wide range of size distribution. Additionally, the termination process may result in the formation of unreactive polymers.

Reversible Deactivation Radical Polymerization

It is a controlled/living radical polymerization technology that provides definite control over the molecular weight distribution and structure of polymers. This technology relies on a reversible deactivation process, wherein the polymer chains that are actively progressing are temporarily rendered "dormant" or "inactive" through a reversible reaction. This technique minimizes the incidence of chain termination and enables more precise polymer synthesis. There are three main categories of RDRP techniques: ATRP or metal-catalyzed living radical polymerization, RAFT, and NMP [13]. ATRP is a polymerization technique that employs a metal catalyst, often copper, to facilitate the reversible activation and deactivation of polymer chains.

i) Conventional Free radical polymerization (CFRP)

vinyl chloride → (AIBN) → poly (vinyl chloride)

ii) Reversible addition fragmentation chain transfer (RAFT) polymerization

vinylbenzyl chloride + benzyl ethyl trithiocarbonate (RAFT agent) → (AIBN) → poly (vinylbenzyl chloride)

iii) Radical polymerization of styrene using (TEMPO) Nitroxide-Mediated Polymerization (NMP)

Fig. (4). Reactions of different types of FRP.

Reversible Addition-Fragmentation Chain Transfer (RAFT)

It is a method of forming polymer chains by utilizing chain transfer agents, such as dithiocarbamates, xanthates, trithiocarbonates, or dithiobenzoate to control their production [14]. Hernandez *et al.* in 2024 synthesized a molecularly imprinted polymer (MIP) platform for smartphone-based tartrazine detection in carbonated drinks on polyethylene terephthalate (PET) [15]. The MIP-PET platform was synthesized using RAFT agent cumyl dithiobenzoate, which regulates the growth of the polymer chain, whereas potassium persulphate commences polymerization of the acrylamide monomer, resulting in a material with improved selectivity and high adsorption capacity. The RAFT polymerization system is environmentally friendly, which does not rely on a metal catalyst, and is suitable for the synthesis of polymers for biomedical and electronic materials. Polymerization of vinylbenzyl chloride using AIBN and benzyl ethyl trithiocarbonate (BET) with controlled molecular weight was obtained using RAFT polymerization. Refer to Fig. (**4**) ii for details. This method allows for the synthesis of complex polymer structures.

Nitroxide-Mediated Polymerization

It is a method that employs stable nitroxide radicals to inactivate the growing radical chain ends reversibly. It helps to control stereochemistry, narrow distribution, and control the MW of the polymer. These methods facilitate the production of polymers with accurate configurations, minimal deviations in molecular weight, and complicated arrangements, all of which are essential for the utilization of sophisticated materials. The initiators include 2,2,6,6-tetramethyl-piperidine-1-oxyl (TEMPO), Styryl-TEMPO, 2,2,5-trimethyl-4-phen-1-3-azahexane-3-oxyl (TIPNO), Styryl-TIPNO, 2,2,5-trimethyl-4-tert-butyl-3-azahexane-3-oxyl (TITNO), N-(tert-butyl)-N-(1-diethylphosphono-2,2-dimethyl-propyl) nitroxide (SG1), Dispolreg 007 (D7) [16]. In 2023, Roggi *et al.* modified commercial poly (styrene-b-butadiene-b-styrene) (SBS) by introducing benzoyl peroxide (BPO) and TEMPO through free-radical activation [17]. Both vinylbenzyl chloride (VBC) and styrene/VBC random copolymer chains were grafted from SBS using the obtained macroinitiator. It was observed that the styrene/VBC-grafted copolymer exhibited greater mechanical resistance compared to the corresponding graft copolymer without the styrene component. NMP gives us a colourless, metal-free polymer that does not need to be purified after synthesis. The main drawbacks of NMP are typically lower polymerization rates, higher temperatures required, and limitations on monomer use. Radical polymerization of styrene using TEMPO and BPO initiator leads to a narrow molecular weight distribution. Refer to Fig. (**4**) iii for details.

Reactive Extrusion Polymerization (REP)

It is a method that enhances reaction rates by increasing the concentration of monomers and initiators or catalysts. It enables polymerization at elevated temperatures without the need to be concerned about the evaporation of solvents or pressure. Screw extruders have been designed to process highly viscous substances, rendering them valuable for continuous polymerization reactions without the requirement of solvents. Nevertheless, screw extruders may lack effectiveness during the first phases of polymerization when the material has not yet attained sufficient viscosity for the screws to provide a consistent flow. For these situations, it is common to utilize a combination of stirred tank reactors and a twin-screw extruder, or many extruders arranged in a sequence. This technique has the ability to improve various polymerization processes that do not require solvents, such as free-radical, anionic, ring-opening polymerization, and polycondensation, while simultaneously addressing the associated challenges [18]. Plastics undergo multiple processing steps when it comes to recycling. The process involves melting and extruding the material multiple times, which can degrade the plastic and affect its quality. To prevent this degradation, specific additives like chain extenders are employed. Those that are designed to cause post-polymerization after processing can be utilized to mitigate material degradation. Schall *et al.* 2023 explored the effectiveness of linear chain extenders in polyamides [19]. They repeated the processing of a polyamide-6 material, and after each cycle, they examined its molar mass distribution. Initially, the polyamide was regranulated five times in a twin-screw extruder without any additives and then again with varying quantities of Bruggolen M1251 chain extender. The findings indicated that incorporating Bruggolen M1251 chain extender in a closed cycle of polyamide material maintained its average molecular weight. The absence of chain extenders showed deterioration of the material at some point. Reactive extrusion involves chemical modification of the polymer during the extrusion process. This technique has been employed to improve the characteristics of oil-based biopolymers such as PLA, polyhydroxyalkanoates (PHAs), and poly (butylene succinate) (PBS).

Ionic Grafting

Ionic grafting (IG) is a polymerization technique in which ions or ion pairs function as the active centers for the formation of polymeric chains. When positively or negatively charged active centers are utilized for polymerization, they are classified as cationic or anionic polymerization, respectively [20]. IG polymerization offers the synthesis of a polymer with a precise molecular weight. However, it necessitates stringent conditions, including an anhydrous environment, the elimination of oxygen, the use of high-quality chemicals, and extremely low temperatures for the reaction. Consequently, a certain number of reports are available that discuss the process of ionic graft polymerization. The polymerization process begins with the formation of reactive ions. In the polymerization process, monomers with electron-releasing groups form stable cations, while monomers with electron-withdrawing groups form stable anions. Unlike radical polymerization, which is mainly determined by monomer chemistry and radical stability, the rate of reaction is significantly influenced by reaction conditions. Impure monomers would lead to early termination, whereas the use of polar solvents greatly affects the rate of reaction. Ion pairs that are loosely coordinated and solvated lead to more reactive, fast-polymerizing chains, but they can also disrupt the polymerization process in other ways, such as by destroying the growing chain or interacting with the initiator ions, making them less commonly used. Ionic polymerization typically uses nonpolar solvents, such as pentane, or moderately polar solvents, like chloroform [11].

Cationic Polymerization

In the process of ionic polymerization, an initiator is typically referred to as a catalyst. Upon polymerization, the initiator molecules or their fragments are integrated into the generated chains. In cationic polymerization, the use of a cocatalyst is mandatory, as a single catalyst alone is insufficient. Strong acids and their salts, including sulfuric acid and perchloric acid, as well as potassium hydrogen sulfate (KHSO4), are used as initiators. However, HCl cannot be used as an initiator because the chloride counterpart is unstable. Lewis acids, including boron trifluoride (BF_3), tin tetrachloride ($SnCl_4$), aluminum chloride ($AlCl_3$), and titanium tetrachloride ($TiCl_4$), are used as initiators. However, they are required to be used with water or alcohol as a co-catalyst [21]. IG polymerizations exhibit significantly higher reaction rates compared to FRP due to the presence of a million times higher concentration of active centers than the concentration of free radicals in radical polymerization. The cationic/anionic active centers in the propagation step are accompanied by a counterion of opposite charge. Therefore, the stereochemistry and rate of polymerization are significantly dependent on the ionic centers and type of counterion. Additionally, the polarity of the solvent and the solvation of the counterion also have a great impact on IG. In the termination step, a reaction between two ionic centers cannot occur due to the repulsion between like charges. Fig. (**5**) summarizes all three steps of cationic polymerization. Consequently, termination occurs through a chain transfer reaction with the monomer or solvent [21].

It is necessary to optimize the amount of protonic cocatalyst in the cationic polymerization that LA is catalyzing. To initiate the polymerization, an optimum level is needed; otherwise, an excess amount leads to a low molecular weight polymer.

Living Cationic Polymerization – It involves precise and controlled initiation and propagation, which minimizes unwanted reactions and termination. It is only possible for a specific subset of monomers. Vinyl ethers, isobutylene, styrene, and N-vinyl carbazole are examples of electron-rich alkenes that can be used as monomers in this process. Generally, it is challenging to produce a stable carbocation for a long time because β-protons may quench the cation in a free monomer. Living polymerization offers a technique for synthesizing extremely high-molecular-weight polymers. Polymerization without termination can be initiated in two ways: through the use of organometallic initiators or by electron transfer. The living polymerization reaction occurs in several ways, including the elimination of chain transfer and termination reactions, and the initiation rate is significantly greater than the propagation rate [23]. The result is that the polymer chain grows steadily, and its length remains relatively similar to traditional chain polymerization, resulting in a very low polydispersity index. Controlled polymerization and live polymerization reactions are sometimes confused or conflated. The definitions of these two polymerization reactions differ, despite their high similarity. Living polymerizations are characterized by polymerizations where chain transfer or termination is eliminated, whereas controlled polymerizations include minimizing termination but not eliminating it by introducing a dormant state in the polymer. Eri Mishima *et al.* in 2011 reported cationic living polymerization using organotellurium compounds as initiators. Organotellurium compounds are effective initiators for living cationic polymerization [24]. In Vinyl ether polymerization, a polymer is synthesized with a preset MW and a narrow MW distribution when a Lewis acid and an organotellurium initiator are present. Through the use of vinyl ethers in end-group transformation and cationic polymerization, block copolymers can be produced in a single pot by adding monomers sequentially without requiring the macroinitiator to be purified.

Step I- Initiation

$$BF_3 + H_2O \rightleftharpoons H^+ BF_3OH^-$$

A total dissociation of catalyst is rare. It exists an equilibrium.

$$[H^+ BF_3OH^-] = K_{eq} [LA] [H_2O]$$

Where, K_{eq} = equilibrium constant, LA = Lewis acid. The ideal ratio of catalyst/cocatalyst depends on the system used and the solvent used.

$$H^+ (BF_3OH)^- + CH_2=CR_1R_2 \longrightarrow CH_3-C^+R_1R_2(BF_3OH)^-$$

Where H^+ = electrophile & $(BF_3OH)^-$ = counterion

Step II- Propagation

$$CH_3-C^+R_1R_2(BF_3OH)^- + CH_2=CR_1R_2 \longrightarrow CH_3-CR_1R_2-CH_2-C^+R_1R_2(BF_3OH)^-$$

Step III- Termination

Termination occur by unimolecular rearrangement of ion pairs

$$CH_3-C^+R_1R_2(BF_3OH)^- \longrightarrow CH_2=CR_1R_2-CH_2 \ H^+(BF_3OH)^-$$

$$CH_3-C^+R_1R_2(BF_3OH)^- + H_2O \longrightarrow CH_2=CR_1R_2 + H^+(BF_3OH)^-$$

Termination occurr through chain transfer to monomer

$$CH_3-C^+R_1R_2(BF_3OH)^- + CH_2=CR_1R_2 \longrightarrow CH_2=CR_1R_2 + CH_3-C^+R_1R_2(BF_3OH)^-$$

Fig. (5). Cationic polymerization (Initiation, propagation, termination) [22]

Anionic Polymerization/Anionic Living Polymerization

The prerequisite for preparing anionic polymers is to start a reaction with a monomer composed of an electron-withdrawing group, such as a nitrile, carboxyl, diene, or phenyl. It is a chain growth process in which the active sites to which monomers are sequentially added are negative ions combined with positive counterions. The extent of interaction between the macromolecular anion and its counterion relies on the specific ions and the medium in which the polymerization occurs [22]. While FRP and cationic polymerizations include termination steps, anionic polymerization is uniquely characterized by the absence of a termination step, as shown in Fig. (**6**). In anionic polymerization, the rearrangement of the ion pair does not occur as it requires the elimination of a hydride ion, which is highly unfavourable during the termination process. In the propagating step, the active carbanionic centres are present at the end groups. At this point, if more monomers are added, the propagation reaction continues until all the monomers are consumed completely. The term "living polymers" refers to a reaction that exhibits an identical reaction each time a monomer is added to the anionic polymer. HMWP is synthesized using the living method, and during the process of styrene polymerization, a potassium amide and ammonia solution initiates the reaction. The initiation reaction starts with the dissociation of potassium amide. Following this, the amide ion adds to styrene. The termination process occurs when a proton is abstracted from ammonia, which is a reaction that transfers the chain to the solvent. This chain transfer reaction generates an amide ion, which has the potential to initiate a new chain.

Anionic polymerization can be initiated by acrylates, butadiene, hexamethyl cyclotrisiloxane, lactones, methacrylate, and styrene monomers. Christoph Hahn *et al.* in 2024 demonstrated the quantitative conversion of ferulic acid to produce 4-vinylguaiacol (VG) and innovative hydroxyl-protected styrene monomers obtained from biobased resources [25]. They utilized silyl and acetal protective groups to safeguard the phenols of VG. By subjecting these protected phenolic styrene derivatives to living anionic polymerization in tetrahydrofuran (THF) at low temperatures, the researchers successfully generated the polymers with controlled and narrow molecular weights [26].

Physical Grafting

Sometimes polymers are difficult to functionalize *via* chemical methods due to insufficient functional groups and reactivity issues. Consequently, physical methods are employed for safer, selective grafting, reducing chemical use, and preventing toxic byproducts. Physical methods of grafting involve surface modification and the grafting of polymers without the use of chemicals, achieved through physical activators such as plasma irradiation, UV, and photochemical radiation.

Initiation -

$$KNH_2 \rightleftharpoons K^+ + N^-H_2$$

$$N^-H_2 + CH_2=CHPh \xrightarrow{Ki} H_2NCH_2-C^-HPh$$

Where, Ki = initiation rate constant

Propagation -

$$H_2N-(-CH_2-CHPh-)_{n-1}-CH_2-CHPh + CH_2=CHPh$$
$$\downarrow Kp$$
$$H_2N-(-CH_2-CHPh-)_n-CH_2-CHPh$$

Kp= propagation rate constant

Chain Transfer -

$$H_2N-(-CH_2-CHPh-)_n-CH_2-CHPh + NH_3$$
$$\downarrow Ktr$$
$$H_2N-(-CH_2-CHPh-)_n-CH_2-CH_2Ph + N^-H_2$$

Ktr= transfer rate constant

Fig. (6). Anionic polymerization (Initiation, propagation, chain transfer process) [28].

Photochemical Grafting

The grafting process starts with the generation of radiation, leading to the excitation of macromolecules and the formation of free radicals. In most cases, the absorption of radiation is carried out by the chromophore group, causing the dissociation of chemical bonds due to excitation. Photochemical grafting employs low-energy radiation, specifically using UV and visible light. However, in some instances, a photosensitizing agent is necessary to initiate the photochemical process. Compared to traditional thermal polymerization, this technique is more environmentally friendly as it reduces volatile organic compound emissions, requires less energy and heat, and allows for rapid polymerization. In photopolymerization grafting, liquid monomers or oligomers are converted into solid materials through light irradiation, which includes UV-visible rays or near-IR rays. Oligomers are HMWP that serve as the backbone in the grafting reaction. The photoinitiator or photoactivator is a crucial component for initiating polymerization as it is responsible for generating active species when exposed to light. When irradiated with light, the photoinitiator absorbs light and undergoes an electronic transition, creating free radicals that react with the monomeric and oligomeric units to initiate polymerization [27]. Table 1 illustrates examples of photoinitiators and photosensitizers. Photosensitizers are unique materials known for their exceptional light absorption properties. When excited, photosensitizers undergo electron or energy transfer reactions with photoinitiators, leading to the generation of active species that kick-start the photopolymerization grafting process. Photochemical polymers can be prepared with or without using photosensitizers [11]. Photochemical reactions also occur in the absence of a photosensitizer. The polymerization process occurs by dissociation of the C-C single bond, which generates free radicals. Then these free radicals interact with free radicals of the individual unit to produce a grafting unit. In photosensitizers, it generates free radicals that can diffuse, removing hydrogen atoms from the base polymer, which creates radical sites necessary for the grafting process [28]. On the other hand, photoactivator induces the creation of free radicals, which in turn drive the abstraction of a proton from the polymer bearing donor functional groups. Such polymers often contain amino, thiol, or alcohol groups with α-hydrogen [29].

Table 1. Examples of photoinitiators and photosensitizers used in photochemical grafting.

-	Photoinitiator [30]	Photosensitizers [31, 32]
I.	UV-based photo initiator: Benzyl, benzoin's and oxime's ester, methyl benzoyl formate, 2,2-dimethoxy-2-phenylacetophenon	Organometallic photosensitizers: Chlorophyll A, Tris(2-phenylpyridine) iridium, chromium acetylacetonate, Ruthenium(2,2-bipyridine)
II.	Cationic photoinitiator (photo acid generator):	Organic photoinitiators: Benzophenone, methylene blue, rose Bengal, Flavins, pterin
III.	Ionic type: Aromatic diazonium salts, diaryliodonium salt, triaryl sulfonium salt Non-ionic type: triazine, sulphonic ester	Nanomaterial: Quantum dots, nanorods
IV.	Visible light photoinitiator: Coumarin-based oxime esters, naphthylamide oxygen ether, anthraquinone derivatives, cyclohexane derivative	Thioxanthene derivatives: (2,4-diethylthixanthane, isopropyl thioxanthane).

Photochemical grafting employs low-energy radiation, specifically UV and visible light. Ngo *et al.* in 2016 grafted acrylic acid (AA) onto composite film of polyamide (TFC-PA) through photo-induced graft polymerization [33]. UV light was used as a photoinducer source that absorbed on the TFC-PA membrane. The hydrogen from the TFA-PA membrane gets separated and forms a free radical. The initiated free radicals help to graft AA onto the TFA-PA membrane. The grafted polymer, which contains a carboxylic acid functional group, contributed to the hydrophilicity property of the photochemically synthesized graft polymer and was used to reduce the fouling properties of the polymer. In 2021, Whba *et al.* performed a radical photopolymerization process to graft acrylonitrile onto epoxidized natural rubber (ENR-25) [34]. The procedure involved the grafting of acrylonitrile onto the isoprene units of ENR-25, resulting in increased grafting efficiency and enhanced dielectric characteristics. The photoinitiator utilized to commence the reaction was 2,2-dimethoxy-2-phenyl-acetophenone (DMPA). During the initiation stage, DMPA decomposes upon UV exposure, leading to alpha cleavage and producing acetal and benzoyl free radicals. These free radicals underwent further fragmentation, resulting in the formation of methyl radicals. These methyl radicals then reacted with the double bonds present in the 1,4-ci--isoprene units of ENR-25. This mechanism resulted in the dissociation of a hydrogen atom from the isoprene unit, producing a free radical. During the chain propagation stage, the free radical generated underwent a reaction with additional monomer molecules, such as acrylonitrile, leading to grafting. Ultimately, during the chain termination stage, the developing radical terminated either by combining with another radical or by undergoing disproportionation, that involves the transfer of hydrogen from one radical to another. This process resulted in the acrylonitrile-g-ENR by photopolymerization grafting reaction.

Plasma Induced Polymerization Technique

Plasma is the fourth state of matter, which consists of a gaseous mixture of ions and electrons. The plasma state exhibits electric conductivity, whether or not an electric field is applied. When a polymer backbone comes into contact with gases that are in an ionized state, the movement of charged electrons creates places where a substrate can be attached. Plasma polymerization grafting is performed on the surface of a polymer using electron-induced excitation, ionization, and dissociation. By utilizing highly accelerated electrons, it is easier to cleave the chemical bonds on the polymer surface to initiate grafting. The degree of grafting and desired surface modification is achieved by altering the external factors of plasma. In the plasma state, gases like carbon dioxide, oxygen, and ammonia dissociate and generate carboxy, hydroperoxide, and amino functional groups [35]. Generation of these functionalities makes the surface of polymers "bioreceptive". In the case of inert gases like Argon (Ar), which generate radical sites on the polymer backbone, in the presence of oxygen, these sites are transformed into polar functionalities. These polar sites act as "anchorage sites," leading to the attachment of monomers or biological entities [31]. There are multiple established techniques for grafting monomers using plasma. Grafting done *via* a liquid medium involves treating an activated polymer backbone with a liquid phase that contains a reactive precursor. In the gas-phase grafting technique, the activated polymer backbone is directly exposed to the vapour precursor. This technique is referred to as post-irradiation grafting. When monomers are directly constituted in the plasma gas phase, high-energy particles present in air interact with the polymer backbone to create a monomer coating on the polymeric surface. Fragmentation of monomeric unit, its recombination, crosslinking and insertion on surface of substrate occur in single step. This technique is influenced by several operating factor including power supply to generate plasma, flow rate of monomer, temperature in reaction chamber, activated species in gas phase. Polymers are grafted *via* direct or indirect method by treatment with plasma. The direct method, also referred to as plasma deposition or plasma polymerization method, involves exposing the monomer to a gentle plasma environment, and deposition of activated monomer on the polymer surface. Polymer's surface properties can be altered by reacting monomers in the vapor phase. Deposition can be achieved using aliphatic and fluorinated monomers. The chemical composition of monomers and exposure conditions of the plasma are the primary controlling factors for the grafting process. Indirect method also referred as plasma grafting, involves exposing a polymer to plasma to activate its surface. The activated surface then reacts with a monomer to start the graft polymerization process. For instance, when polyethylene films are treated with oxygen plasma, peroxide and hydroperoxide groups are generated on the polymer surface. Plasma polymerization alters the surface characteristics to a depth of a few manometers, leaving the polymer backbone unaltered. This allows the retention of bulk properties of the polymer while enhancing surface morphology through the introduction of polar functionalities. Plasma-induced polymerization is commonly used for surface grafting to create separators with functional groups, offering benefits such as an eco-friendly process, low production costs, and easy commercialization. Stupavská *et al.* in 2024 developed a technique for attaching acrylic acid monomers with -COOH groups to polypropylene nonwoven fabric through plasma-induced polymerization. This process involves treating the fabric with a pulsed underwater diaphragm electrical discharge, which is produced in an aqueous solution of acrylic acid. The result is a one-step method that activates the surface of the fabric and simultaneously attaches the acrylic acid monomers to the fabric. This research is significant as it offers a method to permanently modify polymer surfaces, improving their ability to absorb liquids and allowing for the incorporation of -COOH groups. This modification is crucial for the attachment of biomolecules or in the application of regenerative medicine [36]. The mechanism behind AA polymerization in the presence of underwater plasma discharge involves the formation of hydroperoxyl radicals on the surface of fabric, which is similar to the reactions that occur at the water-plasma interface in solution [37]. However, the process of underwater plasma discharge is complex due to the fragmentation and rearrangement of acrylic acid monomers during the polymerization, making it challenging to predict the exact polymerization process [38]. Tu *et al.* employed atmospheric pressure plasma as an eco-friendly and straightforward method to combine the advantageous properties of PCL, chitosan, and polyethylenimine (PEI) to form a composite membrane capable of adsorbing copper ions, which is then utilized as a filtration membrane for the elimination of heavy metal pollutants from industrial wastewater. The impact of atmospheric pressure nitrogen plasma on chitosan-PCL and the process of PEI grafting onto chitosan-PCL were also explored for the adsorption of copper ions in water treatment [39].

Radiation Grafting

In this technique, high-energy radiation is used for polymer grafting, *i.e.*, gamma irradiation, ion beams, X-ray, electron beam, *etc*. The radiation is bombarded onto a macromolecule to produce highly efficient free radicals. The radical forms a covalent bond with the monomer, which requires less energy and does not necessitate the use of catalysts or additives [40]. This process can improve the properties of polymers by changing their nano-dimensional morphology, solubility, and compatibility, among other factors. Baranowska *et al* in 2024, explored the application of radiation-induced polymer grafting [41]. Bitumen is a mineral used as a binder in road construction to mix with asphalt, and it is also used in the chemical industry as a roofing and coating agent, among other applications. Due to the grafting process, the efficiency of bitumen as a binder increases. Poly(styrene-butadiene-styrene) graft with modified bitumen, irradiated with ^{60}Co γ-rays, is used as a binder, which improves its performance. Radiation-induced methods are classified as individual radiation and Mutual or Simultaneous radiation techniques. The following section elaborates on the methods using suitable examples.

Individual Radiation Technique

The polymers are exposed to radiation individually, and then they react with the monomer.

- **Pre-radiation techniques:** Radiation is responsible for generating free radicals by attacking the backbone of the macromolecule. Then it is grafted with a monomer. The advantage is that there is no chance of homopolymer formation, but as a disadvantage, the polymer may degrade due to direct irradiation on the backbone of the polymer. This process occurred in a vacuum or an inert condition.

- **Peroxidation:** In the presence of radiation, air, or oxygen, the polymer forms peroxide or hydroperoxides depending on the compound's nature. The peroxide undergoes decomposition, forming radicals that are then grafted onto the monomer. The advantage of this technique is that the peroxides formed before grafting can be stored for a long time.

Mutual or Simultaneous Method

In this method, both the polymer and the monomer are irradiated to form a free radical. The polymeric substrate is simultaneously irradiated with the monomer. This process occurs in the air or an inert atmosphere. In the mutual method, homo-polymerization leads to the formation of a homopolymer. The homopolymer is soluble in monomer, which increases viscosity and is responsible for decreasing the amount of monomer available for grafting and reducing the degree of radiation grafting. So, adding inorganic salt in polymer grafting reduces the amount of homopolymer radiation that is responsible for forming an ion, *i.e.*, an anion or cation, which is further grafted with a monomer. According to Lorelis et al, the effect of radiation on the polymer depends on the specific monomer that grafts with the cellulosic chain. Electrons from the gamma rays irradiating the glycoside bond of cellulose form a C-centered radical and an alkoxy radical. This ionizing radiation breaks the glycosidic bonds of cellulose [42].

Biological Grafting

The technique of biological grafting involves altering the surface or characteristics of polymers by integrating biological molecules into them. This method enhances the functionality of the polymer, making it suitable for drug delivery and biosensor applications. Utilizing biological activators like enzymes, proteins, and microbes for graft initiation makes this approach more eco-friendly and sustainable. Enzymes possess inherent capabilities to facilitate various chemical reactions like oxidation, reduction, and acyl transfer, resulting in the production of graft polymers with desirable chemical and physical properties.

Enzymatic Grafting

In our search for a green method of polymerization to modify the properties of polymer bulk, enzymatic grafting has proven to be the most effective. It is a synthesis method that is free from chemicals and is environmentally friendly, making it ideal for the non-destructive functionalization of polymers. Enzymatic grafting provides a better alternative to chemical methods, avoiding the creation of unwanted reactive species and the production of toxicity. By controlling the incubation parameters of enzymes, polymer synthesis can be managed, allowing for the achievement of specific crystal structures, graft densities, architectures, and functionalities in polymeric materials through the proper selection of enzymatic catalysts and reaction conditions. Enzyme-based catalysts offer several advantages, such as low environmental impact, sustainable manufacturing, and substrate specificity, surpassing chemical catalysts in these aspects. Table **2** depicts grafting polymer synthesized using enzymes. The use of enzymatic copolymerization has attracted considerable interest because this graft polymerization reaction occurs without interrupting the main structure. Enzymes in polymer synthesis offer several potential advantages by eliminating hazards from reactive reagents [43]. Their environmental benefits include their selectivity, which reduces wasteful protection and deprotection steps [44].

Table 2. Polymer graft modification using enzymes.

Enzymes	Polymer grafting modification	References
Laccase	Grafting of acrylic acid onto lignin.	[45]
Laccase	Pulp bleaching	[46]
Horseradish Peroxidase	Starch-poly(methyl acrylate) (PMA) was synthesized using horseradish peroxidase	[47]
Peroxidases	Dodecyl gallate (DDG) grafting onto the polysaccharide chitosan.	[48]
Lipase	ε-caprolactone onto hydroxyethyl cellulose	[49, 50]
Tyrosinase	Poly(sulfobetaine-co-tyramine) grafted onto polyurethane *via* tyrosinase-mediated reaction.	[51]
Tyrosinase	Grafting of caffeic acid to chitosan	[52]

Enzymes play a crucial role as catalysts in many reactions due to their catalytic rates, specificity, and ability to function under mild conditions. Enzymes within living cells catalyze the formation of polysaccharides, proteins, polypeptides, and nucleic acids with unmatched selectivity, a feat that synthetic methods cannot replicate. Chen *et al.* in 2000 converted phenols into reactive o-quinones using the tyrosinase enzyme, then it reacts with chitosan [53]. The study demonstrated that in aqueous methanolic solutions, tyrosinase catalyzes the oxidation of hexyloxyphenol. The reactions were conducted in both heterogeneous and homogeneous conditions using chitosan films and aqueous methanolic mixtures, respectively capable of dissolving chitosan and hexyloxyphenol. Hexyloxyphenol-modified chitosan changes physicochemical properties, such as increasing the hydrophobicity of the chitosan surface. It also alters the functional properties of chitosan. Schroeder *et al.* in 2008 grafted methacrylate monomers with different amine groups onto polypropylene surfaces using plasma pretreatment and laccase and guaiacol sulfonic acid (GSA), then studied the efficiency of enzymatic covalent bonding using [54]. Enzymatic grafting produces a unique functionalization of inert polymer surfaces without damage and the use of extensive chemicals under mild conditions. An increase in sulfur ten times greater than that of the control indicates that GSA is grafted in the presence of laccase. The benefits of the technique include strong covalent bonding of functional groups to synthetic polymer surfaces, allowing for customized modifications using enzymes with specific substrate preferences and regional selectivity. Enzymatic reactions are not limited to polymers with existing functional groups, whether natural or synthetic, but can be applied to inert polymers like polyolefins as well.

FACTORS CONTRIBUTING TO GRAFT POLYMERIZATION

Nature of Backbone

There are two ways in which the backbone polymer might influence the grafting and crosslinking kinetics: 1) How easily can reactive sites be created on the main polymer through reactions with the initiator? 2) How well do the reactive sites that are produced react with monomer to form grafting or with other reactive sites to form crosslinking? Grafting will occur more easily if the generated radicals are highly reactive with the monomer; crosslinking is more likely to occur if they are not reactive. Grafting involves the chemical bonding of a monomer to an existing polymer chain [55]. The parent/backbone polymer is essential in graft copolymerization because it forms covalent bonds with the grafted monomer. An example of this is chitosan, a water-insoluble polymer. However, its derivatives can become water-soluble through structural modifications. For instance, grafting chitosan with other polymers or monomers makes it hydrophilic, such as carboxymethyl chitosan. In this derivative, a carboxymethyl group is introduced onto chitosan at the C6-OH position, the C_2-NH_2 position, or both positions to enhance water solubility [56]. Functional groups on the backbone also affect the grafting process. Cellulose acetate-p-nitrobenzoate demonstrates high efficiency in grafting with styrene, suggesting the pendant aromatic nitro group plays a significant role in the formation of the graft copolymer.

Effect of Monomer

Reactivity of monomers depends upon the nature of the monomers, their concentration, and activation. For grafting and crosslinking reactions, the relative reactivity of a monomer toward both its polymer and other radicals produced in the system will be crucial. The reactivity of a monomer is affected by its polarity, the strength of its chemical bonds, and its overall chemical composition, among other factors. The monomer concentration and nature of the solvent have a significant impact on the degree of grafting. Rate and effectiveness of grafting are influenced by the reactivity of a monomer, which is dependent on both the process kinetics and the diffusion of the reactant to the polymer surface [57] Vinyl monomers, such as MMA, glycidyl methacrylate, acrylamide (AAm), AA, N-vinylpyrrolidone, and NIPAAm, are the most commonly utilized monomers in graft polymerization. Using the direct radiation grafting approach, Ghaffar *et al.* in 2016 studied the graft reactions between carboxymethyl cellulose (CMC) and AAm as well as MAA at varying concentrations to form CMC hydrogels [58]. With an increase in monomer concentration, grafting yield and grafting ratio also increase. Additionally, it is observed that Poly (CMC/AAm) has a greater graft ratio and graft yield than Poly (CMC/MAA) due to enhanced hydrogen bonding between the amide functional group of AAm with the hydroxy functional group of CMC. Also, MAA structure exerts a greater steric effect than AAm, leading to lower diffusion of MAA than that of AAm. As a result, Poly (CMC/AAm) has greater grafting ratio and grafting yield.

Effect of Solvent

The solvent acts as a carrier in grafting methods, facilitating the breakdown of monomers near the backbone polymer. Solvent selection plays a significant role in polymer grafting; it can affect the propagation rate of various monomers. The degree of grafting of the same monomer can vary significantly depending on the solvent used during the polymerization reaction. The selection of a solvent is crucial for grafting, as it affects the effectiveness and uniformity of the grafted chains. Factors like dosage, reaction environment, temperature, monomer solubility, swelling properties of the backbone, compatibility with multiple solvents, and generation of free radicals all play a role in the process [57, 59]. The solvent compatibility study of poly (dimethylsiloxane) (PDMS) with organic solvents was investigated by Ng Lee *et al.* in 2003. Propanol, glycerol, water, *etc.*, are examples of low solubility (weakly swelling) solvents in which PDMS is poorly soluble. These solvents are optimal for conducting organic reactions in microfluidic channels fabricated from PDMS. On the other hand, pentane, hexane, and heptane are examples of high-solubility (highly swelling) solvents for PDMS that are effective in extracting impurities from large quantities of the material and modifying its surface characteristics. The study reveals that PDMS can effectively resist swelling when combined with solvents with different solubility levels, preventing the polymer from absorbing highly soluble solvents. The solvent compatibility study of PDMS explores the potential applications of PDMS-based microfluidic devices in organic reactions. For the grafting process, it's crucial to produce free radicals from the solvent. Additionally, the formation of free radicals on the monomer and backbone is also significant. For this reason, selecting the right solvent is important. Jun *et al.* have studied the effect of pure solvents and mixed solvents on the graft polymerization of NIPAAm onto cotton cellulose using the γ-preirradiation method [60]. In a water-methanol mixture with a 4:1 ratio, an increase in grafting yield was observed, but this yield dropped to nearly zero when pure methanol was used. In systems including water and organic solvents, the grafting yields exhibited a progressive decline as the solvent transitioned from pure water to pure organic solvents, ultimately reaching nearly zero yields in pure organic solvents. In general, alcohols exhibited varying levels of inhibition on the grafting of NIPAAm, but acetone had the least impact.

Effect of Initiator

All polymer grafting approaches need an initiator to initiate the reaction, except radiation-induced graft polymerization, in which radiation is enough to initiate the reaction. The concentration and solubility of the initiator in solution have a great impact on the rate of polymer grafting. If the reaction proceeds in an organic solvent, the initiator must dissolve in the solvent. Also, the selection of the initiator depends on its decomposition temperature. The decomposition temperature must be at or below the solvent's boiling point; for example, BPO, 2,2'-Azobis(2-methylpropionitrile). Water-soluble photoinitiators include Irgacure 2959, eosin Y, and camphorquinone. High-temperature polymerization initiators, such as 1,1'-azobis (cyclohexanecarbonitrile), are used if polymerization occurs at or below 20 °C. Water-soluble initiators, such as KPS and azo compounds like AIBN, are commonly used in industrial and laboratory applications. In industrial and laboratorial applications, thermal initiators are used to initiate the reaction, such as azo compounds AIBN [61]. Zhang *et al.* in 2019 studied the grafting of PS on cellulose nanocrystals *via* surface-initiated ATRP using a macroinitiator, using ethyl α-bromoisobutyrate (EBiB) as a sacrificial initiator [62]. The molar ratio between the monomer and sacrificial initiator affects the chain length of the grafted polymer. Pourmahdi *et al.* in 2023 demonstrated the effect of varying initiator concentrations on the synthesis of (PAAm) grafted onto lignin [63]. The grafting was carried out using tert-butyl hydroperoxide and cerium (IV) sulfate as initiators for the radical polymerization process. Their studies demonstrated that the grafting percentage is not dependent on the initiator concentration. They concluded that increasing the amount of initiator does not necessarily result in a higher grafting percentage.

Effect of Temperature

Temperature is a crucial factor in controlling the rate, kinetics, and yield of graft polymer. Yield of graft polymers increases as the temperature is lowered, facilitating diffusion into the backbone. The efficiency of initiators is defined by the decomposition rate of the initiator at that temperature. As the temperature increases, initiators produce macro radicals that facilitate polymer grafting. The rate of grafting initially increases after irradiating the polymer backbone due to peroxide decomposition, which leads to radical generation that facilitates polymer grafting. Initially, decomposition of peroxides occurs rapidly, so the rate of grafting increases due to irradiation of the polymer, as enough radicals are available for grafting. With increasing temperature, molecular motion within the polymer increases, leading to an overall reduction in grafting at a certain temperature. Near the glass transition temperature (Tg), the maximum graft yield is observed. Below Tg, radicals formed within polymer chains are unable to react, which causes decreased diffusion into monomers. As the Tg increases, the number of free radicals available for grafting decreases. Consequently, an increase in concentration of monomer radicals leads to a reduced graft yield [64]. Grafting yield can be increased by selecting a solvent system that prevents the formation of unwanted radicals and oxidation. The rate of grafting reactions initially slows as the initiator decomposes, but increases over time. The thermal properties of the solvent are crucial, as its boiling point should be higher than the polymerization temperature, depending on the initiator temperature condition. While grafting with Iron (II)–hydrogen peroxide (Fenton reagent), the temperature should be kept low. Required temperatures for initiator to start radical formation include 40°C for APS, 80°C for KPS, and 60°C for both BPO and AIBN. Mehra *et al.* 2020, synthesized microwave-assisted soy protein isolate-graft-[acrylic acid-co-4-(4-hydroxyphenyl) butanoic acid] hydrogel using N, N-methylene-bis-acrylamide as crosslinker and KPS initiator. The optimum temperature for KPS is 50°C. They maintained a reaction condition between 40 and 100°C and analyzed the temperature effect. Grafting percentage (GP%) increased significantly up to 80 °C. The grafting percentage (GP%) increased significantly up to 80 °C, but decreased beyond this point. At temperatures below 50°C, the redox reaction between KPS and SPI proceeded slowly, contributing to the lower GP%. Similarly, at higher temperatures above 600°C, the redox reaction processes faster, and GP% increases. However, above 80°C, the reaction is terminated because radical formation and homopolymerization chances increase instead of graft polymerization [65]. The schematic presentation of factors contributing to graft polymerization is depicted in Fig. (7).

RECENT APPLICATIONS IN POLYMER GRAFTING TECHNIQUES

Surface Grafting Polymer

Polymer brushes, often referred to as surface-grafting polymers, have emerged as a significant method for functionalizing or altering the surface of materials to enhance their performance [66]. By altering the parameters of the grafting process, it is possible to graft different functionalities onto the surfaces of polymers [67]. The grafted polymers for covalent bonds between the polymer brushes lead to improved stability, durability, and enhanced pattern of polymer formation. Surface grafting of polymers has played a significant role in advancing electronic devices, including organic solar cells (OPV), organic light-emitting diodes (OLED), organic field-effect transistors (OFET), and others [68]. Adhered surface grafting is another technique used for altering surfaces or engineering interfaces, and has attracted interest in organic electronics due to its unique characteristics. This approach has effectively been utilized in various parts of devices, including insulating layers, conducting layers, and semiconducting layers [69]. Surface grafting offers a significant advantage over other surface processing techniques, such as modification of surface properties and surface energy by grafting polymers with specified functionalities onto substrates [70]. It is also effective for modification of polymer brushes thickness and the grafts density by adjusting parameters. For example, using polymer brushes with a high density for insulating purposes reduces the current leakage through pinholes in the polymer layer compared to conventional approaches such as spin coating, scrape coating, inkjet printing, *etc*. It improves injecting charge by precisely modifying energy of the device-interface surface [71]. In "grafting to" method, a chemical group is first added to the substrate. Then, a preformed polymer is chemically attached to surface by reacting polymer's end group with reactive functional group on the substrate. This approach forms polymer brushes without directly polymerizing monomers on the surface. Instead, it relies on reactions like condensation or chemical coupling to create the bond [72, 73]. The "Grafting from" method starts by attaching an initiator to the substrate surface through chemical interactions like condensation or coupling reactions. Polymer chains then grow directly from the surface when a precursor solution is added, resulting in densely packed polymer brushes. Hence, this method is preferred for creating high-density polymer brushes [74]. For example, PS brushes can be created using "grafting to" approaches: 1) Condensation reaction of PS containing hydroxyl groups (PS-OH) at the end with silicon dioxide. 2) a covalent bonding reaction between PS (PS-Si $(CH_3)_2Cl$), terminated with dimethyl chlorosilane and hydroxyl functionalized surface after exposure to UV, ozone, or oxygen plasma. Both methods result in a strong covalent attachment to the active site of the backbone. However, the challenge with the "grafting to" strategy is the steric effect, where the previously grafted polymers introduce steric hindrance for the attachment of additional side chains on the surface. This steric hinderance makes it complicated to achieve polymer brushes with high density using "grafting to" technique [75, 76].

Conducting Polymer

Conducting polymers (CP) are fourth-generation polymers, characterized by their ability to conduct electricity. For almost a decade, we have utilized polymers as insulators for cable wires, electrical plugs, sockets, and other similar applications. CPs are a class of stimuli-responsive polymers that exhibit changes in charge, conductivity, and mechanical and physical properties when oxidized or reduced.

Hackett *et al.* have described various conducting polymer materials for controlled drug delivery applications [77]. CPs are those polymers that have π electrons in their backbone with alternating double or triple bonds, which intrinsically conduct electricity. Some examples of CPs are mentioned in Fig. (**8**) [78].

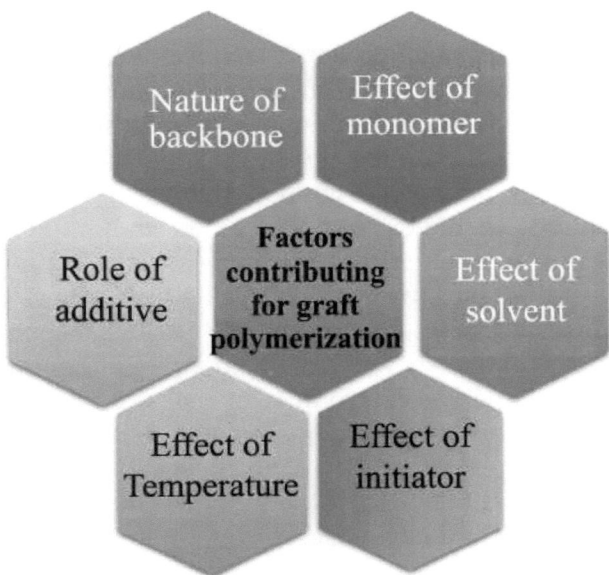

Fig. (7). Factors contributing to graft polymerization (created by using SmartArt).

There are two main types of CPs:

- **Intrinsically conducting polymers (ICP):** These polymers can conduct electricity on their own. They consist of hydrocarbon chains with alternate single and double bonds, such as polyethylene and polyacetylene.
- **Extrinsically conducting polymers (ECP):** In these polymers, conductivity is achieved by adding conductive materials like carbon black (graphite) to a polymer matrix, such as polyethylene. Conductivity is determined by the proximity of the graphite particles. If the particles are close enough, the polymer conducts electricity; if not, it acts as an insulator. Conductivity also changes with temperature: as the temperature increases, the polymer expands, pulling the particles apart and reducing conductivity, and vice versa. ECPs are used in self-regulating heater cables and polyswitch resettable circuits.

The physicochemical properties of CPs are influenced by numerous reasons, including the MW of the monomeric unit, side chains, functional groups attached, and the polymeric backbone. Side chain attachment in polymers mainly acts as a solubilizing agent, but it also directly affects the backbone's electronic, structural, optical, and electrical properties. Key factors in advancing CP technology include processing ability, morphological parameters, sustainability, and longevity. These characteristics are affected by the chemical composition of the backbone, along with functional group attachment and side chain composition, which also contribute to the same [79]. To adjust the physicochemical properties of CPs, the chemical nature of the polymer is modified while preserving the backbone to retain its properties. However, modifying these polymers can sometimes be more challenging than creating them from scratch. Novel methods for polymer modification, such as covalent and non-covalent grafting, are explored, with covalent grafting being the more commonly used approach [80]. An inherent limitation of CP synthesis following functionalization is the presence of large molecular components that remain in the final product. Functionalization typically alters surface properties, and bulk properties remain untouched [81]. CPs are often soluble in a limited range of solvents, such as PANI, which can hinder their functionalization. As a result, the surface can be modified without affecting core characteristics like conductivity when only surface changes are desired [82].

Fig. (8). Structures of conducting polymer.

Conductive Polymer Covalent Grafting

It is a standard technique for manufacturers of graft polymers, which includes several chemical reactions such as electrophilic substitution, bromination, electrophilic aromatic substitution, coupling with diazonium, and nucleophile addition on quinonimine units. Zhu *et al.* 2021 created a comprehensive polymer aqueous proton battery that outperforms other electrochemical systems by incorporating catechol molecules onto the poly(3,4-ethylenedioxythiophene) (PEDOT) polymer backbone to form both the cathode and the anode [83]. By adding catechol groups to the PEDOT chain, they achieved a high potential cathode, which significantly enhances electron movement. By merging an aqueous proton battery with a capacitive-type cathode along with a diffusion-type anode, which has obtained material with high voltage capacity, outstanding energy masses, and an impressive rate capacity. Ayed *et al.* have performed oxidative polymerization of aniline to obtain polyaniline-grafted chitin nanocrystals (PANI-g-ChNCs) nanohybrids *in situ*. Nanocrystals of chitin are used as a backbone, and ammonium persulfate acts as an oxidant to activate the amino groups present on the surface of ChNCs. PANI-g-ChNCs is a hybrid nanofiller used to conduct electricity and act as a wood adhesive. This polymeric assembly is also applicable for manufacturing sensors, biomedical instruments, anticorrosion materials, and capacitors [84].

Conducting Polymer Noncovalent Grafting

Scientists have discovered that the dynamic nature of non-covalent bonds can be utilized to create self-assembly polymers, offering a way to address synthetic challenges in polymer synthesis without relying on traditional approaches. Noncovalent interactions are considered a more versatile approach as they help overcome several problems in covalent grafting, such as irreversible attachments, complex modification techniques, and unwanted side reactions [85]. These traits can be changed by employing non-covalent bonding in a polymer. Noncovalent graft polymers, also known as supramolecular polymers, are formed through self-assembly *via* noncovalent interactions. In the natural world, DNA serves as a prime illustration of polymers with noncovalent bonding. DNA involves hydrogen bonding between nucleotides due to weak dipole-dipole interactions. There are two main types of supramolecular polymers: side-chain and main-chain polymers. Polymers that are connected by noncovalent interactions within the polymer backbone. Side-chain supramolecular polymers have a polymer backbone that is covalently linked and contains molecular recognition units on the side chain, which can be further utilized for non-covalent interactions that lead to grafting and self-assembly [86]. Supramolecular interaction forces involve non-covalent bonds between molecules, which, despite being weaker than covalent bonds, significantly influence material properties. By integrating these non-covalent bonds into covalently crosslinked CPs, new functionalities can be introduced, such as reversibility that allows dynamic self-repair upon damage. This incorporation also enhances the controllability and selectivity of CPs, enabling precise tailoring of physical and chemical properties. Common strategies for preparing CPs include hydrogen bonding, electrostatic interactions, host-guest interactions, and coordination bonding. Jiang *et al.* (2022) developed a systematic approach to incorporating poly(3,4-ethylenedioxythiophene) conducting polymer with polystyrene sulfonate (PEDOT: PSS) into polyrotaxanes to improve the mechanical strength and stretchability of the inherently soft polymer [87]. This polymer is used in wearable, surface-mounted, and implanted bioelectronic devices. On the other hand, Sun *et al.* (2024) reviewed different assemblies based on supramolecular strategies, which are mainly focused on wearable electronic sensors [88].

Graft Polymerization *via* Click Chemistry

In 2001, Barry Sharpless introduced the concept of "click chemistry," which encompasses simple, versatile, high-yielding, and stereospecific reactions. These reactions, often described as "spring-loaded," facilitate the formation of C–C or C–N bonds with minimal side reactions and without the need for extensive purification steps. The high thermodynamic driving force of these reactions contributes to their stereospecificity and regioselectivity. A prime example of click chemistry is the formation of 1,2,3-triazoles through the 1,3-dipolar cycloaddition of azides and alkynes. This method is valued for its ease of preparation and purification, making it suitable for a wide range of applications, including new developments. Click chemistry is especially useful in creating polymeric soft materials with precise architectural and functional control. Additionally, integrating the efficiency of click chemistry with the chain-end functionality control of living free radical polymerization opens up further possibilities for the design of advanced materials [89]. Click chemistry encompasses several highly effective reactions for polymerization and polymer modification, including azide-alkyne cycloaddition (AAC), Diels-Alder reactions, thiol-X reactions, carbonyl-based additions, and SuFEx. These techniques have significantly advanced the ability to synthesize copolymers with innovative structures such as heterolayer dendrimers, dendron-polymer conjugates, and nanoscale constructs. Furthermore, these methods enable the attachment of polymers to a wider array of biopolymers and integration into complex living systems. Reports also indicate that functional cyclic backbones and cyclic-graft polymers can be modified by grafting linear polymers, expanding their utility and applications [90]. Ahmed Eisa and colleagues have developed a versatile approach for functionalizing and grafting 2-hydroxyethyl cellulose (HEC) using click chemistry. Their method enables the incorporation of both neutral and ionic functionalities into HEC. The study also explores a sequential click reaction process to create both neutral and ionic forms of HEC, employing PDMS as a key component. This technique was utilized to graft PLA and PEG segments onto HEC, resulting in the formation of grafted polymer HEC-g-PLA [91].

FUTURE PERSPECTIVES AND CHALLENGES

Graft polymerization is a pivotal process for enhancing sophisticated characteristics and properties of polymeric materials. This approach potentially expands polymer properties, such as charge-carrying capabilities, solubility, homo- or heterogeneous polymer dispersibility, and biocompatibility *etc*. By altering the arrangement and composition of various polymerization reactions, the versatile applications of specific polymers can be expanded and improved. One of the major challenges in graft polymerization is achieving biodegradability while ensuring the material's suitability for various industrial applications, including textiles, plastics, rubber, and the automotive sector. Additional challenges include ensuring the stability and durability of grafted polymers, particularly under harsh environmental conditions, and maintaining the right balance between newly introduced properties and the original characteristics of the base polymer. Another concern is optimizing the compatibility of grafted polymers with existing manufacturing processes and meeting regulatory requirements. Furthermore, radiation sources used in photopolymerization methods are often prohibitively expensive, limiting the widespread adoption of these techniques. Managing the high costs associated with industrial-scale polymer grafting, especially in complex and energy-intensive processes, remains a significant hurdle. The scalability of these processes is another critical issue, as controlling uniformity and efficiency in large-scale production can be difficult. Despite these challenges, grafting holds great potential to enhance the biodegradability of plastics, which is essential in the effort to reduce environmental pollution. As research and technology in polymer grafting continue to advance, the field is expected to offer promising solutions, enabling the replacement of traditional polymers with more efficient and sustainable grafted alternatives. Enzymatic techniques give us a green and sustainable method for the production of polymers, but these methods are underdeveloped and more complex as they involve sophisticated enzyme engineering and require precise control over enzyme activity and reaction conditions. Click chemistry opens up the door for the creation of complex polymers with distinct macromolecular architectures. Future opportunities include controlling polymer tacticity, synthesizing precision oligomers and copolymers with unconventional architectures like dendrimers, polymer hybrids, nanostructures, and developing functionalized sequence-controlled macromolecules. Advancements in new click chemistries will facilitate one-pot polymerizations and allow non-experts access to diverse libraries of functional macromolecules in complex media. Graft polymerization *via* click chemistry holds promise for creating advanced materials with precise molecular architectures, enhancing functionality in fields like drug delivery and nanotechnology. Future advancements could lead to more efficient and scalable processes, as well as novel applications in smart materials and biomedical devices.

CONCLUSION

Polymer grafting imparts advanced characteristics to the polymer backbone, including changes in stability, solubility, nanoscale structure, cell compatibility, and conductivity, which significantly alter polymer bulk properties. It allows modifications in polymer chemistry that expand and strengthen the applicability of specific polymers across a wide range of fields. As discussed in this chapter, several approaches are available for polymer grafting, such as the "grafting-to" method, which enables the coupling of monomers as side chains to increase chain length, and the "grafting-from" technique, which incorporates multiple monomers into the polymer backbone with the help of an initiator. Additionally, "grafting-through" techniques allow the combination of macromonomers on the polymer backbone. Grafting is a precise technique that allows the control of the grafting rate, MW of the polymer assembly, surface and chemical properties, density of graft polymers, and grafting ratio. Several factors influence these parameters, including solvent composition, monomer concentration, temperature, additives, initiator decomposition rate, side-chain length, and surface exposure time to radiation. Various methods, such as surface treatment, plasma treatment, polymer control, thermal methods, enzyme-mediated grafting, microwave-assisted techniques, and click chemistry reactions, offer a wide range of grafting possibilities. By using different inhibitors and additives, one can develop new polymer materials with enhanced biocompatibility, specifically tailored to meet human needs in industries such as textiles, plastics, agriculture, waste disposal, and sewage management. Consequently, grafting techniques help meet the growing market demand by utilizing the latest non-conventional, advanced grafting approaches.

REFERENCES

[1] Choudhary S, Sharma K, Sharma V, Kumar V. Grafting Polymers. In: Gutiérrez TJ, Ed. Reactive and Functional Polymers Volume Two. Cham: Springer International Publishing 2020; pp. 199-243. [Internet]
[http://dx.doi.org/10.1007/978-3-030-45135-6_8]

[2] Hadjichristidis N, Pispas S, Pitsikalis M, Iatrou H, Lohse DJ. Graft Copolymers. In: Mark HF, Ed. Encyclopedia of Polymer Science and Technology. 3rd ed., Wiley 2002. [Internet]
[http://dx.doi.org/10.1002/0471440264.pst150]

[3]   Hadjichristidis N, Pitsikalis M, Iatrou H, Driva P, Chatzichristidi M, Sakellariou G, et al. Graft Copolymers. In: Mark HF, Ed. Encyclopedia of Polymer Science and Technology. 3rd ed., Wiley 2010. [Internet]
[http://dx.doi.org/10.1002/0471440264.pst150.pub2]

[4]   Nakagawa Y, Ushidome K, Masuda K, et al. Multi-Armed Star-Shaped Block Copolymers of Poly(ethylene glycol)-Poly(furfuryl glycidol) as Long Circulating Nanocarriers. Polymers (Basel) 2023; 15(12): 2626.
[http://dx.doi.org/10.3390/polym15122626] [PMID: 37376272]

[5]   Vega-Hernández MÁ, Cano-Díaz GS, Vivaldo-Lima E, et al. A Review on the Synthesis, Characterization, and Modeling of Polymer Grafting. Processes (Basel) 2021; 9(2): 375.
[http://dx.doi.org/10.3390/pr9020375]

[6]   Goddard JM, Hotchkiss JH. Polymer surface modification for the attachment of bioactive compounds. Prog Polym Sci 2007; 32(7): 698-725.
[http://dx.doi.org/10.1016/j.progpolymsci.2007.04.002]

[7]   Pitsikalis M. Ionic Polymerization. In: Reference Module in Chemistry, Molecular Sciences and Chemical Engineering [Internet]. Elsevier; 2013 [cited 2024 Aug 8]. p. B9780124095472054196. Available from: https://linkinghub.elsevier.com/retrieve/pii/B9780124095472054196

[8]   Lacerda PSS, Gama N, Freire CSR, Silvestre AJD, Barros-Timmons A. Grafting Poly(Methyl Methacrylate) (PMMA) from Cork via Atom Transfer Radical Polymerization (ATRP) towards Higher Quality of Three-Dimensional (3D) Printed PMMA/Cork-g-PMMA Materials. Polymers (Basel) 2020; 12(9): 1867.
[http://dx.doi.org/10.3390/polym12091867] [PMID: 32825164]

[9]   Pizzi D, Humphries J, Morrow JP, et al. Poly(2-oxazoline) macromonomers as building blocks for functional and biocompatible polymer architectures. Eur Polym J 2019; 121: 109258.
[http://dx.doi.org/10.1016/j.eurpolymj.2019.109258]

[10] Savaş B, Çatıker E, Öztürk T, Meyvacı E. Synthesis and characterization of poly(methyl methacrylate-g-α-methyl-β-alanine) copolymer using "Grafting Through" method. J Polym Res 2021; 28(5): 194.
[http://dx.doi.org/10.1007/s10965-021-02551-9]

[11] Bhattacharya A. Grafting: a versatile means to modify polymersTechniques, factors and applications. Prog Polym Sci 2004; 29(8): 767-814.
[http://dx.doi.org/10.1016/j.progpolymsci.2004.05.002]

[12] Betraoui A, Seddiki N, Souag R, et al. Synthesis of New Hydrogels Involving Acrylic Acid and Acrylamide Grafted Agar-Agar and Their Application in the Removal of Cationic Dyes from Wastewater. Gels 2023; 9(6): 499.
[http://dx.doi.org/10.3390/gels9060499] [PMID: 37367168]

[13] Ouchi M, Sawamoto M. Sequence-controlled polymers via reversible-deactivation radical polymerization. Polym J 2018; 50(1): 83-94.
[http://dx.doi.org/10.1038/pj.2017.66]

[14] Mayadunne RTA, Rizzardo E, Chiefari J, Chong YK, Moad G, Thang SH. Living Radical Polymerization with Reversible Addition−Fragmentation Chain Transfer (RAFT Polymerization) Using Dithiocarbamates as Chain Transfer Agents. Macromolecules 1999; 32(21): 6977-80.
[http://dx.doi.org/10.1021/ma9906837]

[15] Hernández CJ, Medina R, Maza Mejía I, et al. Preparation of a Molecularly Imprinted Polymer on Polyethylene Terephthalate Platform Using Reversible Addition-Fragmentation Chain Transfer Polymerization for Tartrazine Analysis via Smartphone. Polymers (Basel) 2024; 16(10): 1325.
[http://dx.doi.org/10.3390/polym16101325] [PMID: 38794519]

[16] Cameron NR, Lagrille O, Lovell PA, Thongnuanchan B. Solution homopolymerizations of n-butyl acrylate and styrene mediated using 2,2,5-trimethyl-4-tert-butyl-3-azahexane-3-oxyl (TITNO).

Polymer (Guildf) 2014; 55(3): 772-81.
[http://dx.doi.org/10.1016/j.polymer.2013.12.060]

[17] Roggi A, Guazzelli E, Resta C, Agonigi G, Filpi A, Martinelli E. Vinylbenzyl Chloride/Styrene-Grafted SBS Copolymers *via* TEMPO-Mediated Polymerization for the Fabrication of Anion Exchange Membranes for Water Electrolysis. Polymers (Basel) 2023; 15(8): 1826.
[http://dx.doi.org/10.3390/polym15081826] [PMID: 37111973]

[18] Li TT, Feng LF, Gu XP, Zhang CL, Wang P, Hu GH. Intensification of Polymerization Processes by Reactive Extrusion. Ind Eng Chem Res 2021; 60(7): 2791-806.
[http://dx.doi.org/10.1021/acs.iecr.0c05078]

[19] Schall C, Altepeter M, Schöppner V, Wanke S, Kley M. Material-Preserving Extrusion of Polyamide on a Twin-Screw Extruder. Polymers (Basel) 2023; 15(4): 1033.
[http://dx.doi.org/10.3390/polym15041033] [PMID: 36850316]

[20] Penczek S, Moad G. Glossary of terms related to kinetics, thermodynamics, and mechanisms of polymerization (IUPAC Recommendations 2008). Pure Appl Chem 2008; 80(10): 2163-93.
[http://dx.doi.org/10.1351/pac200880102163]

[21] Obi BE. Polymer Chemistry and Synthesis. Polymeric Foams Structure-Property-Performance. Elsevier 2018; pp. 17-40. Internet
[http://dx.doi.org/10.1016/B978-1-4557-7755-6.00002-1]

[22] Zhu S, Hamielec A. Polymerization Kinetic Modeling and Macromolecular Reaction Engineering. Polymer Science: A Comprehensive Reference. Elsevier 2012; pp. 779-831.
[http://dx.doi.org/10.1016/B978-0-444-53349-4.00127-8]

[23] Guo X, Choi B, Feng A, Thang SH. Polymer Synthesis with More Than One Form of Living Polymerization Method. Macromol Rapid Commun 2018; 39(20): 1800479.
[http://dx.doi.org/10.1002/marc.201800479] [PMID: 30238698]

[24] Mishima E, Yamada T, Watanabe H, Yamago S. Precision synthesis of hybrid block copolymers by organotellurium-mediated successive living radical and cationic polymerizations. Chem Asian J 2011; 6(2): 445-51.
[http://dx.doi.org/10.1002/asia.201000402] [PMID: 21254423]

[25] Hahn C, Becker S, Müller AHE, Frey H. Anionic polymerization of ferulic acid-derived, substituted styrene monomers. Eur Polym J 2024; 211: 113004.
[http://dx.doi.org/10.1016/j.eurpolymj.2024.113004]

[26] Rudin A, Choi P. Ionic and Coordinated Polymerizations. The Elements of Polymer Science & Engineering. Elsevier 2013; pp. 449-93. Internet
[http://dx.doi.org/10.1016/B978-0-12-382178-2.00011-0]

[27] Husár B, Hatzenbichler M, Mironov V, Liska R, Stampfl J, Ovsianikov A. Photopolymerization-based additive manufacturing for the development of 3D porous scaffolds. Biomaterials for Bone Regeneration. Elsevier 2014; pp. 149-201.
[http://dx.doi.org/10.1533/9780857098104.2.149]

[28] He X, Zang L, Xin Y, Zou Y. An overview of photopolymerization and its diverse applications. Appl Res 2023; 2(6): e202300030.
[http://dx.doi.org/10.1002/appl.202300030]

[29] Sakhare MS, Rajput HH. Polymer grafting and applications in pharmaceutical drug delivery systems – a brief review. Asian J Pharm Clin Res 2017; 10(6): 59.
[http://dx.doi.org/10.22159/ajpcr.2017.v10i6.18072]

[30] Hayat U, Tinsley A, Calder M, Clarke D. ESCA investigation of low-temperature ammonia plasma-treated polyethylene substrate for immobilization of protein. Biomaterials 1992; 13(11): 801-6.
[http://dx.doi.org/10.1016/0142-9612(92)90022-G] [PMID: 1391403]

[31] Gupta B, Anjum N. Plasma and Radiation-Induced Graft Modification of Polymers for Biomedical

Applications. In: Kausch H, Anjum N, Chevolot Y, *et al.* eds. Radiation Effects on Polymers for Biological Use. Adv Polym Sci. Berlin, Heidelberg: Springer 2003; pp. 35-61.
[http://dx.doi.org/10.1007/3-540-45668-6_2]

[32] Zhong M, Wang L, Yan W, et al. Light-controlled radical polymerization: mechanisms, methods, and applications. Chem Rev. 2022;122(15):12345-12400.
[http://dx.doi.org/10.1021/acs.chemrev.1c01023]

[33] Hong Anh Ngo T, Tran DT, Hung Dinh C. Surface photochemical graft polymerization of acrylic acid onto polyamide thin film composite membranes. J Appl Polym Sci. 2017 Feb 5;134(5):app.44418.

[34] Whba R, Su'ait MS, Tian Khoon L, Ibrahim S, Mohamed NS, Ahmad A. Free-Radical Photopolymerization of Acrylonitrile Grafted onto Epoxidized Natural Rubber. Polymers (Basel) 2021; 13(4): 660.
[http://dx.doi.org/10.3390/polym13040660] [PMID: 33672185]

[35] Haddadi-Asl V, Roghani-Mamaqani H. In situ controlled radical polymerization: a review on synthesis of well-defined nanocomposites by grafting from polymers. Iran Polym J. 2023;32(4):455-478.
[http://dx.doi.org/10.1007/s13726-023-01112-5]

[36] Mohammad AW, Teow YH, Ang WL, Chung YT, Oatley-Radcliffe DL, Hilal N. Nanofiltration membranes review: Recent advances and future prospects. Desalination 2015; 356: 226-54.
[http://dx.doi.org/10.1016/j.desal.2014.10.043]

[37] Joshi RP, Thagard SM. Streamer-Like Electrical Discharges in Water: Part II. Environmental Applications. Plasma Chem Plasma Process 2013; 33(1): 17-49.
[http://dx.doi.org/10.1007/s11090-013-9436-x]

[38] Kováčik D, Šrámková P, Multáňová P, *et al.* Plasma-induced Polymerization and Grafting of Acrylic Acid on the Polypropylene Nonwoven Fabric Using Pulsed Underwater Diaphragm Electrical Discharge. Plasma Chem Plasma Process 2024; 44(2): 983-1001.
[http://dx.doi.org/10.1007/s11090-024-10454-y]

[39] Tu SL, Chen CK, Shi SC, Yang JHC. Plasma-Induced Graft Polymerization of Polyethylenimine onto Chitosan/Polycaprolactone Composite Membrane for Heavy Metal Pollutants Treatment in Industrial Wastewater. Coatings 2022; 12(12): 1966.
[http://dx.doi.org/10.3390/coatings12121966]

[40] Saito K, Fujiwara K, Sugo T. Fundamentals of Radiation-Induced Graft Polymerization. Innovative Polymeric Adsorbents. Singapore: Springer Singapore 2018; pp. 1-22.
[http://dx.doi.org/10.1007/978-981-10-8563-5_1]

[41] Baranowska W, Rzepna M, Ostrowski P, Lewandowska H. Radiation and Radical Grafting Compatibilization of Polymers for Improved Bituminous Binders—A Review. Materials (Basel) 2024; 17(7): 1642.
[http://dx.doi.org/10.3390/ma17071642] [PMID: 38612155]

[42] List R, Gonzalez-Lopez L, Ashfaq A, Zaouak A, Driscoll M, Al-Sheikhly M. On the Mechanism of the Ionizing Radiation-Induced Degradation and Recycling of Cellulose. Polymers (Basel) 2023; 15(23): 4483.
[http://dx.doi.org/10.3390/polym15234483] [PMID: 38231912]

[43] Kumar G, Smith PJ, Payne GF. Enzymatic grafting of a natural product onto chitosan to confer water solubility under basic conditions. Biotechnol Bioeng 1999; 63(2): 154-65.
[PMID: 10099592]

[44] Jayakumar R, Prabaharan M, Reis RL, Mano JF. Graft copolymerized chitosan—present status and applications. Carbohydr Polym 2005; 62(2): 142-58.
[http://dx.doi.org/10.1016/j.carbpol.2005.07.017]

[45] Yu C, Wang F, Fu S, Liu H, Meng Q. Laccase-Assisted Grafting of Acrylic Acid onto Lignin for its Recovery from Wastewater. J Polym Environ 2017; 25(4): 1072-9.

[http://dx.doi.org/10.1007/s10924-016-0846-8]

[46] Arias ME, Arenas M, Rodríguez J, Soliveri J, Ball AS, Hernández M. Kraft pulp biobleaching and mediated oxidation of a nonphenolic substrate by laccase from Streptomyces cyaneus CECT 3335. Appl Environ Microbiol 2003; 69(4): 1953-8.
[http://dx.doi.org/10.1128/AEM.69.4.1953-1958.2003] [PMID: 12676669]

[47] Wang S, Wang Q, Fan X, *et al.* Synthesis and characterization of starch-poly(methyl acrylate) graft copolymers using horseradish peroxidase. Carbohydr Polym 2016; 136: 1010-6.
[http://dx.doi.org/10.1016/j.carbpol.2015.09.110] [PMID: 26572441]

[48] Vachoud L, Chen T, Payne GF, Vazquez-Duhalt R. Peroxidase catalyzed grafting of gallate esters onto the polysaccharide chitosan. Enzyme Microb Technol 2001; 29(6-7): 380-5.
[http://dx.doi.org/10.1016/S0141-0229(01)00404-5]

[49] Cheng HN, Gu QM. Enzyme-Catalyzed Modifications of Polysaccharides and Poly(ethylene glycol). Polymers (Basel) 2012; 4(2): 1311-30.
[http://dx.doi.org/10.3390/polym4021311]

[50] Li J, Xie W, Cheng HN, Nickol RG, Wang PG. Polycaprolactone-Modified Hydroxyethylcellulose Films Prepared by Lipase-Catalyzed Ring-Opening Polymerization. Macromolecules 1999; 32(8): 2789-92.
[http://dx.doi.org/10.1021/ma981816b]

[51] Kwon HJ, Lee Y, Phuong LT, *et al.* Zwitterionic sulfobetaine polymer-immobilized surface by simple tyrosinase-mediated grafting for enhanced antifouling property. Acta Biomater 2017; 61: 169-79.
[http://dx.doi.org/10.1016/j.actbio.2017.08.007] [PMID: 28782724]

[52] Liu Y, Zhang B, Javvaji V, *et al.* Tyrosinase-mediated grafting and crosslinking of natural phenols confers functional properties to chitosan. Biochem Eng J 2014; 89: 21-7.
[http://dx.doi.org/10.1016/j.bej.2013.11.016]

[53] Chen T, Kumar G, Harris MT, Smith PJ, Payne GF. Enzymatic grafting of hexyloxyphenol onto chitosan to alter surface and rheological properties. Biotechnol Bioeng 2000; 70(5): 564-73.
[PMID: 11042553]

[54] Schroeder M, Fatarella E, Kovač J, Guebitz GM, Kokol V. Laccase-induced grafting on plasma-pretreated polypropylene. Biomacromolecules 2008; 9(10): 2735-41.
[http://dx.doi.org/10.1021/bm800450b] [PMID: 18771316]

[55] Russell KE. Free radical graft polymerization and copolymerization at higher temperatures. Prog Polym Sci 2002; 27(6): 1007-38.
[http://dx.doi.org/10.1016/S0079-6700(02)00007-2]

[56] Farag R, Mohamed R. Synthesis and characterization of carboxymethyl chitosan nanogels for swelling studies and antimicrobial activity. Molecules 2012; 18(1): 190-203.
[http://dx.doi.org/10.3390/molecules18010190] [PMID: 23262448]

[57] Nasef M. Preparation and applications of ion exchange membranes by radiation-induced graft copolymerization of polar monomers onto non-polar films. Prog Polym Sci 2004; 29(6): 499-561.
[http://dx.doi.org/10.1016/j.progpolymsci.2004.01.003]

[58] Abdel Ghaffar AM, El-Arnaouty MB, Abdel Baky AA, Shama SA. Radiation-induced grafting of acrylamide and methacrylic acid individually onto carboxymethyl cellulose for removal of hazardous water pollutants. Des Monomers Polym 2016; 19(8): 706-18.
[http://dx.doi.org/10.1080/15685551.2016.1209630]

[59] Wang YM, Kálosi A, Halahovets Y, Beneš H, de los Santos Pereira A, Pop-Georgievski O. Solvent effects on surface-grafted and solution-born poly[N -(2-hydroxypropyl)methacrylamide] during surface-initiated RAFT polymerization. Polym Chem 2024; 15(20): 2070-80.
[http://dx.doi.org/10.1039/D4PY00177J]

[60] Jun L, Jun L, Min Y, Hongfei H. Solvent effect on grafting polymerization of NIPAAm onto cotton

cellulose *via* γ-preirradiation method. Radiat Phys Chem 2001; 60(6): 625-8.
[http://dx.doi.org/10.1016/S0969-806X(00)00375-3]

[61] Bahrani S, Aslani R, Hashemi SA, Mousavi SM, Ghaedi M. Introduction to molecularly imprinted polymer. In: Ramezani G, editor. Interface Science and Technology. Vol 32. Amsterdam: Elsevier; 2021. p. 511-6.
[https://www.sciencedirect.com/science/chapter/bookseries/abs/pii/B9780128188057000060?via%3Dihub]

[62] Zhang Z, Wang X, Tam KC, Sèbe G. A comparative study on grafting polymers from cellulose nanocrystals *via* surface-initiated atom transfer radical polymerization (ATRP) and activator regenerated by electron transfer ATRP. Carbohyd Polym 2019; 205: 322-9.
[http://dx.doi.org/10.1016/j.carbpol.2018.10.050] [PMID: 30446111]

[63] Pourmahdi M, Mohsenpour M, Abdollahi M. Synthesis and characterization of lignin-graft-polyacrylamide copolymers: effect of type and concentration of initiator and co-initiator, monomer concentration, and reaction temperature and time on efficiency of graft copolymerization. Wood Sci Technol 2023; 57(5): 1099-123.
[http://dx.doi.org/10.1007/s00226-023-01485-3]

[64] Gürdağ G, Sarmad S. Cellulose Graft Copolymers: Synthesis, Properties, and Applications. In: Kalia S, Sabaa MW, Eds. Polysaccharide Based Graft Copolymers. Berlin, Heidelberg: Springer Berlin Heidelberg 2013; pp. 15-57.
[http://dx.doi.org/10.1007/978-3-642-36566-9_2]

[65] Mehra S, Nisar S, Chauhan S, Singh V, Rattan S. Soy Protein-Based Hydrogel under Microwave-Induced Grafting of Acrylic Acid and 4-(4-Hydroxyphenyl)butanoic Acid: A Potential Vehicle for Controlled Drug Delivery in Oral Cavity Bacterial Infections. ACS Omega 2020; 5(34): 21610-22.
[http://dx.doi.org/10.1021/acsomega.0c02287] [PMID: 32905438]

[66] Azzaroni O. Polymer brushes here, there, and everywhere: Recent advances in their practical applications and emerging opportunities in multiple research fields. J Polym Sci A Polym Chem 2012; 50(16): 3225-58.
[http://dx.doi.org/10.1002/pola.26119]

[67] Gao T, Ye Q, Pei X, Xia Y, Zhou F. Grafting polymer brushes on graphene oxide for controlling surface charge states and templated synthesis of metal nanoparticles. J Appl Polym Sci 2013; 127(4): 3074-83.
[http://dx.doi.org/10.1002/app.37572]

[68] Wang S, Wang Z, Li J, Li L, Hu W. Surface-grafting polymers: from chemistry to organic electronics. Mater Chem Front 2020; 4(3): 692-714.
[http://dx.doi.org/10.1039/C9QM00450E]

[69] Alonzi M, Lanari D, Marrocchi A, Petrucci C, Vaccaro L. Synthesis of polymeric semiconductors by a surface-initiated approach. RSC Advances 2013; 3(46): 23909.
[http://dx.doi.org/10.1039/c3ra43680b]

[70] Chen T, Amin I, Jordan R. Patterned polymer brushes. Chem Soc Rev 2012; 41(8): 3280-96.
[http://dx.doi.org/10.1039/c2cs15225h] [PMID: 22234473]

[71] Song Y, Yan L, Zhou Y, Song B, Li Y. Lowering the Work Function of ITO by Covalent Surface Grafting of Aziridine: Application in Inverted Polymer Solar Cells. Adv Mater Interfaces 2015; 2(1): 1400397.
[http://dx.doi.org/10.1002/admi.201400397]

[72] Koutsos V, van der Vegte EW, Pelletier E, Stamouli A, Hadziioannou G. Structure of Chemically End-Grafted Polymer Chains Studied by Scanning Force Microscopy in Bad-Solvent Conditions. Macromolecules 1997; 30(16): 4719-26.
[http://dx.doi.org/10.1021/ma961625d]

[73] Ito Y, Ochiai Y, Park YS, Imanishi Y. pH-Sensitive Gating by Conformational Change of a

Polypeptide Brush Grafted onto a Porous Polymer Membrane. J Am Chem Soc 1997; 119(7): 1619-23.
[http://dx.doi.org/10.1021/ja963418z]

[74] Ge F, Wang X, Zhang Y, *et al.* Modulating the Surface *via* Polymer Brush for High-Performance Inkjet-Printed Organic Thin-Film Transistors. Adv Electron Mater 2017; 3(1): 1600402.
[http://dx.doi.org/10.1002/aelm.201600402]

[75] Ribelli TG, Lorandi F, Fantin M, Matyjaszewski K. Atom Transfer Radical Polymerization: Billion Times More Active Catalysts and New Initiation Systems. Macromol Rapid Commun 2019; 40(1): 1800616.
[http://dx.doi.org/10.1002/marc.201800616] [PMID: 30375120]

[76] Barbey R, Lavanant L, Paripovic D, *et al.* Polymer brushes *via* surface-initiated controlled radical polymerization: synthesis, characterization, properties, and applications. Chem Rev 2009; 109(11): 5437-527.
[http://dx.doi.org/10.1021/cr900045a] [PMID: 19845393]

[77] Hackett AJ, Malmström J, Travas-Sejdic J. Functionalization of conducting polymers for biointerface applications. Prog Polym Sci 2017; 70: 18-33.
[http://dx.doi.org/10.1016/j.progpolymsci.2017.03.004]

[78] Pal T, Banerjee S, Manna PK, Kar KK. Characteristics of Conducting Polymers. In: Kar KK, editor. Handbook of Nanocomposite Supercapacitor Materials I [Internet]. Cham: Springer International Publishing; 2020 [cited 2024 Aug 20]. p. 247–68. (Springer Series in Materials Science; vol. 300).
[http://dx.doi.org/10.1007/978-3-030-43009-2_8]

[79] K N, Rout CS. Conducting polymers: a comprehensive review on recent advances in synthesis, properties and applications. RSC Advances 2021; 11(10): 5659-97.
[http://dx.doi.org/10.1039/D0RA07800J] [PMID: 35686160]

[80] Maity N, Dawn A. Conducting Polymer Grafting: Recent and Key Developments. Polymers (Basel) 2020; 12(3): 709.
[http://dx.doi.org/10.3390/polym12030709] [PMID: 32210062]

[81] Mishra AK. Conducting Polymers: Concepts and Applications. Journal of Atomic, Molecular, Condensate and Nano Physics 2018; 5(2): 159-93.
[http://dx.doi.org/10.26713/jamcnp.v5i2.842]

[82] Abel SB, Frontera E, Acevedo D, Barbero CA. Functionalization of Conductive Polymers through Covalent Postmodification. Polymers (Basel) 2022; 15(1): 205.
[http://dx.doi.org/10.3390/polym15010205] [PMID: 36616554]

[83] Zhu M, Zhao L, Ran Q, *et al.* Bioinspired Catechol-Grafting PEDOT Cathode for an All-Polymer Aqueous Proton Battery with High Voltage and Outstanding Rate Capacity. Adv Sci (Weinh) 2022; 9(4): 2103896.
[http://dx.doi.org/10.1002/advs.202103896] [PMID: 34914857]

[84] Ben Ayed E, Ghorbel N, Kallel A, Putaux JL, Boufi S. Polyaniline-Grafted Chitin Nanocrystals as Conductive Reinforcing Nanofillers for Waterborne Polymer Dispersions. Biomacromolecules 2022; 23(10): 4167-78.
[http://dx.doi.org/10.1021/acs.biomac.2c00635] [PMID: 36082444]

[85] Dai S, Feng Y, Wang P, *et al.* Highly conductive copolymer/sulfur composites with covalently grafted polyaniline for stable and durable lithium-sulfur batteries. Electrochim Acta 2019; 321: 134678.
[http://dx.doi.org/10.1016/j.electacta.2019.134678]

[86] Qin B, Yin Z, Tang X, *et al.* Supramolecular polymer chemistry: From structural control to functional assembly. Prog Polym Sci 2020; 100: 101167.
[http://dx.doi.org/10.1016/j.progpolymsci.2019.101167]

[87] Jiang Y, Zhang Z, Wang YX, *et al.* Topological supramolecular network enabled high-conductivity, stretchable organic bioelectronics. Science 2022; 375(6587): 1411-7.

[http://dx.doi.org/10.1126/science.abj7564] [PMID: 35324282]

[88] Sun Z, Ou Q, Dong C, *et al.* Conducting polymer hydrogels based on supramolecular strategies for wearable sensors. Exploration 2024; 4(5): 20220167.
[http://dx.doi.org/10.1002/EXP.20220167] [PMID: 39439497]

[89] Kolb HC, Sharpless KB. The growing impact of click chemistry on drug discovery. Drug Discov Today. 2003 Dec 15; 8(24): 1128-37.
[http://dx.doi.org/10.1016/S1359-6446(03)02933-7]

[90] Opsteena JA, van Hest JCM. Modular synthesis of block copolymers via cycloaddition of terminal azide and alkyne functionalized polymers. Chem Commun 2005; 1(1): 57-9.
[http://dx.doi.org/10.1039/b412930j]

[91] Eissa AM, Khosravi E, Cimecioglu AL. A versatile method for functionalization and grafting of 2-hydroxyethyl cellulose (HEC) *via* Click chemistry. Carbohydr Polym 2012; 90(2): 859-69.
[http://dx.doi.org/10.1016/j.carbpol.2012.06.012] [PMID: 22840013]

CHAPTER 3

Principles of Green Chemistry in Grafting of Polymers

A.R. Chabukswar[1,*], S.C. Jagdale[1], Yash D. Kale[1] and S.A. Polshettiwar[1]

[1] *Department of Pharmaceutical Sciences, School of Health Sciences and Technology, Dr. Vishwanath Karad MIT World Peace University, Pune, Maharashtra 411038, India*

Abstract: The polymer system and its chemistry have their roots in the systematic study of applied chemistry. Despite its origins in the last century's changes in climate and environmental pollution, polymer and green chemistry have become a hot topic for chemists. The polymer industry plays a significant role in applied chemistry as polymers have become omnipresent. Polymer grafting is a unique technique that enhances the chemical, morphological, biocompatibility, and physical properties, thereby improving the potential of polymers in terms of conduction and various properties beyond charge transport. A polymer's base activity can be enhanced by further conjugation with copolymers or by grafting. The grafting procedure can enhance the desired properties of the parent polymer and the activity of natural or synthetic polymers without altering their core nature. By this technique, the polymer backbone gains new properties from grafting different monomers like improved elasticity, ion exchange, hydrophobic/hydrophilic character, heat resistance, absorption of water, chemosensitivity, pH sensitivity, antibacterial effect, dye adsorption capabilities, *etc*. This chapter also covers the effectiveness and advantages of the polymer grafting approaches with their various applications in different fields. For the environment's safety, the solutions used in the solution polymerisation technique often face issues due to their viscosity and heat transfer. The central issue faced is removing solvent from the polymer, which requires intensive energy consumption methods like distillation. Since not all polymerization techniques can be performed using greener solvents, chemists are working to find alternative routes and safer solvents to minimize waste and help save the environment for such polymers.

Keywords: Biocompatibility, Grafting, Green chemistry, Monomers, Pollution, Polymer.

---

[*] **Corresponding author A.R. Chabukswar:** Department of Pharmaceutical Sciences, School of Health Sciences and Technology, Dr. Vishwanath Karad MIT World Peace University, Pune, Maharashtra 411038, India; E-mail: anuruddha.chabukswar@mitwpu.edu.in

Kuldeep Vinchurkar, Satish Polshettiwar, Nilesh Mahajan & Yogeshwar Bachhav (Eds.)
All rights reserved-© 2026 Bentham Science Publishers

INTRODUCTION

Overview of Green Chemistry

As the world has expanded its reach in science and technology, this has significantly boosted the economy and development, yet also led to environmental degradation [1]. Such sudden progress has manifested a high risk of climate change, ozone holes, and the accumulation of harmful, destructive organic pollutants in the biospheres [2]. Considering all these aspects, it is necessary to balance using natural resources and conserving the environment for a better future for the next generations [3]. Since green chemistry has already been practised over the last two decades, the environment has benefited a lot, and the harmful gas evolution rate has decreased over these years [4]. Due to this fact, new rules and regulations have been put forth to enhance the protection of the ecosystem from harmful chemicals and the synthesis of new compounds or moieties using much greener solvents and reagents. This is a step towards green chemistry [5].

Green synthesis/chemistry is a new branch of chemistry that involves using techniques or developing methods to help chemical engineers create chemical compounds and their synthetic routes in a much more straightforward manner, thereby eliminating harmful chemicals. This step towards eco-friendly and much more efficient products will lead to a brighter future for this ecosystem. Green chemistry has become a crucial tool in the fields of synthetic and applied chemistry [6].

Definition and Goals

According to the Environmental Protection Agency, green chemistry is defined as a chemistry that designs chemical products and processes that are environmentally harmless. Chemical products should be formulated in a manner that ensures they do not persist in the environment after application and break down into components that are harmless to the environment.

Historical Context and Development

Green chemistry was introduced by Poul T. Anastas in 1991 at the US Environmental Protection Agency's particular program launching event to provide a roadmap to a sustainable development method in chemistry, for industries working on chemical technology, academia, and government. In 1995, the annual US Presidential Green Chemistry Challenge was broadcast to the public. In 1996, the International Union of Pure and Applied Chemistry created a separate working party. In 1990, the first book and journal concerning green chemistry were published by the Royal Society of Chemistry [7].

The goals concerning green chemistry include:

- Clean chemistry
- Environmentally benign chemistry
- Atom economy

To reduce the use of dangerous or harmful substances in synthetic and applied chemistry, Paul Anastas developed twelve principles of green chemistry. However, when designing any reaction, it is very difficult to follow all twelve principles of the process simultaneously; however, it is recommended to apply as many principles as possible while carrying out synthesis.

Introduction to Polymer Grafting

Polymer grafting is used to enhance the polymer's shape, chemical composition, and physical characteristics. The current conduction and non-charge characteristics of polymers could be enhanced by this method. Transport improves the solubility of nano-dimensional traits such as morphology, biocompatibility, and biocommunication, and others of the parent polymer as illustrated in Fig. (**1**). The physicochemical characteristics of a polymer can be altered, considerably more by grafting or copolymerising with another polymer.

The significance of various chemical methods for polymer grafting—such as alkylation, click reactions, amide production, and free radical polymerization—has been addressed here, along with the process's significance and a thorough scientific justification. The efficiency of graft-to techniques and their use in a variety of domains are also covered in this review, which will give the learner a glimpse of polymer grafting [8]

Definition and Significance

Graft polymers are segmented copolymers comprising one composite's linear backbone and another composite's randomly distributed branches. The "graft polymer" in Fig. (**2**) illustrates the covalent link between the grafted chains of species B and polymer species A. Despite their structural differences, the individual grafted chains can be either homopolymers or copolymers. For the creation of stable blends or alloys, graft polymers, which have been manufactured for many years, are particularly utilized as impact-resistant materials, compatibilizers, thermoplastic elastomers, or emulsifiers. High-impact polystyrene, which has grafted polybutadiene chains over a polystyrene backbone, is one of the more well-known examples of a graft polymer [8].

Graft copolymers are branched copolymers in which the side chains' constituents differ structurally from the main chain's constituents. Due to their limited and compact fold structures, graft copolymers with more side chains can exhibit a worm-like conformation, a compact molecular dimension, and significant chain end effects. Graft copolymer preparation has been practised for many years. Graft copolymers can be created with general physical properties using any synthetic method. They can be applied to impact-resistant materials and are frequently utilised in the creation of stable alloys or blends as compatibilizers, emulsifiers, or thermoplastic elastomers. Materials produced using grafting techniques for copolymer synthesis are often more thermally stable than those produced by homopolymer synthesis [8].

To create a graft polymer, three synthesis techniques are utilized: grafting onto, grafting from, and grafting through.

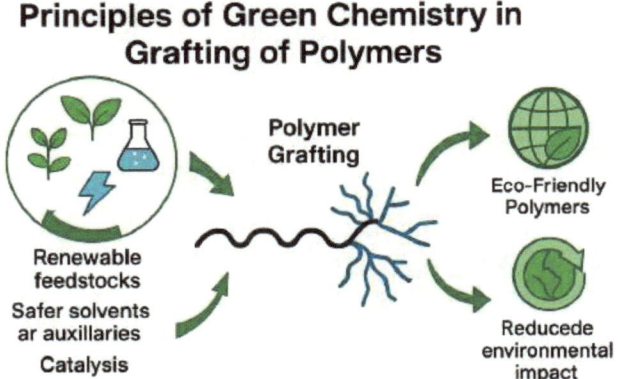

Fig. (1). Principles of green chemistry in grafting of polymers [1].

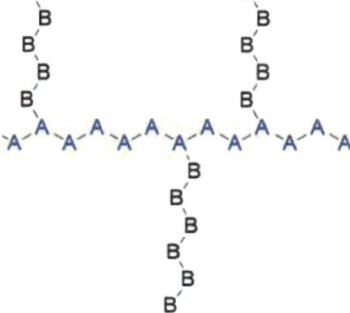

Fig. (2). Graft copolymer consist of a main polymer chain of (A) covalently bonded to the side chains (B) [8].

Types of Grafting Techniques

- **Grafting-Onto Method:**

The grafting-onto method involves using a backbone chain that has functional groups A randomly distributed along its length. The coupling reaction between the reactive end groups of the branches and the functional backbone leads to the development of the graft copolymer. Chemical modifications to the backbone enable these coupling processes. These copolymers are frequently created through reaction mechanisms such as anionic polymerization, atom-transfer radical polymerization, free-radical polymerization, and living polymerization. Anionic polymerisation processes are commonly employed in preparing copolymers derived from the grafting-onto approach. This process leverages a coupling interaction between the propagation site of an anionic living polymer and the electrophilic groups of the backbone polymer as shown in Fig. (3). Without creating a reactive group-containing backbone polymer, this technique would not be feasible. The popularity of this technique has increased as click chemistry has become popular. In the grafting-onto method of polymerization, atom transfer nitroxide radical coupling chemistry is a high-yield chemical reaction [8].

- **Grafting-from method:**

The grafting-from approach chemically alters the macromolecular backbone to introduce active sites that can start functioning. The starting sites may already be present in the polymer, be inserted *via* copolymerization, or be incorporated in a post-polymerization step. The number of active sites can regulate the number of chains grafted on the macromolecule if the active sites along the backbone contribute to forming a single branch. The lengths of each grafted chain may vary, despite the controllable number of grafted chains, due to steric and kinetic hindrance effects. "Grafting-from" reactions involving polyethylene, polyvinyl chloride, and polyisobutylene have been carried out. Atom-transfer radical polymerisation, free-radical polymerisation, anionic grafting, and cationic grafting have all been employed in the synthesis of grafting from copolymers. The grafting-from approach uses graft copolymers that are typically synthesised using anionic and cationic grafting procedures together with ATRP processes [8].

- **Grafting-Through Method:**

The macromonomer approach, commonly referred to as grafting-through, is a straightforward technique for creating graft polymers with well-defined side chains. Generally, a macromonomer functionalised with acrylate is copolymerised with a lower molecular weight monomer using free radicals. The number of grafted chains is determined by the copolymerisation behavior of the monomers and macromonomers as well as the ratio of their molar concentrations. The monomer-to-macromonomer concentrations fluctuate during the reaction, resulting in graft copolymers with varying branch counts and randomly arranged branches. Depending on the reactivity ratio of the macromolecule's terminal functional group to the monomer, this technique permits the addition of branches either homogeneously or heterogeneously. Variations in the graft distribution significantly impact the physical characteristics of the grafted copolymer. Macromonomers such as polyethylene, polysiloxanes, and poly(ethylene oxide) have been integrated into polystyrene or poly(methyl acrylate) frameworks as mentioned in (Scheme 1). Any known polymerisation method can be used with the macromonomer or grafting *via* approach. Particular control over the molecular weight, molecular weight distribution, and chain-end functionalisation is provided by living polymerisations [8].

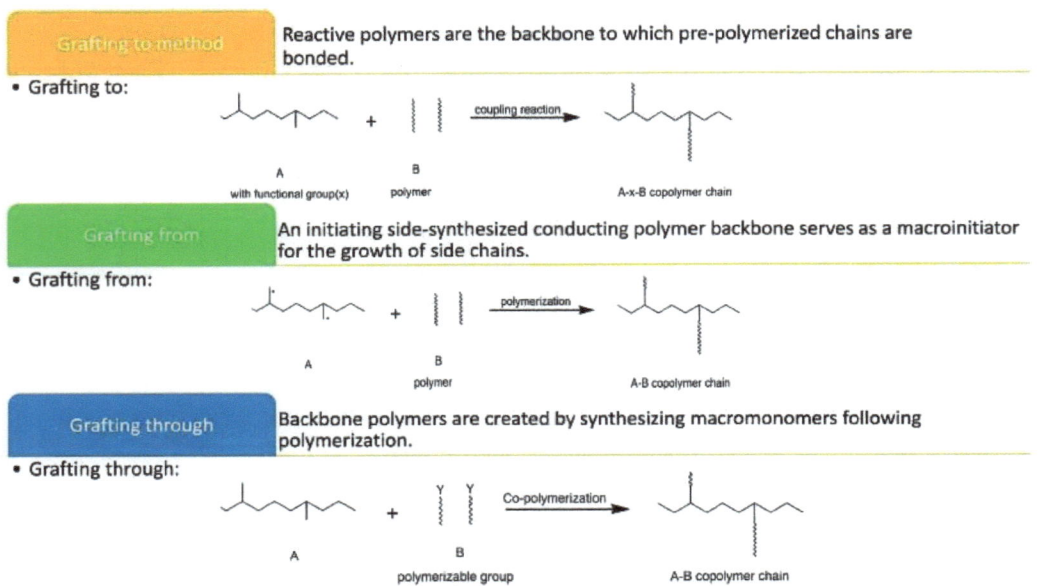

Scheme 1. Types of polymer grafting methods with pictorial representation [8].

Importance of Green Chemistry in Polymer Grafting

Since they have more applications than other copolymers, such as alternating, periodic, statistical, and block copolymers, which typically feature linear chains, graft copolymers have attracted considerable research attention [9].

- **Graft copolymers are often used in the following applications:**
  - Membranes for the separation of gases or liquids
  - Hydrogels
  - Drug delivery systems
  - Thermoplastic elastomers
  - Compatibilizers for polymer blends
  - Polymeric emulsifiers
  - Impact-resistant plastics

Environmental Impact

Green chemistry is the design (or redesign) of products and industrial processes to minimise their negative effects on human health and the environment [9]. The foundation of the GC concept is sustainability, which calls for reducing environmental effects and preserving natural resources for future generations. Although many of the concepts of "green chemistry" are not novel, the degree to which they have been utilized and organized into a coherent framework has increased interest in this field among regulatory, business, and academic institutions [4, 5]. Green chemistry has advanced to the point where academia and industry now view it as a regular procedure. Green chemistry is the creation of chemical products and processes that reduce or eliminate the need for hazardous chemicals and the processes that generate them. Thus, regarding both economic and environmental benefits, green chemistry techniques are intended to benefit society. Students today are eager to learn about sustainability and green chemistry concepts as public awareness of environmental challenges increases [10]. Because of this, numerous colleges have started incorporating green chemistry principles into their research projects or curricula [10].

Green chemistry ideas have demonstrated succesl in general chemistry and organic chemistry courses that are currently being offered. Because green chemistry is interdisciplinary, it may be taught alongside other topics, and several new courses in the field have been developed and added to the curriculum. While some of these courses were designed with science or chemical majors in mind, others were intended for non-science majors [11]. An undergraduate course in green chemistry at Westminster College now includes a community-based service-learning project. As part of this project, students designed lab activities related to green chemistry that a nearby high school will utilise [12].

There have been few documented attempts to incorporate community-based service learning into green chemistry classrooms, even though the concepts of green chemistry are strongly related to environmental and socioeconomic challenges [12]. The concept of green and sustainable chemistry, which first surfaced in the early 1990s, only started gaining momentum and recognition at the turn of the millennium [13]. The development of methods and tools that lead to more effective chemical processes that generate less waste and emit fewer pollutants than conventional chemical reactions is the focus of green and sustainable chemistry [13]. All facets and varieties of chemical processes that are less detrimental to the environment and public health than current best practices are collectively referred to as "green chemistry." The concepts and recommendations of Green Chemistry for any chemical process aim to achieve the following objectives.

• Increase the number of resources that are accessible for the development of a chemical process [14].

• Minimize the amount of waste produced when handling and preparing chemicals.

• Better methods should be employed to produce materials that have fewer detrimental effects on the environment.

• Substitute hazardous materials and goods with less environmentally damaging options with comparable features and uses.

• Cut down on the energy required to produce the compounds of interest by employing considerably faster methods or switching to renewable energy sources with higher efficiency and lower energy prices.

• Lessen both the compound's and its constituent material's toxicity [15].

• Chemists can significantly lessen environmental and human health risks by reducing or eliminating the use of potentially harmful compounds associated with a certain synthesis or process.

Green chemistry offers a useful concept for environmentally sound conservation supported by science. It is both a necessity of the present and the future. When designing reaction processes and selecting catalysts, scientists, researchers, and pharmaceutical companies must apply the principles of green chemistry [16]. By using green chemistry techniques, we can reduce waste, restrict the use of dangerous chemicals, maintain the atom economy, and protect the environment, which isour next generation's heritage. Integrating environmental preservation and technological advancement is one of the challenges of the new millennium. Chemists will largely determine the conditions for long-term development, and green chemistry may be the secret to their success. Green chemistry seeks to solve these problems by creating new synthetic schemes and equipment that streamline chemical manufacturing processes, creating novel reactions that maximise the desired products while minimizing by-products, and searching for greener solvents that are ecologically friendly [17].

Economic and Safety Benefits

- **Economic benefits:**

By utilizing safer materials and implementing waste reduction strategies, green chemistry can reduce expenses associated with disposal and mitigate health risks. Additionally, it can reduce petroleum product consumption, conserve energy and water, and increase the plant's capacity. Green chemistry also has the potential to increase chemical reaction yields by using less feedstock in those reactions [17].

- **Safety benefits:**

Green chemistry can contribute to a safer workplace by reducing the amount of hazardous chemicals that employees and customers are exposed to. Additionally, it can aid in averting chemical-related mishaps before they occur. The manufacture of goods and procedures with minimal usage and the creation of toxic materials is known as "green chemistry." It aims to stop pollution at the molecular level while considering economic and environmental goals [17].

PREVENTION OF WASTE IN POLYMER GRAFTING

Efficient Grafting Techniques

The backbone, monomer, initiator, and solvent of the system, as well as their interactions with one another, are key chemical elements that affect polymer grafting procedures and the reactions they entail. Several distinct polymer grafting factors are considered, including temperature and the use of additives along with the graft polymerisation synthetic pathway . Among the activators, there are several fascinating and adaptable pathways available for polymer modification [18, 19].

Grafting Through Chemical

There are two primary methods of chemical grafting: ionic initiator grafting and free radical grafting. Because it outlines the path of the grafting method, the initiator's role is crucial.

Free-radical Grafting

Free radicals are produced in these kinds of chemical reactions and are then transferred to the substrate to react with monomers to form graft copolymers. Different polymerization techniques, such as polymer grafting, free radical processes (such as FRP, RDRP, and REX), *etc.*, are utilized. In the grafting process, the polymer can be created by radiation, which causes homolytic fission in macromolecules. The behaviour of free radicals determines their lifespan. Polymer backbone is a crucial component in forming stable Graft polymers. Table 1 shows types of free- radical polymerisation methods. Several variables affect polymer grafting by FRP and other reactions, including the composition of the system's elements (backbone, monomer, initiator, and solvent) and their interactions, as shown in Fig. (3) [20, 21].

Table 1. Types of free-radical polymerisation methods [20, 21].

Conventional Radical Polymerization(FRP)	Reversible Deactivation Radical Polymerization(RDRP)	Reversible Deactivation Radical Polymerization(RDRP)
Step 1- Initiation, in which free radicals are formed. Step-2- Propagation, in which free radicals react with the monomer. Step 3 - Polymer radicals are eliminated by combining and disproportionating to small molecules. These reactions occur simultaneously and produce a broad range in the molar mass distribution.	Controls the growth of the polymer molecule while it is polymerizing. Controllers that function through a few unique chemical routes can reversibly deactivate polymer radicals. It is a reversible catalytic process where an alkyl halide macromolecule reacts with the catalyst to produce a radical that spreads till it reacts with the catalyst once more.	Reactive polymerizations occur in the melt phase and are mostly carried out in extruders through five main types: bulk polymerization, polymer grafting, polymer functionalization, controlled degradation, and reactive blending. Polyolefin and starch additions are examples of REX polymer grafting.

Atom Transfer Radical Polymerization (ATRP)

Through the catalysis of metal complexes, or metal complex-mediated living radical polymerisation, the reversible activation/deactivation process was further accomplished. It utilizes halogen atoms to shield the inactive polymer chains, primarily bromine or chlorine atoms, as shown in Scheme **2**. Atom transfer radical polymerization (ATRP) is the name given to this technique, which evolved from atom transfer radical addition (ATRA) in organic synthesis. A halogen atom is simultaneously abstracted from the carbon-halide link, and the metal complex, MX_n/L_y (such as copper(I) bromide), is subjected to one-electron oxidation. This radical (P•) is then generated to proliferate monomers. To return to the dormant state of P-X (the polymer chain end-capped with a halogen atom), the radical can, however, also react with MX_{n+1}/L_y generated at the higher oxidation state (for example, copper(II) dibromide). The vast range of monomers that can be successfully processed with this process includes styrene, acrylates, methacrylates, acrylamides, and acrylonitrile in bulk, suspension, solution (with organic or water-based solvent), and supercritical carbon dioxide [22].

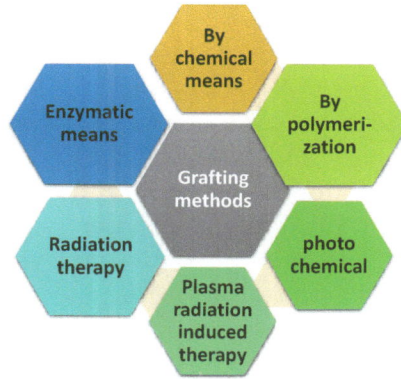

Fig. (3). Grafting with its type of methods [19].

$$MX_n/L_y + P-X \rightleftharpoons MX_{n+1}/L_y + P^\bullet$$

M - Metal atom
X - Br, Cl
L - Ligand

kp
Monomer

Scheme 2. Atomic transfer radical polymerization representation (ATRP) [22].

Table 2. The ATRP systems [22].

Metal	ligand	Initiator
CuBr, CuCl	2,2-Bipyridine	(1-phenylethyl halide structure)
RuBr$_2$	Multidentate amine	(methyl 2-halopropanoate structure)
FeBr$_2$	PPh$_3$ + Al(OiPr)$_3$	
NiBr$_2$	PPh$_3$	(methyl 2-halopropanoate structure)

(Table 2) cont.....

Metal	ligand	Initiator
PdBr$_2$	PPh$_3$	phenyl-SO$_2$Cl

An initiator, a metal halide (in a low oxidation state) complexed with ligand(s), and a monomer make up a typical ATRP system. Cu(I), Ni(II), Ru(II), Fe(II), and Rh(II) elements have provided the basis for the development of several catalysts (Table 1); however, Cu(I) and Ru(II) systems have received the greatest attention. Bipyridine or multidentate amines are the typical ligands for copper-based catalysts, while PPh3 is the typical ligand for other catalysts. The ligand system type, which encompasses electronic, steric, and solubility features, has a significant impact on catalyst activity and the management of polymer molecular weight. To polymerize a monomer, the initiator and the monomer must be compatible. The initiators utilised in ATRP are alkyl halides with activating groups such as aryl, carbonyl, or nitrile groups on the a-position; examples of these initiators include I-phenyl ethyl bromide, CCl4, 2-chlorobutyrate or 2-bromobutyrate, methyl a-bromophenyl acetate, and arene sulfonyl chloride as shown in Table **2**. These initiators differ structurally from those used in conventional free-radical polymerization [22].

Reversible Addition-Fragmentation Chain Transfer (RAFT) Polymerization

Reversible addition-fragmentation chain transfer, or RAFT polymerization, is a more adaptable technique for imparting radical polymerization. It is a type of reversible deactivation radical polymerisation (RDRP). It has been explained how RAFT polymerisation has evolved historically at CSIRO. The following are the characteristics of RAFT.

- The capacity to control the polymerisation of the majority of monomers polymerisable by radical polymerisation is one benefit of RAFT polymerisation. Vinyl monomers, styrene, dienes, acrylonitrile, (meth)acrylates, and meth(acrylamides) are a few examples.
- Monomer and solvent (*e.g.*, OH, NR$_2$, COOH, CONR$_2$, SO$_3$H) tolerance for unprotected functionality. Aqueous or protic medium can be used for polymerisations.
- Harmony with the reaction circumstances (bulk, organic or aqueous solution, emulsion, mini-emulsion, suspension, *etc*.
- Simple to use and affordable in comparison to other technologies.

In a perfect living polymerization, there is no irreversible chain transfer or termination; instead, all chains are initiated at the beginning of the process, grow constantly, and survive the polymerization. Chains can be extended by introducing more monomers to the reaction if initiation and propagation proceed quickly. This will result in a highly narrow molecular weight distribution. All chains cannot be active simultaneously during a radical polymerisation. These characteristics are demonstrated in RDRP, such as RAFT polymerization, when reactant conditions allow for a quick equilibrium between the active and dormant chains, and when reactants can reversibly deactivate propagating radicals, thereby maintaining the majority of living chains in a dormant form (Fig. **4**).

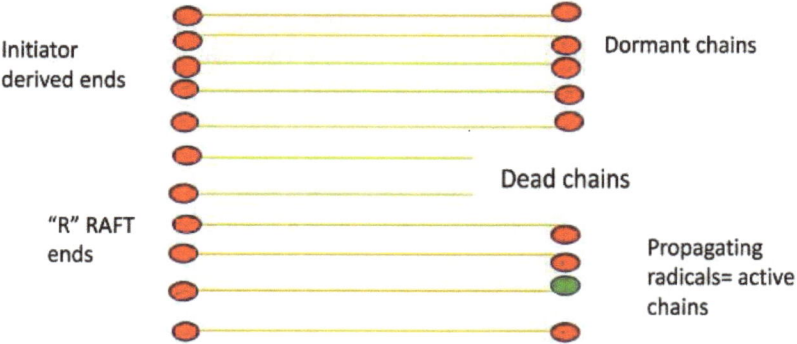

Fig. (4). RAFT polymerisation Scheme. The quantity of each chain displayed here does not commensurate with what would be anticipated from a carefully planned experiment. Due to the quick propagation-related equilibration of the dormant and active chain ends, all living chains typically grow synchronously and have identical lengths [23].

Applications

- The process of synthesis of polymers *via* molecular weight under control:

The first application of ATRP is in synthesizing different polymers with extremely low polydispersity and controlled molecular weights. Could the molecular weight distribution of polystyrene (PS) produced by ATRP, for instance, be the same as that of PS produced conventionally using live anionic polymerisation?. Additionally, the theoretical values determined by the [M]o/[I]o ratio and the molecular weights of PS agreed [22, 23]. Low-polydispersity polymer synthesis in an ATRP method depends on the rapid initiation and deactivation of living chains. For all the polymer chains to grow concurrently, the fast initiation necessitates that the initiation reaction occurs far quicker than the propagation [24]. A delayed initiation results in a mixture of long (early propagating) and short (later propagating) chains; hence, low polydispersity is impossible. As a result, the initiator must be compatible with the type of monomer [25, 26].

- The Production of Block Copolymers

The strategy for synthesising well-defined block copolymers is to use a proper order of monomer additions or a halide-exchange method. The former is related to the fast-initiation requirement discussed above [25, 26].

- Synthesis of Star Polymers

Star polymer synthesis is another use for ATRP. The ATRP produces multi-armed star polymers when a molecule with several initiator moieties (carbon-halide bonds) is utilised as a promoter. To make 3-armed star polymers, for instance, Compounds A and B (Scheme 3) can be utilized. In contrast to linear polymers, star polymers have measured molecular weights that are lower than the calculated values due to their smaller molecular hydrodynamic volumes compared to equivalent linear polymers with the same molecular weights [26].

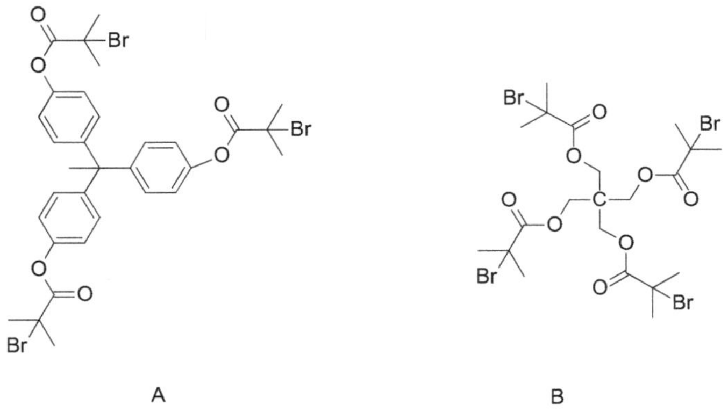

Scheme 3. The initiators for synthesising star polymers [26].

- Synthesis of Hyperbranched Polymers

While they are significantly more straightforward to synthesize, hyperbranched polymers share a similar branching pattern with dendrimers. ATRP offers a novel and effective method for creating hyperbranched polymers from vinyl monomers. Every unit added to the polymer chains during the ATRP polymerisation of a monomer with a carbon-halide group (Scheme 4) also adds a new branching site. Ultimately, without cross-linking, this type of polymerisation will result in a three-dimensional structure. Compared to linear polymers of the same molecular weight, hyperbranched polymers have substantially smaller gyration volumes due to their highly condensed chain structure [27, 28].

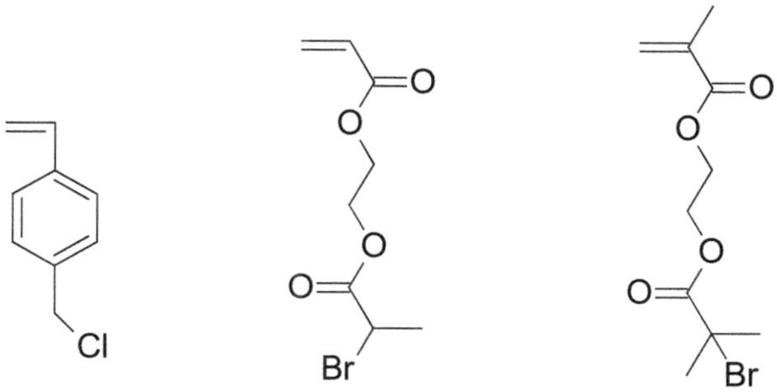

Scheme 4. The monomers for hyperbranched polymers [28].

- Synthesis of End-Functionalized Polymers:

When an alkyl halide compound contains a functional group such as hydroxyl, amino, or vinyl as an initiator, the functional group will be attached to the polymer chain end. These end-functionalized polymers are helpful for supramolecular construction. There were also two reports on the synthesis of end-unsaturated, functionalized polystyrene using ATRP in the literature. Vinyl chloroacetate and allyl bromide were used as initiators to prepare polystyrene macromonomers. The vinyl-terminated polystyrene copolymerized with N-vinylpyrrolidone yielded hydrogels of poly(N-vinylpyrrolidone-g-styrene) [27, 28].

Challenges and Solutions

With ATRP, one can now easily create specialised polymers with smaller molecular weights. Low catalyst efficiency, which often necessitates high catalyst concentrations, is a significant issue for ATRP.. The monomer's metal halide content is between 0.1% and 1% (molar). When this catalyst is used in the product, it contains a significant amount of catalyst residue. The result is potentially poisonous and has a rich colour due to this leftover catalyst. Consequently, post-purification, which typically involves passing the polymer/catalyst mixture through silica or alumina gel or resins, is necessary to remove the catalyst from the product produced by ATRP [27]. As a result, the catalyst is wasted, and this post-treatment is expensive and time-consuming. Thus, creating methods to directly lower the catalyst residue concentration in the finished product to prevent post-purification is a significant problem for ATRP. Catalyst support was created to address this issue. The theory is that the catalyst may be readily extracted from the polymer solution and, ideally once regenerated, it becomes immobile on insoluble particles. By complexing with their ligands grafted onto particles, ATRP catalysts were immobilised onto particles (Scheme 5). Nevertheless, it was discovered that the catalysts based on particles were ineffective at mediating the ATRP process. Using CuBr supported on silica gel or crosslinked polystyrene particles *via* Schiff-base, the molecular weights of polystyrene and PMMA generated by ATRP were significantly greater than that anticipated with high polydispersity [27, 28].

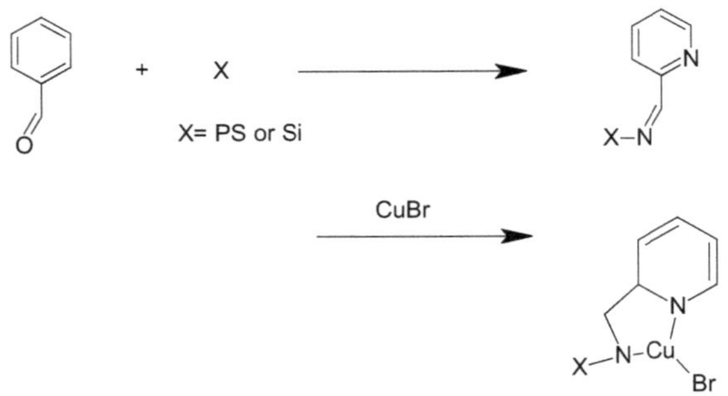

Scheme 5. Catalyst immobilization technique [28].

ATOM ECONOMY IN GRAFTING REACTIONS

Principles of Atom Economy

A chemist finds the yield percentage of a chemical process to be quite significant. It provides a means to compare theoretical and actual product quantities. It assesses reaction efficiency. However, to compare the effectiveness of chemical reactions, the computation of "atom economy" has gained importance recently [29].

To determine the effectiveness of a specific reaction, chemists frequently compute percentage yield in the manner shown below:

$$\% \text{ yield} = \frac{\text{Actual yield}}{\text{Theoretical yield}} \times 100$$

Using this method, a 100% yield of the intended product indicates that a reaction is efficient. However, the amount of undesirable products generated in the reaction pathway is not disclosed by the % yield estimate. Numerous instances of extremely "efficient" chemical reactions in the industry produce trash that is significantly larger in mass and volume than the intended product. The term "atom economy," which was coined by Stanford University Professor Barry Trost, is derived from "green chemistry" ideas. Designing chemical products and processes to eliminate or minimise the usage or production of hazardous compounds is the primary goal of green chemistry. Not every reactant atom will always be included in the final product during the synthesis of a given chemical compound. The atoms not included in the intended product will be used to create the by-product, and waste materials could be dangerous to the environment. Sheldon, a professor at Delft University in the Netherlands, measured the idea of an atom economy. The molecular weight of the intended product divided by the molecular weights of all the products formed in a reaction yielded the following percentage atom utilisation, which he used to determine the quantity of waste product for a given reaction [29].

% Atom utilisition = Formula weight of desired product Formula weight of (the desired product + the waste and by-product) X 100

Strategies for Achieving a High Atom Economy in Grafting

The reagents employed in a given reaction, along with the type of chemical reaction involved, determine the reaction's atom economy. Rearrangement (*e.g.*, migration of an alkyl group), addition (Example 1), substitution (*e.g.*, chlorination of methane), and elimination (*e.g.*, dehydration) can be used to categorise the majority of typical chemical reactions. Since addition and rearrangement reactions only require repositioning reactant atoms within the same molecule or their inclusion into a second molecule, they are atom-economical by nature. However, substitution reactions have an innately low atom economy since they entail substituting one group for another. Because deleted atoms are constantly lost as waste, elimination reactions are intrinsically uneconomical. Therefore, the industrial chemist may favour addition and rearrangement processes over less ecologically friendly substitution and elimination reactions when creating an atom-efficient chemical pathway [29].

Selective Grafting Techniques

While considering both the yield and the atom economy provides a more accurate assessment of a reaction's efficiency and environmental acceptability, other considerations must also be taken into account (Scheme **6**).

A) Rearrangement Reactions:

This involves atoms that form molecules. Hence, the atom economy of these reactions is 100%

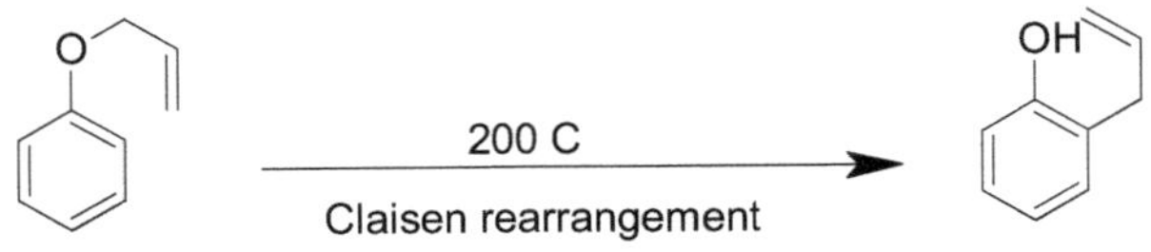

[Claisen rearrangement scheme: allyl phenyl ether → 2-allylphenol at 200 C]

in this rearrangement reaction, %Atom economy= (134.1/134.1)*100=100%

B) Addition reaction:

[2,3-dimethyl-2-butene + Br₂ → 2,3-dibromo-2,3-dimethylbutane]

Addition Reaction- halogenation of an alkene

in this addition reaction, %Atom economy= (215.8/215.8)*100=100%

Scheme 6. Atom economy schematic representation [29].

• Atom Economy in Elimination Reactions:

The remaining discussion on reaction efficiency will exclusively focus on the topic of an atom economy based on reaction stoichiometry. It might also be a good idea to compute the atom economy based on the amounts of reagents used when encountering these reactions in the lab (experimental atom economy). Furthermore, one might want to consider energy consumption, toxicity, the use of auxiliary materials, the distinction between catalytic and stoichiometric reagents, and the choice between sustainable and non-renewable feedstocks.

The atoms of the leaving group (OH) that are being replaced, the counterion (sodium) of our nucleophile (bromide), and the sulfuric acid that is needed for this reaction are all wasted in the formation of unwanted products in this reaction, by the substitution reaction. Elimination reactions are much worse than substitution reactions in terms of their atom economy because they only need the loss of atoms from the reactant, while acquiring none [29, 30].

Take the atoms in the following elimination reaction as an illustration. One popular technique for converting alkyl halides into alkenes through elimination is base-promoted dehydrohalogenation. The reaction between 2-bromo-2-methylpropane and sodium ethoxide in Scheme 7 results in the formation of methyl propene (Table 3). The reactant atoms in this reaction that are integrated into the desired product, and this product, are depicted in green. In contrast, the reactant atoms that are not used and the reactant atoms in the reaction's undesirable products are displayed in brown. The atom economy of this reaction is presented in Table 4, with a calculated atom economy of 27%. Equation 2 depicts the formation of 2-bromo-2-methyl propane [29, 30]

Table 3. Economy is a result not only of the loss of the HBr but also because this is a base-promoted reaction, and all of the atoms of the sodium ethoxide base are found in unwanted side products [29].

Reagents Formula	Reagents FW	Utilized Atoms	Weight of Utilized Atoms	Unutilized Atoms	Weight of Unutilized Atoms
7 C_4H_9Br	137	4C,8H	56	HBr	81
8 C_2H_5ONa	68	-	0	2C,5H,O,Na	68
Total 6C,14H,O,Br,Na	205	4C,8H	56	2C,6H,O,Br,Na	149
%AE =27%	-	-	-	-	-

Equation 1:-

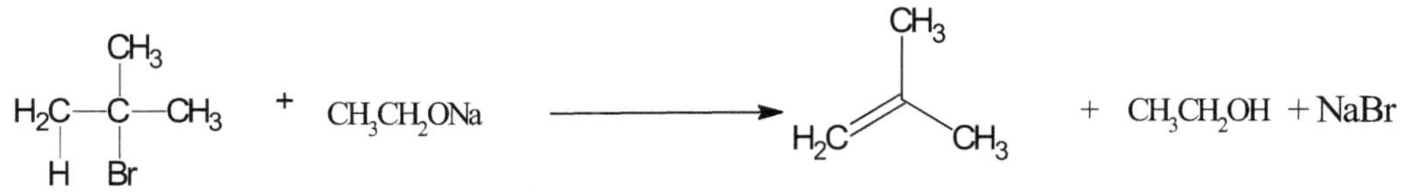

Scheme 7. Formation of methyl propene [29].

• **Atom Economy in Addition Reactions:**

Addition reactions produce excellent atom economy because they typically result in incorporating all the reactant atoms into the final intended products. Therefore, addition reactions are more environmentally friendly than elimination and substitution processes from the perspective of atom economics. Considering the following example of the addition of hydrogen bromide to methyl propene, since every atom in the reactants is used in the ultimate desired product, all the atoms of the reactants are displayed in green in this example (Scheme 8). The exceptional atom economy of this reaction is further shown by the atom economy table (Table 4) and the computation of 100% atom economy [29].

Table 4. Atom economy equation 2 [29].

Reagents Formula	Reagents FW	Utilized Atoms	Weight of Utilized Atoms	Unutilized Atoms	Weight of Unutilized Atoms
9 C_4H_8	56	4C,8H	56	-	0
11 HBr	81	HBr	81	-	0
Total 4C,9H,Br	137	4C,9H,Br	137	-	0
-	-	-	-	-	%AE=100%

Equation 2:-

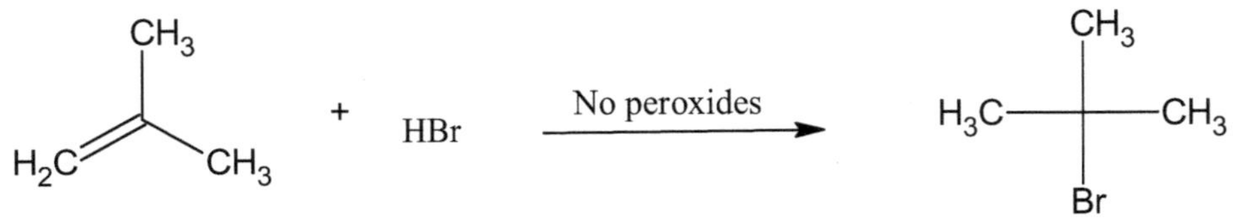

$$\underset{H_2C}{\overset{CH_3}{\diagdown}}{=}\underset{CH_3}{\diagup} + HBr \xrightarrow{\text{No peroxides}} H_3C-\underset{Br}{\overset{CH_3}{\underset{|}{C}}}-CH_3$$

Scheme 8. Formation of 2-bromo-2-methyl propane [29].

• Atom Economy in Rearrangement Reactions:

The atoms in a molecule are rearranged during the rearrangement process. From the perspective of atom economy, rearrangement processes are environmentally favorable because no atom substitution, addition, or elimination occurs in the molecule undergoing the reaction. Consider the acid-catalyzed rearrangement of 2,3-dimethyl-2-butene from 3,3-dimethyl-1-butene as an example. Because they are all integrated into the intended product, the reactant atoms, in this instance, are all displayed as green. The atom economy percentage can be determined by setting up an atom economy table, just as in the other examples. The acid (H+) employed in this reaction is only used in catalytic levels; as a result, it is indicated in black (rather than green or brown) and is not considered in the atom economy table or calculation, even though it is not integrated into the target product. For rearrangements, the reaction's % atom economy is 100%, as was previously predicted in Table **5** (Scheme **9**).

Table 5. Atom economy equation 3 [29].

Reagents Formula	Reagents FW	Utilized Atoms	Weight of Utilized Atoms	Unutilized Atoms	Weight of Unutilized Atoms
12 C_6H_{12}	84	6C,12H	84	-	0
Total 6C,12H	84	6C,12H	84	-	0
	-				%AE=100%

Equation 3:-

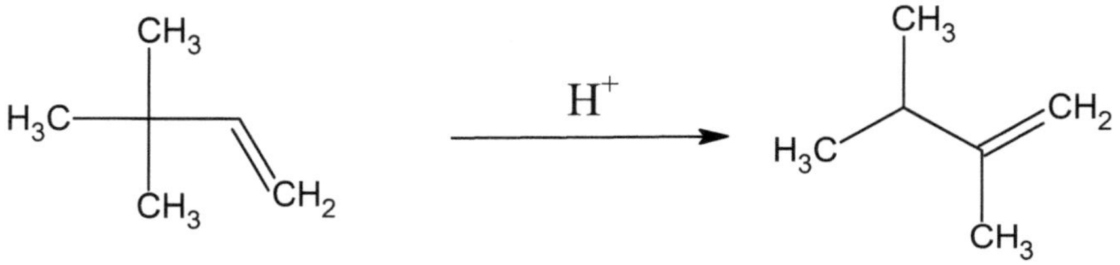

Scheme 9. Formation of 2,3-dimethyl-2-butene [29]

Controlled Polymerization Methods

• Nitroxide Mediated Polymerization

Stable free radicals, including nitroxides, are employed in NMP systems as reversible terminating agents to regulate the polymerization process. Reversible deactivation of the developing chains occurs through the establishment of covalent bonds, resulting in the generation of dormant chains. The bond experiences homolytic cleavage at high temperatures, which yields the nitroxide radical and the active growth chain. A brief deactivation process occurs immediately after activation, adding a few monomer units to the propagating chain [30, 31].

• Atom Transfer Radical Polymerization

Atomic Switching Radical polymerisation, first described by Matyjaszewski and Sawamoto, is based on a continuous and reversible halogen transfer (pseudo halogen) between a transition metal (such as Cu), complexed by a ligand (such as bipy), in its lower oxidation state, and a dormant propagating species (Pn). A one-electron oxidation of the transition metal occurs concurrently with halogen transfer, and propagation is accomplished by adding monomers to the active chains. The fast, simultaneous initiation and quick deactivation of the developing chains regulate the homogeneity of the polymeric chains. For this control to be successful, the deactivation rate needs to be higher than the propagation rate. Later phases of termination and chain transfer processes are suppressed by low concentrations of active centers, which are maintained at low levels by deactivation [31].

USE OF RENEWABLE FEEDSTOCKS

Renewable and Natural Polymers

There is a growing concern that producing polymers heavily reliant on fossil fuels is not sustainable. Inevitably, polymer manufacture will completely shift to renewable resources. Biodegradable polymers derived from renewable resources encompass cellulose, starch, chitosan, lignin, proteins, oils, common commodity polymers (such as polyethene and polyethene terephthalate), and microbiological poly(ester)s (Fig. **5**). It is a significant endeavor to conduct fundamental research on the synthesis, modification, property enhancement, and new applications of these materials [32].

Cellulose, Starch, and Chitosan as Grafting Substrates

A review of renewable resource-related polymers now reveals a plethora of worldwide activity in both the academic and industrial domains. The preparation, processing, characteristics, and performance of several polymers derived from renewable resources are discussed here.

Cellulose

Wood, used to make a wide range of fundamental furniture and construction materials, mostly consists of cellulose. Typically, plants that utilize different pulping techniques grow to produce clean cellulose. Pulp fibres are used in printing, industrial, and everyday home items, including paper towels, handkerchiefs, facial tissues, napkins, and toilet paper. It is this structure that gives cellulose its unique properties. The hydroxyls in cellulose have equatorial locations, which allow them to protrude laterally along the stretched molecule and make them easily accessible for hydrogen bonding as shown in Fig. (**6**). The chains assemble into a highly organized structure because of these hydrogen bonds. Due to the strong interchain hydrogen bonds present in the crystalline areas, the resulting structure exhibits high surface energy, good strength, and a high hydrophilic character. It is insoluble in most solvents. Additionally, they prevent cellulose from melting, making it a non-thermoplastic. The three OH groups that are present in each anhydrous glucose unit (AGU) are what give cellulose its notable reactivity [33, 34].

Because of the species and the processes used in their isolation and purification, most celluloses have a high degree of polymerization. These latter procedures frequently add functions to the macromolecular backbone, such as carboxyl and carbonyl groups.

Starch

The primary type of carbohydrates that green plants store is starch. It is the main ingredient in roots, tubers, and seeds. Commercial starch is produced from various ingredients, including potatoes, sago, tapioca, wheat, and corn. Glucopyranose, a six-membered ring glucose unit, is found in starch. The molecular structure of starch can be divided into two main categories. Amylopectin is a structure in which the starch molecules are highly branched. It is the predominant component of starch, making up as much as 100% of waxy starches, 72% of regular maize starch, and 80% of potato starch. The crystalline structure of starch is attributed to the double helix a-1-6 links that branch off from chains of glucose units connected by a-1-4 linkages to form amylopectin molecules [35, 36].

Amylose is the name for the other type of starch structure, which is made up of repeated units of 1-4-a-glucose with relatively few branches. The figure displays the structures of both amylopectin and amylose. In addition to its use in food applications, starch is also used extensively as a glue in the corrugated board sector and as a surface-modifying agent in the paper and textile industries. Laundry, baby powder, oil manufacturing, and the creation of bioplastics or biodegradable polymers are among the additional uses for starch [37, 38].

Chitosan

The fully or partially N-deacetylated derivative of chitin is called chitosan (Fig. **6**). Acetic acid and other organic solvents can dissolve chitosan. In many applications, it is superior to chitin. Because chitin and chitosan have a higher nitrogen content (6.89%) than synthetically replaced cellulose (1.25%), they are of commercial interest. Since chitin and chitosan include amino functionality, they can be changed appropriately to impart desirable qualities and unique biological roles, such as solubility as shown in Fig. (**7**) [39, 40].

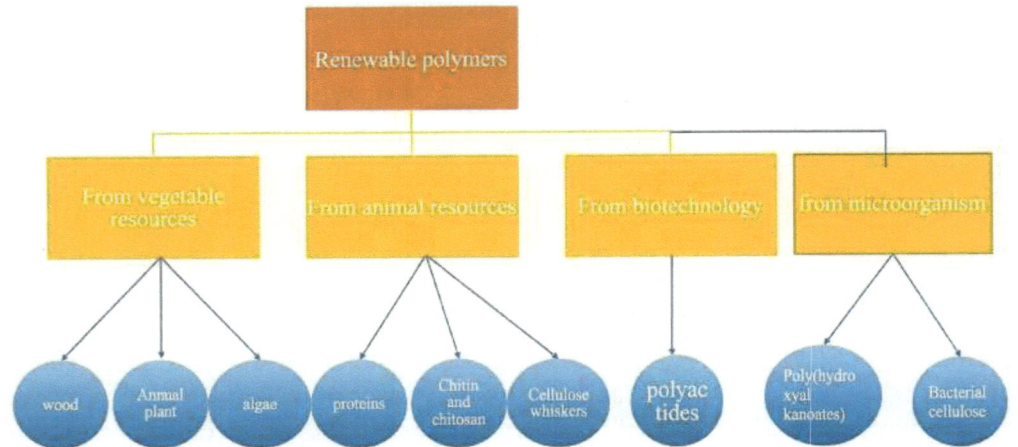

Fig. (5). Renewable resources classification [32].

Fig. (6). Chemical structure of cellulose [34].

Fig. (7). Typical chitosan structure [40].

Benefits and Challenges of Using Renewable Feedstocks

Although using renewable feedstock in manufacturing has many advantages, there are also drawbacks. To fully realize the potential of renewable feedstock, supply chain challenges and technology constraints must be resolved.

Overcoming Supply Chain Issues

Supply chain management is essential for consistently available and high-quality renewable feedstock. The secret is to establish dependable supply chains that can manage a diverse range of feedstock materials. Cooperation between farmers, processors, and manufacturers is essential to maximize the flow of renewable feedstock from the field to the factory and overcome logistical obstacles [40 - 43].

Addressing Technological Limitations

There may be technological barriers to using renewable feedstock, particularly regarding scalability and compatibility with current industrial methods. Ongoing research and development initiatives, however, are focused on creating novel solutions and increasing process efficiency. By investing in technological advancements, manufacturers can overcome these constraints and fully capitalize on renewable feedstock [40 - 43].

In brief, renewable feedstock offers several advantages for environmentally friendly production. We can recognize renewable feedstock's potential to lower carbon emissions and preserve natural resources by understanding its various forms and applications in production. The benefits to the economy and technological advancements emphasise the significance of renewable feedstock in promoting environmentally friendly production methods. Even though there are obstacles to overcome, such as supply chain problems and technological constraints, cooperation and continuous progress can help. Renewable feedstock has a bright future in sustainable manufacturing, and a more sustainable and greener world depends on its broad adoption [44, 45].

Case Studies and Industrial Applications

Biopolymers have garnered significant interest in various applications that require sustainable and biodegradable solutions. Treatment of disease still largely depends on developing drug delivery systems to increase the activity of bioactive chemicals, and tremendous progress has been made in this area. In this sense, the creation of drug delivery systems often utilizes synthetic, natural, and semi-synthetic polymers. Several environmental issues arise from the everyday use of artificial and chemical-based polymers by the food and medical industries. The development of packaging materials based on biopolymers is primarily motivated by the need to reduce pollution, manage municipal solid waste, and promote environmental awareness toward sustainability. By using biopolymers, carbon dioxide emissions, municipal solid waste, and dependency on resources derived from petroleum have reduced [46, 47].

Applications of Biopolymers

Biopolymers for Medical Applications

The key to increasing life expectancy is science and technology. This has led to decreased morbidity and death rates through the development of innovative techniques and new technology. One crucial strategy for treating diseases is the use of drug delivery techniques to boost the efficacy of bioactive substances, and significant progress has been made in this field. In this regard, synthetic, semi-synthetic, and natural polymers are frequently used to develop drug delivery systems [48]. Biopolymers have many potential medical uses, including tissue guidance, adhesion, fixing, suturing, covering, occlusion, isolation, contact inhibition, and cell proliferation. To address and investigate the significant functional, architectural and compositional aspects of natural tissues, several biomaterial production techniques and technologies have been created in recent decades. The quest for more superior and tissue-focused implanted devices has expanded the knowledge of biomaterials and sparked interest in multimodal scaffolds with distinctive geometries and physical-chemical characteristics. Because these scaffolds integrate several topographies that are not typically seen in each material, they have multifunctional or multimodal properties, which increase their potential importance in regenerative medical approaches [49]. The development of three-dimensional templates and the production of artificial extracellular matrix (ECM) environments for tissue regeneration are made possible by polymers. Biopolymers can be produced artificially or from natural resources. Since every group has unique advantages and limitations, various composite materials and interpenetrating networks have been developed to achieve the necessary goals [50].

Biopolymers for Industrial Applications

Over the years, there has been a steady increase in industrial interest in biopolymers. The need for novel materials from future biopolymer producers is enormous [51]. Nonetheless, as the material is specifically made available for sustainable development, its cost-effectiveness must increase. Traditional polymers and bio-based polymers are now more similar than ever [52]. Nowadays, bio-based polymers are frequently found in various applications, from consumer items to high-tech ones, thanks to expanded biotechnology research and development and increased public awareness. Food packaging is crucial for preserving food's quality, purity, and safety throughout its entire shelf life, as well as shielding it from outside contamination [53]. The food business mostly uses synthetic polymers for packaging because of these materials' low cost, low manufacturing difficulty, affordability, flexibility, and lightweight nature.

However, these artificial polymers are not biodegradable, and most plastic waste and scraps severely contaminate the environment. Biodegradable polymer compounds must be developed and used to address these environmental issues. Examples of biopolymers include pullulan, hemicellulose, starch, xanthan gum, pectin, carboxymethyl cellulose, and other biopolymers derived from renewable resources [54, 55]. There is considerable promise for alginate, guar gum, gum karaya, agar, gellan, and other materials to replace conventional petroleum-based food packaging materials. Applications of biopolymers, such as chitosan and PLA silk, in medicine are being increasingly investigated. Biopolymers' special qualities, such as their biocompatibility and biodegradability, offer several advantages and raise the possibility of using them in implanted medicinal applications. Since synthetic materials do not satisfy the needs of biological systems, these novel materials are crucial to medicine [56, 57]. Thus, recent research has shown that biopolymers combined with synthetic materials have the potential to change medicine [58].

SAFER SOLVENTS AND REACTION CONDITIONS

Environmentally Benign Solvents

The primary source of auxiliary waste in polymer research is the use of solvents. The use of organic solvents has been reduced or replaced by the investigation of alternative reaction media, driven by the growing importance of sustainable chemistry in polymer research. In polymer chemistry, water, supercritical carbon dioxide, and ionic liquids are the most often utilised green solvents.

Green chemistry utilizes chemical solvents that are environmentally friendly, referred to as "green solvents." They gained notoriety in 2015 when the UN recognised the need for green chemistry and green solvents for a more sustainable future and established a new development plan with a sustainability focus based on 17 sustainable development goals. As an alternative to petrochemical solvents, green solvents are being developed that are more environmentally friendly and derived from processing agricultural commodities or other sustainable techniques. Low toxicity, ease of recycling, and ease of biodegradation are a few expected qualities of green solvents [59, 60].

The following can be classified as green solvents according to their production methods or the raw resources they use to make them:

Water

Water does not qualify as an organic solvent because it is carbon-free. Its chemical makeup makes it a polar protic solvent. It is renewable and non-toxic.

Many ions and proteins are dissolved in water in living things. It is the most affordable and plentiful solvent for a wide variety of industrial chemistry reactions and procedures [60].

Supercritical Fluids

When brought to temperatures and pressures above their critical points, some compounds in the gas phase at room temperature and pressure can function as solvents. At this stage, a supercritical fluid, a mixture of liquid and gas states existing in one phase, has characteristics halfway between a liquid and a gas, such as gas mobility and liquid dissolving power [60].

At a pressure of 22.05 MPa and a temperature of 374.2°C, supercritical water (SCW) is produced. It exhibits the characteristics of dense gas and dissolves just as well as low-polarity organic solvents. On the other hand, inorganic salt solubility is drastically decreased in SCW (Scheme **10**). SCW is used as a reaction medium, particularly in oxidation procedures, to eliminate hazardous materials like those present in industrial aqueous effluents. Two major technological obstacles to the utilisation of supercritical water are salt deposition and corrosion [60].

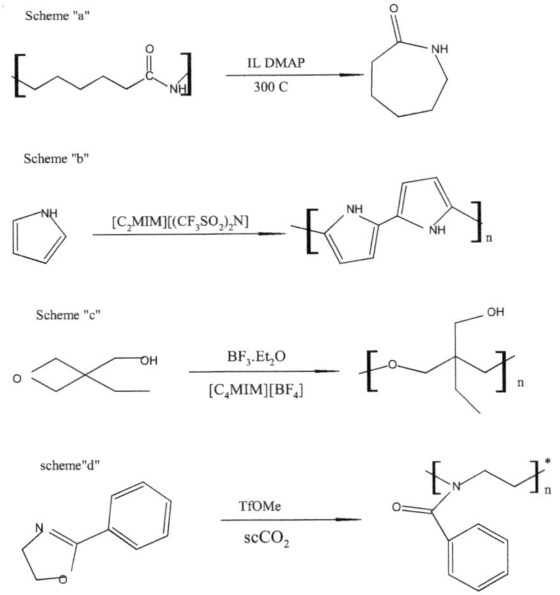

Scheme 10. a) Depolymerization of nylon-6 in ionic liquids catalyzed by DMPA at 300°C
b) Polymerization of pyrrole in 1-ethyl-3-methylimidazolium bis(trifluoromethanesulfonyl)amide
c) Cationic ring-opening polymerization of 3-hydroxymethoxetane in [C4MIM][BF4]
d) Cationic ring-opening polymerization of 2-phenyl-oxazoline initiated by methyl triflate in $scCO_2$ [60].

Because of its generally accessible conditions, supercritical carbon dioxide (CO_2) is the most often employed supercritical fluid. It becomes supercritical at temperatures above 31°C and pressures exceeding 7.38 MPa, at which point it functions effectively as a nonpolar solvent [60].

Mild Reaction Conditions:

Examples of Greener Solvents in Polymer Grafting

- **Solvents derived from carbohydrates:**

After water, ethanol is the most used solvent. As a result, it is utilised in coatings, some cleansers, cosmetics, and toiletries. Research indicates that simple alcohols like methanol and ethanol are better for the environment than formaldehyde or dioxane. This means that mixes of methanol and water or ethanol and water are better for the environment than mixtures of pure alcohol or propanol and water [61].

- **Solvents derived from lipids:**

- Fatty acid methyl ester synthesis: Triglycerides, or lipids, can be utilized as solvents by themselves, although they are mostly degraded to produce fatty acids and glycerol (glycerine). Fatty acid esters, such as FAMEs (fatty acid methyl esters), can be produced by esterifying fatty acids with alcohol and methanol. The methanol needed to create FAMEs is often made from natural gas or petroleum, although it can also be obtained through alternative processes, such as the gasification of biomass and hazardous waste from homes. In synthetic chemistry, glycerol from lipid hydrolysis and some of its derivatives can be employed as solvents [62].

- **Deep eutectic solvents:**

The phrase "deep eutectic solvent" (DES) refers to a mixture whose melting point is lower than the melting points of its parts. Many solid substances can be blended in this fashion to create liquids employed as solvents, particularly when the melting-point depression is very significant. Choline chloride, an ammonium salt, is one of the materials most frequently utilised to produce DES. A mixture of urea (melting point: 133 °C) and choline chloride (melting point: 302 °C) in a 2:1 molar ratio has a melting point of 12 °C [63].

- **Terpenes:**

Terpenes are a broad family of naturally occurring chemicals that can be extracted from specific plant parts to produce solvents. With the gross formula (C5H8)n, all terpenes are structurally given as multiples of isoprene. Among the most well-known solvents in this class are turpentine and the monoterpene D-limonene. Although turpentine is obtained from pine trees (sap, stump), it is also obtained from citrus peels. Turpentine is a by-product of the Kraft paper-producing process. The mixture of terpenes that makes up turpentine varies depending on its origin and the method of production. Mass concentrations of 20 to 35% β-pinene, 2 to 20% d-limonene, and 40 to 65% α-pinene are observed throughout the US and Canada. When extracting vegetable oil, α-pinene can replace n-hexane as a solvent. It can also be used to extract compounds like carotenoids that are utilised as food additives [64, 65]. Turpentine was once utilized in organic coatings, but petroleum hydrocarbons have mostly replaced it. These days, its primary purpose is to provide a source for its constituents, such as α and β-pinene [65].

- **Ionic liquids:**

Molten organic salts, typically fluid at ordinary temperatures, are known as ionic liquids. Among the commonly utilised cationic liquids are phosphonium, imidazolium, pyridinium, and ammonium. Halides, tetrafluoroborate, hexafluorophosphate, and nitrate are examples of anionic liquids. Ionic liquids are thermally, chemically, and electrochemically stable and non-flammable. Because of these characteristics, ionic liquids can be utilised as green solvents since, in comparison to traditional solvents, their low volatility reduces VOC emissions [65].

Comparative Analysis with Traditional Solvents

While conventional and eco-friendly solvents perform identical tasks, there are some notable distinctions between them. Conventional solvents, for instance, usually have detrimental effects on the environment due to their involvement in pollution and groundwater contamination. Green solvents, on the other hand, are made more environmentally friendly by lowering pollutants and being biodegradable [66].

Conventional solvents present a serious toxicity issue as well. They risk the environment, animal life, and human health. Conversely, green solvents are a far better choice because they provide less risk to people, animals, and the environment. Additionally, there are numerous sources of solvents. Green solvents are produced from renewable resources, as opposed to traditional solvents derived from fossil fuels. Sustainability is further promoted by the general initiative to lessen dependency on fossil fuels. Another significant factor in the manufacture of solvents is energy efficiency. Conventional solvents typically have higher energy requirements and produce more greenhouse gases as a side effect. Green solvents are less energy-intensive to create and use, and since they're eco-friendly, there are no greenhouse gas emissions [66, 67].

Regulation and compliance, in addition to cost, are other factors that are called into question when comparing standard solvents with green solvents. If you have any experience with conventional solvents, you are likely aware of the stringent regulations your sector has around their usage and disposal. Complying with these requirements may potentially lead to higher costs for your company. On the other hand, green solvents are more approachable because they naturally adhere to the majority of rules, making the process of fulfilling regulatory obligations easier. Therefore, even though green solvents may initially cost more, they are significantly more economical [68, 69].

ENERGY EFFICIENCY IN GRAFTING PROCESSES

Importance of Reducing Energy Consumption

Reducing energy consumption in green chemistry assays is essential for several key reasons:

- **Environmental Impact**: Lower energy consumption directly reduces the carbon footprint associated with chemical processes. Many traditional chemical processes rely on high temperatures and pressures, which require significant energy, often derived from fossil fuels. By minimising energy use, green chemistry aims to lessen the environmental impact and contribute to the reduction of greenhouse gas emissions [70].
- **Cost Efficiency**: Energy costs can constitute a substantial portion of the operational expenses in chemical manufacturing and research. Companies can lower their operational costs by developing assays and processes that use less energy, making the processes more economically viable and potentially more competitive [70].
- **Resource Conservation**: Energy is a finite resource, and reducing its consumption helps conserve it. Green chemistry promotes the use of more efficient processes that require less energy, aligning with broader sustainability and resource conservation goals [70].
- **Process Optimization**: Green chemistry often involves designing more efficient processes. Lower energy consumption is usually a byproduct of improved process design, which can lead to more effective use of raw materials and

reduced waste production [70].
- **Regulatory Compliance**: Many regions are implementing stricter environmental regulations and standards to reduce energy consumption and emissions. Adopting energy-efficient practices in green chemistry can help ensure compliance with these regulations, avoiding potential fines and legal issues [70].

- **Public Perception and Responsibility**: Demonstrating a commitment to energy efficiency and sustainability can enhance a company's reputation and appeal to consumers increasingly concerned about environmental issues. It reflects a company's commitment to corporate social responsibility and ethical practices [70].

In summary, reducing energy consumption in green chemistry assays is crucial for minimising environmental impact, lowering costs, conserving resources, optimising processes, meeting regulatory requirements, and improving public perception.

Microwave-Assisted Grafting

- **Microwave Synthesis Apparatus**

Microwave ovens with single and multimode settings are among the equipment used in microwave-assisted synthesis [71].

- **Single-mode microwave apparatus**

A single-mode apparatus's capacity to produce a standing wave pattern sets it apart. This interface produces an array of nodes with zero microwave energy intensity and a variety of antinodes with the maximum magnitude of microwave energy. One drawback of single-mode equipment is that only one vessel can be exposed to radiation at a time. Still, the device is easy to operate. One benefit of single-mode devices is their rapid heating rate. This is because the sample is always positioned in the field's antinodes, where microwave radiation intensity is at its peak. Under sealed-vessel conditions, these devices can handle volumes between 0.2 and approximately 50 mL; under open-vessel conditions, amounts of about 150 mL. Currently, single-mode microwave ovens are used in small-scale drug discovery, automation, and combinatorial chemical applications [71].

- Multi-mode microwave apparatus:

One essential element of multi-mode equipment is deliberately avoiding the production of a standing wave pattern. The goal is to wreak as much havoc as possible inside the apparatus. Since radiation spreads more broadly at higher levels of chaos, the area inside the device that may effectively heat increases with the degree of disorder. As a result, unlike single-mode microwave heating devices, which can only heat one sample at a time, multi-mode microwave heating devices can heat many samples at once. Thanks to its functionality, a multimode heating device can be used for the extraction process, ashing, and other chemical analysis processes, as well as bulk heating. A large multimode filter can be used to handle several litres of the reaction mixture with both open and closed vessel setups. The inability to effectively manage sample heating due to temperature uniformity is a significant drawback of the multi-mode apparatus [71].

Ultrasound-Assisted Grafting

Because ultrasound-assisted extraction is simple to apply to a wide range of matrices and can be employed with a wide range of solvents, it is a popular technique for removing organic pollutants from solid samples. Sonication and microextraction techniques have been used in numerous procedures for liquid samples in recent years. Due to the tiny vapour pressure of conventional organic solvents, there is a growing interest in either minimising their usage or substituting them with less hazardous solvents, such as ionic liquids, as a result of advancements in extraction techniques. However, these solvents' low biodegradability and some toxicity to aquatic life prevent them from being completely green. The use of more ecologically friendly solvents, like deep eutectic solvents, or innovative sorbents, like magnetic nanoparticles, for the extraction of organic pollutants has been noticed because of the increased interest in the so-called "Green Chemistry". [72]

Ultrasound-assisted Liquid-liquid Microextraction

An aqueous phase, a dispersion solvent, and an extraction solvent comprise the ternary solvent system used in DLLME (dispersive liquid-liquid microextraction). When the extraction solvent or a dispersing agent mixture is quickly injected into the sample, a hazy solution is formed due to the extraction solvent's microdroplets dispersing throughout the aqueous sample, which produces considerable turbulence. Analytes can be quickly partitioned from the aqueous phase into the extraction solvent thanks to the high contact area between the two phases. Centrifugation separates the two phases of the extraction process once it is complete, and the sedimented or floating analyte-enriched extraction solvent is collected for further examination. Solidification can help in the collection of solvents that are lighter than water [72].

An enhanced version of the traditional DLLME procedure called UA-DLLME uses ultrasonic energy to enhance emulsification and the migration of analytes from the aqueous phase to the extraction solvent. The number of solvents used, the sample size, the ionic strength, the pH, and the extraction and disperser solvents used all affect the effectiveness of the extraction. Except for sonication, which enables the extraction solvent to disperse in the sample without the need for a dispersing solvent, USAEME (ultrasound-assisted emulsification microextraction) is a technique very similar to DLLME. Two primary groups can be distinguished based on the type of solvent employed in the proposed ultrasonic-assisted liquid-liquid microextraction (UA-LLME) methods for the identification of organic contaminants: those using new green solvents and those utilizing standard organic solvents [72].

Case Studies and Practical Applications

A mixture of sodium salt (10 mol%) of 1-hexenesulfonic acid, aldehyde, and 1H-indole was dissolved in as little water as possible while constantly stirring. Additionally, the reaction mass was exposed to ultrasonic radiation for the necessary time at room temperature. Clean conversion is provided by this catalyst (94%) [73].

Clean and effective chemical conversions in the absence of solvents and water are made possible by the special surroundings produced by sonochemical processes. Cavitational collapse is the basis for the effective agitation, dissolution, mass and heat transfers, and reagent sonolysis that ultrasound uses to initiate and enhance both physical and chemical changes. The recent advancements in sonochemical engineering, which have powerful flow reactors and enhanced energy yields in ultrasonic generators, present excellent prospects for the industrialisation of these and other effective green synthetic processes [73].

REDUCTION OF DERIVATIVES IN GRAFTING

Covalently grafted polymer films are produced *via* the graftfast process, a grafting technique. It depends on chemically reducing diazonium salts by reducing agents, either in the presence or absence of a vinylic monomer. Unlike electro-induced methods that yield firmly grafted and stable polymer films, such as surface electro-initiated emulsion polymerization (SEEP) or cathodic electro grafting (CE) of vinylic monomers (which requires significant experimental conditions), the Graftfast procedure produces extremely grafted polymer films on any material (conductors, semiconductors, and insulators). Furthermore, compared to the slower ATRP-based techniques, it is a faster one-step reaction that occurs in water at room temperature, atmospheric pressure, and ambient air [74, 75].

Direct Grafting Methods

The process of polymer grafting enhances the polymer's shape, chemical composition, and physical characteristics. By enhancing the current conduction and properties of polymers beyond charge transport, this approach can improve the parent polymer's solubility, nano-dimensional shape, biocompatibility, and biocommunication, among other qualities. By copolymerising a second polymer or by grafting, a polymer's physicochemical qualities can be further altered [75].

- Graft copolymer using click chemistry:

The linear backbone of the graft copolymers is paired with randomly spaced branches. It is widespread for well-known graft copolymers to develop with desired functional groups, regulated lengths, and contents of side and back chains. The three graft copolymer synthetic techniques are graft-to, graft-from, and graft through. These tactics reduce unreacted side chains and increase grafting efficacy. The utilisation of CuAAC in the synthesis of graft copolymers was initially studied by Matyjaszewski and colleagues, as shown in Fig. (**8**) below. Here, PHEMA (Poly(2-hydroxyethyl methacrylate)) with alkynyl side groups was used to prepare CuAAC in a range of azido-terminated polymers, including PEO-N3, PSN3, PnBA-N3, and PS-b-PnBAN3 (Scheme **11**). Due to the steric hindrance of the linked side chains, graft densities were found below 50% for PS, PnBA, and PS-b-PnBA. Consequently, whereas grafting densities for less bulky PEO side chains reached up to 88% [75],

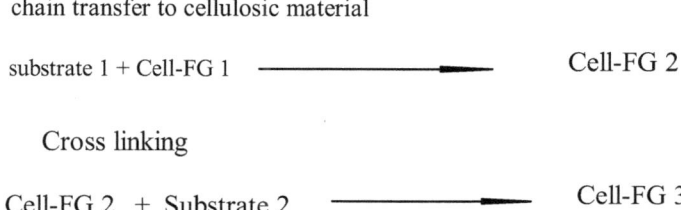

Fig. (8). Overview of chain transfer for cellulose grafting(FG: functional group, Cell: Cellulose) [78].

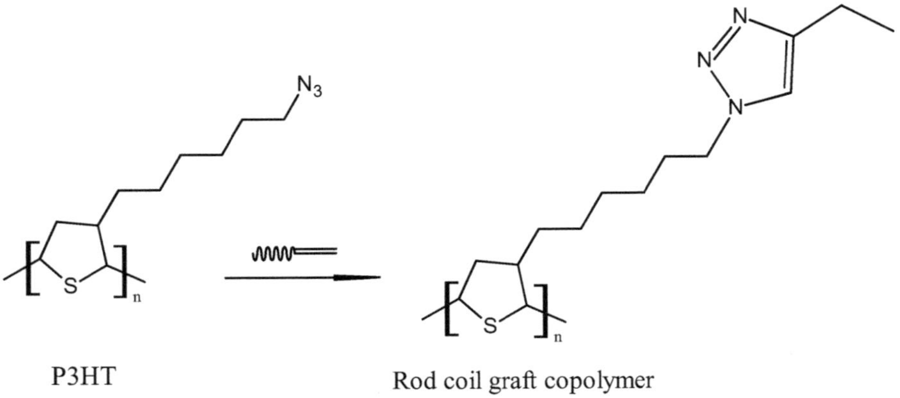

Scheme 11. Procedure for constructing poly(3-hexylthiophene-)-related rod coil graft copolymer through CuAAC [75].

Benefits of Reducing Derivatization Steps

The main objectives of surface modification are to enhance the mechanical properties, wettability, biocompatibility, and other properties of a surface polymer. To give the modified material unique qualities such as improved thermal stability, multiphase and physical responses, compatibility, flexibility, and rigidity, modification of the polymer is necessary. An insoluble polymer can be changed into a soluble polymer, and vice versa. Additionally, it enhances the environment for processing polymers. The most popular techniques for modifying polymers include grafting, mixing, cross-linking, and composite forms. Graft polymerization is a compelling and appealing approach that modifies the rheological properties, hydrophilic potential, polymer charges, molecular chain, aggregation state, and complexing ability of the parent polymer by chemically attaching one or more side chains to the main polymer chain through covalent bonds [76].

Examples and Case Studies

• Bromination

Since polystyrene bromination was created by thermally polymerizing styrene around 70°C and was believed to have little to no branching, this procedure was utilized to ensure that replacement occurred preferentially in the unsaturated region of the molecule. The radical termination approach (Fig. **9**) is the sole way to escape the technique's restriction, which is that terminating developing chains would cause cross-linking and the creation of an intractable network throughout polymerisation. In this process, initial disproportionation is crucial. The image depicts a traditional instance of bromination to obtain an unsaturated link necessary for further polymerisation (Scheme **12**). The rate of reaction, along with other kinetic factors, which are often present in the polymerisation of styrene, regulates the chain length and branching [76].

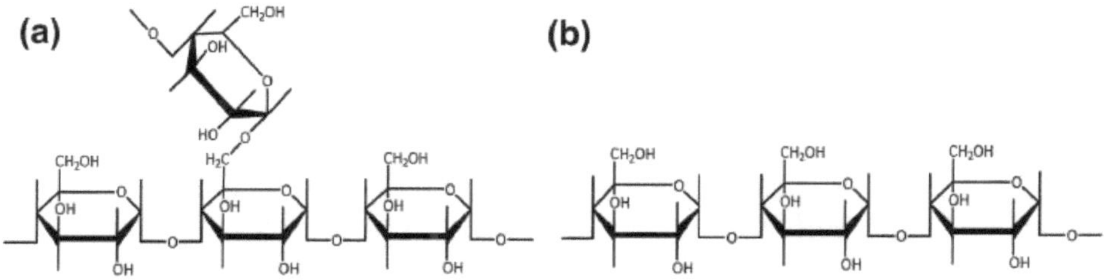

Fig. (9). Amylopectin (a) and amylose structure (b) [38].

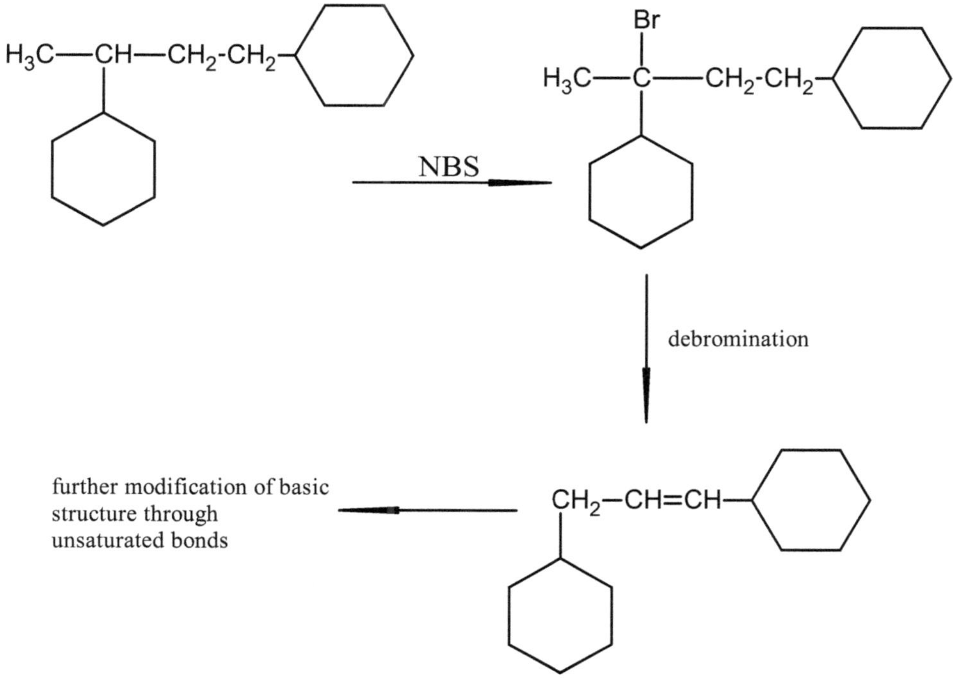

Scheme 12. Polymer grafting through bromination [76].

- **Esterification**

When an alcohol combines with an electron-deficient carbonyl in the presence of an acid, it creates an ester bond. This process is known as esterification. The polymer gains a new property from this connection without losing its intrinsic properties. It has been reported that cellulose nitrate is produced when nitric acid is used to esterify cellulose in the presence of sulfuric acid, phosphoric acid, or acetic acid. Furthermore, it has been discovered that cellulose esters, such as cellulose acetate, cellulose acetate propionate, and cellulose acetate butyrate, are not only beneficial but also economically useful [77].

- **Chain transfer**

The removal of hydrogen atoms from the cellulose molecule stops the polymerization process in cellulose chain transfer, which results in the formation of radicals from the cellulose backbone. This reaction termination is a cause for concern in cellulose grafting, where the chain transfer reaction is widely employed. Since purely radical transfer graft copolymerization does not result in higher graft yields, thiol or xanthate ester groups are often introduced. However, substances with higher chain transfer action, such as thiol groups, can be incorporated into the cellulose molecules prior to grafting. During potassium persulfate initiation, the activating species are generated on the swollen cellulose substrate backbone [78].

DESIGN FOR DEGRADATION

Principles of Degradable Polymers

Biodegradable materials are used in a variety of industries, including packaging, medical, and agriculture. In recent years, there has been an increased focus on biodegradable polymers. Natural and synthetic polymers are the two categories of biodegradable polymers that can be distinguished. Certain polymers are produced with feedstocks derived from petroleum or renewable biological resources, which are classified as non-renewable resources. Natural polymers typically offer fewer benefits than their manufactured counterparts.

Due to their many special qualities, polymer materials are used in orthopaedics, dentistry, hard and soft tissue replacements, medication delivery, and cardiovascular devices, among other biomaterial applications. In actuality, most materials utilized in medicine belong to the polymer class. This section covers the fundamentals of polymer science, shows how materials made of polymers can be specifically created to meet the needs of the biomaterials industry, and gives examples of the various applications this group of materials is seeing today in the medical profession [79].

- **Molecular Structure of Single Polymer Molecules**:

° Polymer molecules are characterised by their large molecular mass. Compared to a water molecule with a molecular mass of 18Da, a single polymer molecule may have a molecular mass of 2,000,000Da. Alterations in the chain sequence of monomer addition can also result in random or gradient copolymers due to differences in monomer reactivity. Branched structures can occur when a primary polymer backbone has shorter chain extensions extending from it. Branches can be intentionally added to the molecular framework or result from unintended side reactions during synthesis. The kind and degree of branching alter a polymer system's characteristics [79, 80].

- **Chemical Structure of Single Polymer Molecules:**

º The same fundamental structure would repeatedly appear if you could see the constituent atoms of a polymer molecule. The repetition unit of a polymer molecule is the name given to this structure. The word polymer derives its etymology from the Greek words "poly," which means many, and "meros," which means part—many parts. The repeat unit is known as the "mer" in polymer molecules [79, 80].

- **Determination of Chemical Composition:**

º The chemical structure of polymers and the composition of copolymer systems are frequently verified by researchers. Nuclear magnetic resonance (NMR) and infrared (IR) spectroscopy are two common methods scientists use. Using the magnetic moments associated with isotopes that have an odd number of protons and neutrons, nuclear magnetic resonance (NMR) analysis can extract information. These atoms have a nonzero nuclear spin or an intrinsic atomic magnetic moment and angular momentum, although all nuclides with even quantities of both have an altogether zero spin [79, 80].

- **Tacticity:**

º The repetition units in polymer chains' stereochemistry are characterised by their tacticity. Let us examine a polypropylene (PP) molecule to demonstrate the tacticity issue stretched into a flat zigzag pattern [81].

- **Connecting Physical Behaviour With Chemical Characteristics**

º The main properties of polymer molecules—tactility, molecular mass, chemical composition, and molecular architecture—have been covered. We will now relate these molecular features to macroscopic qualities and show how these features can be tuned to provide the desired behaviour in a polymer system. We shall concentrate on polymer systems' hydrophilicity, biodegradability, and tensile characteristics [82].

- **The Rubbery State:**

Amorphous polymers are used in rubber. Nonetheless, there is sufficient thermal energy in the random coils for rotation around single bonds. If you could observe polymer molecules in their melted condition, you would notice that every random coil is in a state of constant transformation. As the system's thermal energy rises, the intensity of this molecular mobility increases. Because the molecules can move, these materials are extensible, soft, and flexible on a macroscopic level [82, 83].

- **The Glassy State:**

Because a chain segment must travel (rotate) past a nearby segment, energy barriers are formed that impede rotation around bonds when the polymer system cools down. The rate of segmental motion in a polymer chain decreases with decreasing temperature, resulting in an increasing stiffness of the chain. The interwoven random coils freeze in place when the system gets closer to the Tg [83].

- **The Semicrystalline State:**

At low enough temperatures, glasses are formed by all polymer systems. On the other hand, certain polymers can form stable crystalline domains by packing into a regular lattice as a melt solidifies. Whereas the crystalline chains in syndiotactic poly (vinyl chloride) (PVC) have a spiral shape, the stable crystalline domains in PE are generated by chains in the planar zigzag conformation. This state is known as semicrystalline because only a section of the lengthy polymer chains can crystallize as some segments do not fit into the crystallites [83].

Grafting Techniques for Biodegradable Polymers

A polymer scientist with the knowledge of structure-property interactions can create molecules of polymers tailored to a specific use. In this section, we discuss the synthesis of polymers.

The two most used techniques for creating polymers are addition polymerization and condensation polymerization. Organic chemistry describes reactions that take place between various functional groups. An amine group and a carboxylic acid group, for example, can react to produce an amide bond and release a water molecule. A few of these key organic processes for bond formation are shown in Scheme **13**. The same processes produce several condensation polymers as those listed in Scheme **14**. The sole distinction is that difunctional monomers, which have two functional groups, are utilized in place of monofunctional molecules, which only have one. Scheme **15** [84].

1-hexenesulfonic acid sodium salt
water

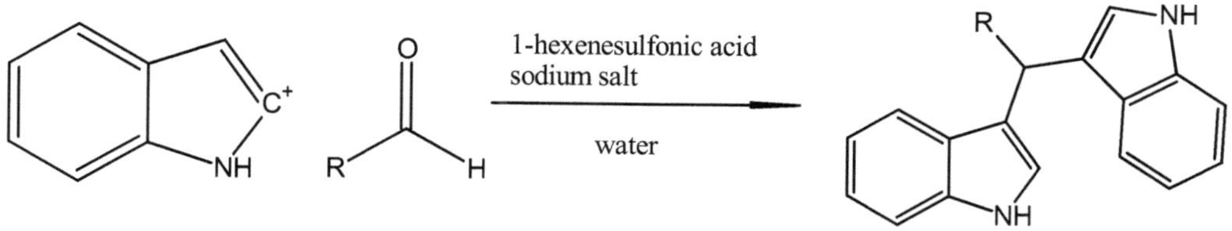

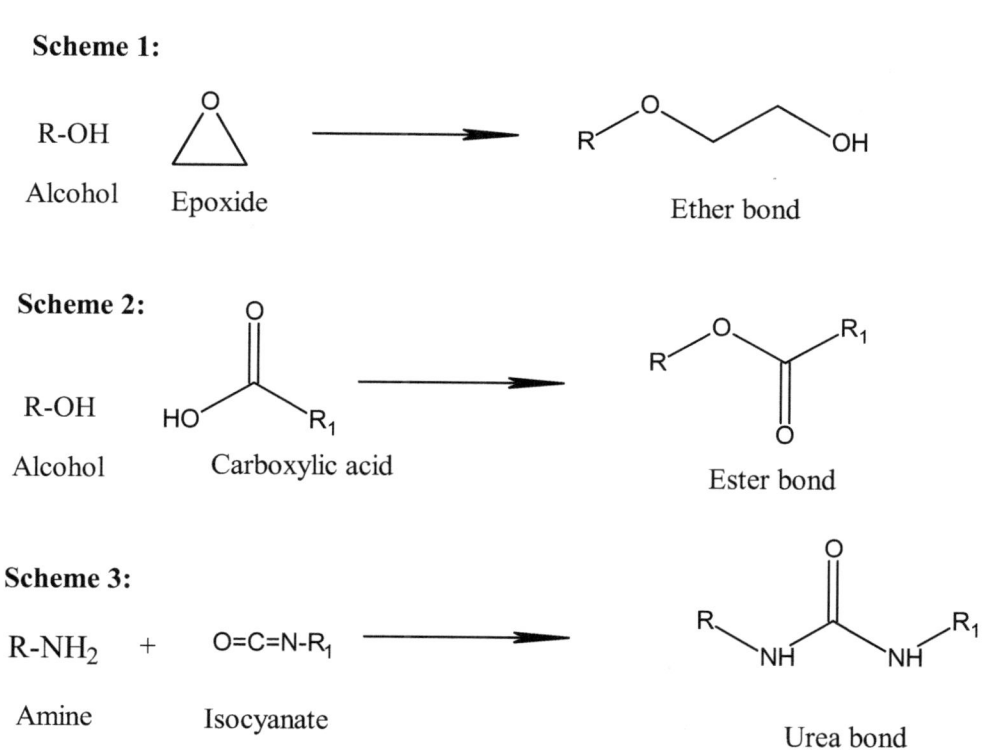

Scheme 13. Synthesis of bis(indoly)methanes [73].

Scheme 1:
R-OH
Alcohol
Epoxide
Ether bond
Scheme 2:
R-OH
Alcohol
Carboxylic acid
Ester bond
Scheme 3:
R-NH2
+
O=C=N-R1
Amine
Isocyanate
Urea bond

Scheme 14. Examples of common organic reactions [84].

Scheme 4

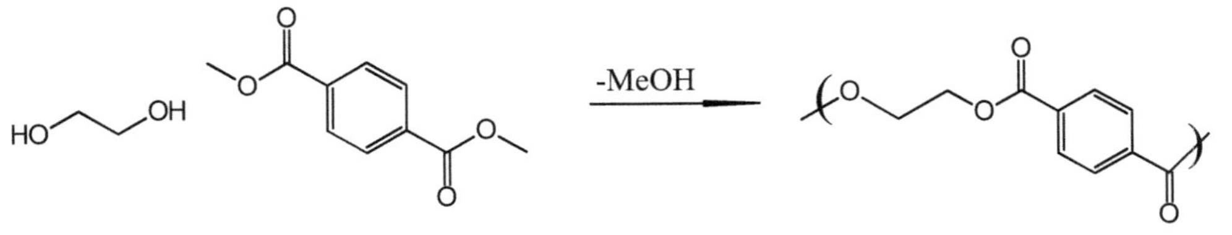

Scheme 15. Condensation reaction of ethylene glycol and dimethyl terephthalate to produce poly(ethylene terephthalate) [84].

Environmental Impact and Applications

Applications of Biodegradable Polymer

With a global market valued at billions of dollars yearly, the usage of biodegradable polymers is expanding quickly. Applications for biodegradable polymers include food packaging, computer keyboards, interior car parts, and medical applications such as tissue engineering, medical delivery, and large implanted devices [85].

Environmental Fate And Assessment

The last 20 years have seen the design and development of biodegradable polymers. They are applied in a manner that facilitates biodegradation. The biological process of biodegradation involves the digestion of dead organic matter by bacteria, which releases inorganic substances as by-products and transforms them into microbial energy and biological mass. Utilizing organic recycling and biodegradation, waste management turns biowaste into compost for soil fertilisation. An additional organic recycling technology that generates biogas and then turns it into compost is anaerobic digesters. Biopolymers are produced to reuse bio-wastes in anaerobic digesters and composting facilities. Biopolymers are also used in agricultural plastics designed to decompose naturally after use and be left in the field. One recurring topic is the impact of compounds created during composting and polymer biodegradation on the environment. These compounds may be directly diffused during their biodegradation in soil, or they may be spread into the environment through compost fertilization [86, 87].

Case Studies of Biodegradable Grafted Polymers

Case Study I: Resorbable Sutures

- What Problems Can Occur?

Stitches are frequently required to keep tissue together following an injury, wound, or surgery. An interior wound sutured with a nonbiodegradable material would need a second surgery to extract the sutures after the wound healed [87].

- What Properties are Required of the Biomaterial?

Adequate tensile force is the most fundamental feature needed in a suture material to prevent the wound from opening again while it heals. Furthermore, as the body heals, the suture materials ought to break down naturally, negating the need for retrieval or removal. Additionally, the degradation products must not be harmful and occur at a suitable rate [87].

- What Polymeric Biomaterial is Used?

Sheep or beef gut was used to make the first resorbable sutures. However, synthetic materials are utilised more frequently because they are more affordable, more accessible to handle, operate consistently, and have a lower risk of spreading illness. A poly(lactide-co-glycolide) copolymer is one substance that is commonly utilized. The material has the high tensile strength and flexibility needed for this application but is permanently amorphous since it is a random copolymer. Additionally, the substance breaks down into lactic and glycolic acids by the hydrolysis of ester linkages in the polymer backbone. Due to their natural occurrence in the body's metabolic process, both substances are safe in tiny doses. It is also possible to customise the pace of deterioration since the ester linkages in the lactide and glycolide repeat units hydrolyse at various rates [87].

CATALYSIS IN GRAFTING REACTIONS

Role of Catalysis in Green Chemistry

Catalysis, or the employment of catalysts to substitute more effective catalytic processes for stoichiometric ones, is the ninth principle of green chemistry. Reagents that participate in a chemical reaction but do not alter once the process is finished are known as catalysts. By engaging with particular sites on the reactants, they lower the energy barrier of a process. Because they can perform a single reaction repeatedly and work well in small quantities, catalysts can help save energy and resources. They are better than overused stoichiometric reagents, which only perform a single reaction [88].

Green catalysts include, for instance, Laccases: Lichens, microbes, plants, and insects all contain these glycoproteins. They are utilised in the pulp and paper, textile dyeing, and printing industries. They can oxidise compounds and create water. Additional catalysts include USY zeolite, palladium, and microwave-assisted synthesis [89].

Enzyme-Catalyzed Grafting

Enzyme-catalyzed reactions are a more environmentally friendly way to functionalize polymers compared to traditional methods. This approach is particularly beneficial because of the enzymes' excellent selectivity, high efficiency, softer reaction conditions, and recyclability as shown in Table **6**. Metal catalysts are not necessary for conducting certain reactions in solventless environments. As a result, this method is gaining popularity in the field of biomedical applications since it eliminates the toxicity caused by solvents and metal catalyst residues.

Here are a few examples of an catalyzed grafting process [89].

Table 6. Examples of enzyme classes and their functions [89].

Enzyme class	Catalytic function	Selected enzymes
Oxidoreductase, EC1	Oxidation-reduction reactions involving intermolecular electron transfer	Laccase, horseradish peroxidase
Transferase, EC2	Transfer of functional groups from donor molecules to acceptor	DNA methyltransferase, Thiaminase
Hydrolase, EC3	hydrolysis	Amylase, pepsin
Lyase, EC4	Non-hydrolytic bond cleavage	Adenylate cyclase, carbonic anhydrase
Isomerase, EC5	Molecular isomerization	Beta-carotene isomerase, photoisomerase
Ligase, EC6	Large molecules coupling	DNA ligase, CTP Synthase

Metal-Free Catalytic Systems

A class of materials where the metal is absent is established because of the aggravation of diminishing the metal loading in catalytic ensembles. Although such metal-free catalysts are not new, their use has increased dramatically in recent years due to factors such as economics, industry, safety, and environmental considerations [90]. Most metal-free catalysts include different types of carbon, which can be bulk or nanostructured, crystalline or amorphous, and chemically or not changed. The carbon material should genuinely contain no metal, or very little metal, to be eligible for this crucial class of catalyst [90]. Since metal impurities are relatively common, it is generally considered acceptable to classify a system as metal-free at levels in the parts per billion range, as long as the characterisation has been done thoroughly and with instruments of sufficient sensitivity [91]. The total absence of metal impurities is challenging to achieve. The notion that a catalyst is metal-free in some circumstances could be misleading, incomplete, and dependent on methods that aren't entirely certain [91].

For instance, reports claiming that X-ray photoemission spectroscopy (XPS) has eliminated all metals may not be clear-cut because this instrument is primarily surface sensitive and typically detects metals within a few nanometres of thickness; it does not reveal information about the material's interior. The majority of reactions take place at the surface. However, it is essential to consider the possibility that some metal atoms deep within the carbon phase may have a functional role [91].

For example, Carbon is a catalyst for gas-phase reactions:

- Aromatic hydrocarbon

Direct dehydrogenation of ethylbenzene yields styrene, a crucial monomer in the petrochemical sector. At temperatures between 560 and 650 °C, multi-promoted iron oxide catalysts mediate the process. Excess steam is used to prevent the development of coke and to keep the activity high for an extended length of time. This results in low energy efficiency (Scheme **16**). Because the oxidative dehydrogenation pathway is intrinsically free of thermodynamic restrictions, it can operate between 300 and 450 °C.

Scheme a:

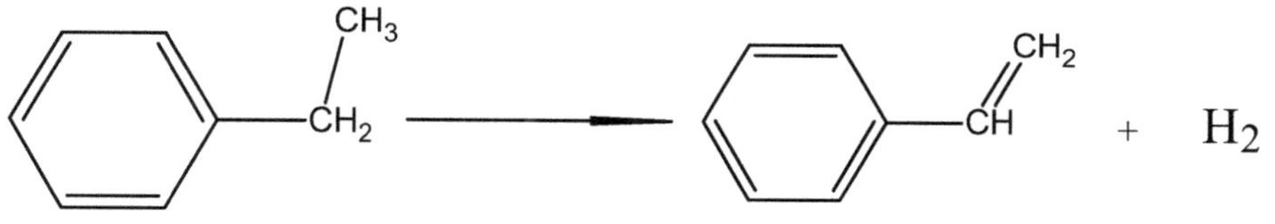

Scheme b:

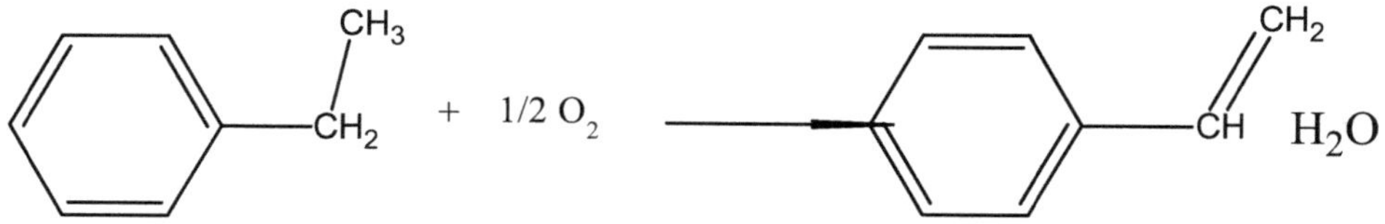

Scheme 16. Direct dehydrogenation of ethylbenzene yields styrene [91].

Case Studies and Examples

- Carbon as a catalyst for liquid-phase reactions

The catalytic activity of nanocarbons is also observed in liquid-phase processes.

For instance, under mild reaction conditions, multi-walled carbon nanotubes (CNTs) have been utilized as a catalyst for the oxidative dehydrogenation (ODH) of dihydroanthracene to anthracene. The transesterification of triglycerides with methanol is thought to be a test reaction for the synthesis of biodiesel. More recently, it was demonstrated that amino-functionalized CNTs are active solid basic catalysts for this reaction. According to the inquiry, since the catalysts have the unique benefit of not having bulk quantities of acidic oxygen atoms, which are invariably present in basic oxide systems, this reaction should also be taken into consideration as a test for the broader use of N-CNT systems in synthetic chemistry [92].

The potential of carbon nanotubes (CNTs) in fluid phase reactions must be completely explored; possible advantages of use over other solid-based systems include wettability and decreasing transport constraints. It has recently been demonstrated that one viable technique for liquid phase reactions is the functionalisation of carbon materials by diazonium chemistry. For example, organised mesoporous carbon can be sulfonated to produce a highly active solid acid. The sulfonated carbons combine hydrophilicity and hydrophobicity into a single substance and may synergistically impact condensation, esterification, and biodiesel synthesis processes [92].

Real-time Analysis for Pollution Prevention

Importance of Real-time Monitoring

The real-time analysis principle primarily emphasises how crucial it is to watch a reaction's progress in real-time. By properly monitoring the reaction using existing analytical instruments, this principle suggests that the creation of hazardous compounds can be controlled or predicted and that preventive measures can be implemented to avoid unfortunate situations. It emphasises the use of cutting-edge analytical techniques such as nuclear magnetic resonance (NMR), infrared (IR), ultraviolet-visible, gas chromatography (GC), gas chromatographic—spectrometry (GC–MS), high-performance liquid chromatography (HPLC), and so on to characterise all potential intermediates and by-products formed during the reaction while also improving understanding of the reaction mechanism. Such real-time analysis would aid in optimising the quantity of their act and other chemicals occasionally utilized excessively [93].

Using this real-time in-process monitoring of the reaction, an effective synthetic pathway with a higher-than-expected product yield, reduced E-factor, and improved atom efficiency might be developed. Various monitoring approaches are available for the correct monitoring of a reaction, including online (sampling of representative aliquots) and in-line (continuous sampling of all material) [93].

Techniques for Monitoring Grafting Reactions

Green chemistry utilizes analytical techniques to monitor and control processes in real-time, preventing the formation of hazardous compounds. These techniques can also be applied to the selection of materials to reduce the likelihood of chemical mishaps such as flames, explosions, and leaks [94].

Here are some techniques for Monitoring Grafting Reactions:

- **Less hazardous chemical syntheses**

To lower the hazards associated with chemical processes, use reagents that are safe for the environment and non-toxic.

- **Reduce derivatives**

When synthesizing the target production, avoid protective groups and needless derivatives. These procedures can result in waste and require extra reagents.

- **Design for degradation**

When a chemical product reaches the end of its useful life, it should be designed to decompose into harmless degradation products and not linger in the environment.

- **Catalysis**

Choose the right catalysts to decrease waste, shorten reaction times, and improve reaction selectivity.

- **Increase energy efficiency**

Make sure you prioritise the use of renewable resources and use as little energy and materials as feasible.

- **Minimize the potential for accidents**

Select substances based on their physical states (solid, liquid, or gas) to reduce the risk of chemical mishaps.

Examples of Implementation and Benefits

Developing safer products and processes through the application of novel concepts from basic research will be the chemical industry's biggest challenge in the future. Moreover, the education and training of a new generation of chemists is critical to the success of green chemistry. It is necessary to introduce the concepts and methods of green chemistry to students at all levels. Here are a few implementations:

- **Next-generation catalyst design**

Opportunities abound to enhance traditional industrial processes. One method to create a novel pathway to the Nylon-6,6 precursor, o-caprolactam, for instance, is to employ nano-porous aluminophosphate catalysts with an acidic and redox-active site distribution (Scheme **17**). Hazardous reagents are eliminated, and less waste byproduct is produced in this one-step, solvent-free synthesis [95].

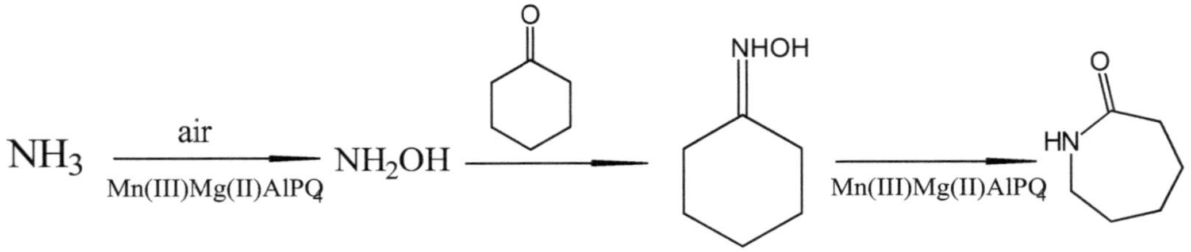

Scheme 17. O-caprolactam synthesis using Nylon-6,6 [95].

- **Sonochemical approach**

The sonochemical approach was created recently to prepare materials. Acoustic cavitations, which involve the development, growth, and implosive collapse of bubbles, occur under ultrasonic irradiation (20 kHz to 10 MHz). The bubbles produce momentary, localised hot patches with a range of physical and chemical properties, including extraordinarily high temperatures, high pressures, and rapid cooling rates. These properties may provide a unique setting for chemical reactions under harsh conditions. Under such severe conditions, chemical bonds break and various materials are created within these quickly collapsing bubbles through the breakdown of volatile precursors. Sonochemical synthesis is used to make pharmaceuticals, dyes, and other products, particularly when combined with other environmentally friendly methods [95].

INHERENTLY SAFER CHEMISTRY FOR ACCIDENT PREVENTION

Designing Safer Grafting Processes

Green chemistry, a science-based concept, aims to create less dangerous and more environmentally friendly chemicals, goods, and processes. It covers a product's whole life cycle, from manufacture to disposal. Green chemistry aims to minimize or eliminate the use of hazardous materials in the creation, manufacturing, consumption, and disposal of chemical products [96].

Here are some ways to select safer chemicals and conditions in green chemistry:

- **Use polar solvents**

At the top of the list of green chemicals are polar solvents, which include ethanol, 1-propanol, acetone, acetonitrile, 2-propanol, and methanol. These solvents are also thought to be safe for the environment [96].

- **Use catalysts**

Catalysts save waste since they work well in small quantities and can repeatedly complete a single reaction. They are better than stoichiometric reagents, which are overused.

- **Design chemicals carefully**

Chemicals and their solid, liquid, or gaseous states should be designed to reduce the risk of chemical accidents, such as fires, explosions, and environmental releases.

- **Apply chemicals in the proper dosage.**

Chemicals should only ever be used within acceptable limits and at the exact dosage that has been prescribed.

Selection of Safer Chemicals and Conditions

A green chemistry concept known as "safer chemistry for accident prevention" calls for the use of safer chemicals in manufacturing and lab settings to lower the possibility of mishaps. Industry and the environment can both gain from this. Here are some more green chemistry guidelines that may assist in averting mishaps:

Less hazardous chemical syntheses

Two benefits of using less hazardous chemicals are that the quantity of hazardous waste produced is reduced along with the chance of unintentional exposure.

Reduce derivatives

Reagent requirements and waste output can be decreased by avoiding derivatives.

Real-time pollution prevention

Through the use of real-time analysis to monitor and regulate chemical processes, chemists can spot problems and take action before they result in waste or contamination.

Use of renewable feedstocks

Switching to renewable energy sources from fossil fuels can decrease mishaps like explosions and spillovers in manufacturing facilities.

Design for degradation

One way to keep chemicals out of the environment is to design them such that, once they are used, they decompose into harmless compounds.

Design for energy efficiency

Reducing energy use can lessen its effects on the environment and the economy. Synthetic procedures can be carried out at room temperature and pressure, and renewable feedstocks should be used whenever feasible [96].

Case Studies of Safety in Polymer Grafting

Choosing the appropriate polymer system is crucial in the dosage form formulation process. The kind of polymer(s) used in pharmaceutical formulations primarily determines the drug's mechanism, release rate, and formulation stability. Short half-lives, low bioavailability, and chemical and physical instability are among the drawbacks of pharmaceutical and biological therapies. Using suitable polymers can successfully deliver medications to the target region at an appropriate concentration for a specific time [97].

To achieve tailored physicochemical properties by grafting of polymers

- While chitosan works best as a flocculant in acidic media, its derivatives with side-chain carboxyl groups exhibited zwitterionic properties. They

demonstrated strong flocculation capabilities in both acidic and basic media.
- Cai *et al.* created hydrogels using irradiation and poly(N-isopropyl acrylamide) (PNIPAAm)-grafted chitosan. The chitosan-g-PNIPAAm hydrogels exhibited increased swelling behaviour proportional to the grafting %, or the quantity of grafted branches.
- There have been numerous reports recently on the potential uses of acrylic acid grafts of chitosan to create hydrophilic and mucoadhesive polymers. In the work of Hu et al., silk peptide as a model drug, CS - poly(acrylic acid) (PAA) graft was demonstrated to be a promising carrier for delivering hydrophilic medicines, proteins, and peptides.
- Succinic acid treatment, acrylamide and 2-acrylamido-2-methylpropane sulfonic acid combinations, and chemical grafting of psyllium with N-hydroxymethyl acrylamide resulted in modified psyllium with changed pH-sensitivity, swelling ability, biodegradability, and drug release mechanism.
- By grafting thermosensitive polymers, such as poly (N-vinyl caprolactam) (PVCL) with poly (ethylene oxide), to create PVCL-graft-C11EO42, Vihola *et al.* were able to stabilise thermally responsive particles and create a thermostable system. These novel polymeric materials give hydrogel particles more thermostable systems that can be employed to regulate the release of drugs [97].

FUTURE TRENDS AND PERSPECTIVES

Emerging Techniques in Green Grafting

To render chemical processes more environmentally friendly and sustainable, innovative technologies have been invented or are currently being developed in the rapidly evolving field of "green chemistry." Among the instances are:

1. **Bio-based feedstocks:** Renewable, plant-based feedstocks are growing in favor as a substitute for petrochemicals. Bio-based plastics and biofuels are two examples of novel materials and chemicals being developed from plants [98].
2. **Low-solvent or solvent-free procedures:** By lowering the usage of volatile or poisonous organic solvents, these processes minimize waste and their negative effects on the environment [98].
3. **Microwave-assisted synthesis:** This method of carrying out chemical processes is quicker and uses less energy. This technology can greatly decrease reaction times, waste, and the energy needed to make chemicals [98].
4. **Flow chemistry:** This method improves efficiency and exact control over reaction conditions by allowing a constant flow of reactants through a reaction chamber. It can lower the usage of dangerous solvents while increasing yields and decreasing waste [98].

5. **Catalysis:** Novel catalysts, such as organocatalysts and biocatalysts, can decrease the energy and resources needed to carry out chemical processes.
6. **Green solvents:** Green solvents, such as ionic liquids, supercritical fluids, and deep eutectic solvents, can lessen the use of dangerous solvents and increase the sustainability of chemical processes [99].
7. **Renewable energy sources:** Chemical processes can have a lower carbon footprint when they use renewable energy sources like solar and wind power.

These are but a handful of the numerous novel innovations in the field of green chemistry. New developments are continually being made in this industry to make chemical processes more ecologically friendly and sustainable [100].

Potential Applications and Innovations

The potential applications and innovations are as follows:

1. **Nano pesticides:**

The population of the planet is still expanding. It is estimated that by 2050, there will be about 10 billion people on the Earth. A significant increase in agricultural output will be necessary to feed such a large population while maintaining sustainable agriculture practices, which include limiting the adverse effects of land usage on the environment, requiring less water, and minimising the contamination of crops by agrochemicals like pesticides and fertilisers. It should come as no surprise that nanotechnology is garnering interest from sectors beyond pharmaceuticals and healthcare. Customised nano-delivery systems have the potential to be an invaluable resource for farmers, enabling them to address the primary issues associated with conventional pesticides, including environmental contamination, bioaccumulation, and the significant rise in insect resistance [101, 102].

2. **Enantioselective organocatalysis:**

Nature has always served as an inspiration to chemists. A few years ago, scientists had an idea for a novel form of catalyst that wouldn't require the usage of pricey metals, just like most natural enzymes. Since its inception in the late 1990s, "organocatalysis" has expanded. Chemists developed methods for forming chiral carbon-carbon bonds with only a tiny portion of the millions of organocatalysis available as they became aware of the industrial ramifications of reducing the catalyst amount. Organocatalysis has led to the development of many other domains, and industries are now using asymmetric organocatalytic processes at scale to produce medications and fine chemicals [103].

3. Solid-state batteries:

Although they were first conceptualized in the 19th century, solid-state batteries began to develop only recently. Compared to lithium-ion batteries, which run modern laptops, tablets, and smartphones, solid-state batteries are lighter, enable more energy storage, and perform better at high temperatures. Moreover, unlike the electrolytes used in lithium-ion technology, solid-state electrolytes are not flammable, which may help to prevent explosions and spontaneous fires. Polymers might be the most sensible and economical choice for numerous additional uses. There is a great deal of research to be done, especially since the constituents of solid-state batteries are intricately linked [104].

4. Flow Chemistry

Achieving some of the Sustainable Development Goals of the United Nations, which aim to create a more sustainable and better future for all by 2030, would need chemistry. Among these, flow chemistry reactions carried out continually in a stream instead of in a batch are especially important for addressing responsible production and consumption. Eventually, flow chemistry procedures reduce the risk of handling hazardous materials and increase output, thereby minimizing damage and reducing environmental impact. Although some view flow chemistry as still in its early stages in small-scale laboratory settings, practical industrial applications are becoming increasingly widespread. It is quite difficult to comprehend how each of them acts [105].

REACTIVE EXTRUSION

Reactive extrusion, which enables entirely solvent-free chemical reactions, is associated with flow chemistry. This procedure is environmentally beneficial because potentially hazardous solvents have been eliminated. The fact that it would necessitate a comprehensive rethinking of current industrial processes presents numerous engineering challenges. Experts in polymers and materials have extensively utilised and studied extrusion procedures. Still, it is only recently that other chemists have begun to delve into their potential for synthesizing organic molecules. While reagents are ground in a ball mill for traditional extrusion processes, more sophisticated screw-based extrusion technologies may even enable solvent-free reactions to occur in flow configurations. The drawback, once more, is effectively scaling up and modifying the systems [106].

Challenges and Opportunities

Green chemistry aims to create and innovate chemical processes and products that minimize or eliminate the use of substances that pose a risk to human health and the environment; an approach to sustainability that is non-regulatory and driven by economics. In the field of green chemistry, there are some obstacles and prospects [107].

Challenges

- Creating economically and environmentally sustainable processes: innovations must adhere to tight safety and environmental regulations while preserving efficiency and cost-effectiveness.
- Locating sustainable and non-toxic raw materials
- Cutting back on waste and energy use

Opportunities

- Creating safer chemicals: As a result of technological advancements, our understanding of chemical toxicity has improved, facilitating the research of chemicals and the development of safer alternatives for the environment and people.
- Real-time analysis to prevent pollution: To prevent environmental build-up, create chemical goods that, when used, decompose into harmless compounds. Incorporate in-process, real-time monitoring and control to reduce or completely eradicate contamination during syntheses.
- Atom economy: Use atom economy techniques to minimise waste generated during a chemical process while maintaining the target product's production.

CONCLUSION

Summary of Key Points

Unquestionably, grafting is an intriguing extension of the conductive polymer family that connects its chemical properties with practical needs. The combination of the flexibility, processability, and biocompatibility of non-conducting conventional polymeric materials with their unparalleled electronic and electrical capabilities and environmental responsiveness can lead to several technological advancements and novel approaches to biocommunication. An excellent extension of the above chemistry might be the association or co-assembly of numerous conducting polymers in the same system, in addition to conducting and non-conducting polymer pairings.

The Role of Green Chemistry in the Future of Polymer Grafting

Polymer reaction engineering is a discipline that has the potential to be transformed towards better sustainability due to its economic significance and its extensive and fascinating technological capabilities. An ecologically friendly polymerisation process can be achieved by coordinating multiple steps; there is no one, unique method for changing polymerisation operations into more sustainable ones. The 12 principles of green chemistry are enabling this shift.

- Prevent waste
- Design safer chemicals and products
- Design less hazardous chemical syntheses
- Use renewable feedstock
- Use catalysts, not stoichiometric reagents
- Avoid chemical derivatives
- Maximize atom economy
- Use safer solvents and reaction conditions
- Increase energy efficiency
- Design for degradation after use
- Analyse in real time to prevent pollution
- Minimize the potential for accidents

As previously mentioned, while there is always space for development, many principles—principles 5, 6, 7, 11, and 12—are essentially being met in contemporary polymer manufacture. While Principles 1, 9, and 10 are well along, they still have significant potential for further advancements. Regarding principle 9, for instance, utilizing photopolymerisation may necessitate more developments in catalysis, whilst employing adiabatic polymerisation methods suggests the need for enhanced polymer property modeling, monitoring, and control as shown in Fig. (**10**). Regarding principle 10, the continuous research and development of biodegradable polymers provides another illustration.

Ultimately, all polymer researchers and engineers should make it a routine practice to monitor all 12 principles of green chemistry. This would substantially simplify the inevitable and required change of polymer synthesis towards a more sustainable future.

Final Thoughts on Sustainability and Innovation

A theory known as "green chemistry" attempts to lessen the harmful effects of chemical processes and products on the environment and human health. It can be an innovative tool to create a more sustainable future. Here are its concluding remarks:

PRINCIPLES OF GREEN CHEMISTRY

Prevention
Waste should be minimized or avoided

Atom Economy
Synthetic methods should maximize incorpotation of all materials used in the process

Less Hazardous Chemical Syntheses
Wherever practicable synthetic methods should be designed to use and generate less toxic substances.

Designing Safer Chemicals
Chemical products should be designed to achieve their desired function while being as non-toxic as possible

Safer Solvents and Auxiliaries
The use of auxiliary substances, should be made unnecessary whenever possible

Design for Energy Efficiency
Energy requirements of processes should be minimized

Use of Renewable Feedstocks
Raw materials should be renewable whenever technically and economically practicable

Reduce Derivatives
Derivatization should be avoided whenever possible

Catalysis
Catalytic reagents are Superior to stoichiometric ones

Design for Degradation
Chemical products should be designed to degrade into innocuous products

Real Time Analysis for Pollution Prevention
Analytical methodologies need to be further developed to allow for real-time monitoring

Inherently Safer Chemistry for Accident Prevention
Substances and their forms should be chosen to minimize the potential for chemical accidents

Fig. (10). Principles of green chemistry.

Economic stability

Green chemistry has the potential to guarantee sustained economic stability by integrating sustainability into commercial operations.

Environmental targets

Green chemistry can help achieve environmental goals by reducing the use of dangerous chemicals and preventing contamination at the molecular level.

Innovation

Green chemistry has the potential to spur innovation and open up new economic opportunities in several sectors.

Human health

Green chemistry can create goods and procedures that pose less risk to people's health.

Social Challenges

A foundation for creating answers to societal problems can be found in green chemistry.

Energy Consumption

One of the main objectives of green chemistry is minimising the amount of energy used in separation processes.

Green chemistry can support the implementation of the 2030 Sustainable Development Agenda.

Impact on the Environment

The ultimate goal of green chemistry is to stop all chemical emissions from entering the environment.

All branches of chemistry can benefit from applying green chemistry, which employs cutting-edge scientific methods to address pressing environmental issuee.

REFERENCES

[1] Purohit P, Bhatt A, Mittal RK, Abdellattif MH, Farghaly TA. Polymer Grafting and its chemical reactions. Front Bioeng Biotechnol 2023; 10: 1044927.
[http://dx.doi.org/10.3389/fbioe.2022.1044927] [PMID: 36714621]

[2] Sheldon RA, Arends I, Hanefeld U. Green Chemistry and Catalysis. Weinheim: Wiley; 2007.

[3] Grewal A, Kumar K, Redhu S, Bhardwaj S. Microwave assisted synthesis: a green chemistry approach. Int Res J Pharm Appl Sci 2013; (3): 278-85.

[4] Lancaster M. Green Chemistry: An introductory text. Cambridge: Royal Society of Chemistry 2010; pp. 1-16.

[5] Joshi UJ, Gokhale KM, Kanitkar AP. Green Chemistry: Need of the Hour. Indian J Pharm Edu Res 2011; 45(2): 168-74.

[6] Clark JH, Macquarrie DJ. Handbook of Green Chemistry and Technology. 1st ed. Wiley 2002; pp. 10-25.
[http://dx.doi.org/10.1002/9780470988305]

[7] Ravichandran S, Karthikeyan E. Microwave Synthesis-A Potential Tool for Green Chemistry. Int J Chemtech Res 2011; 3(1): 466-70.

[8] Krstenansky JLI, Cotterill I. Recent advances in microwave-assisted organic syntheses. Curr Opin Drug Discov Devel 2000; 3(4): 454-61.

[PMID: 19649876]

[9] Hayes BL. Microwave Synthesis: Chemistry at the Speed of Light. Matthews (NC): CEM Pub 2002; p. 11-23.

[10] Sekhon BS. Microwave-Assisted Pharmaceutical Synthesis: An Overview. Int J Pharm Tech Res 2010; 2(1): 827-33.

[11] Rajak H, Mishra P. Microwave assisted combinatorial chemistry: The potential approach for acceleration of drug discovery. J Sci Ind Res (India) 2004; 63(8): 641-54.

[12] Wathey B, Tierney J, Lidström P, Westman J. The impact of microwave-assisted organic chemistry on drug discovery. Drug Discov Today 2002; 7(6): 373-80.
[http://dx.doi.org/10.1016/S1359-6446(02)02178-5] [PMID: 11893546]

[13] Lidström P, Tierney J, Wathey B, Westman J. Microwave assisted organic synthesis—a review. Tetrahedron 2001; 57(45): 9225-83.
[http://dx.doi.org/10.1016/S0040-4020(01)00906-1]

[14] Gabriel C, Gabriel S, Grant EH, Grant EH, Halstead BSJ, Mingos DMP. Dielectric parameters relevant to microwave dielectric heating. Chem Soc Rev 1998; 27(3): 213-24.
[http://dx.doi.org/10.1039/a827213z]

[15] Strauss CR, Trainor RW. Developments in Microwave-Assisted Organic Chemistry. Aust J Chem 1995; 48(10): 1665-92.
[http://dx.doi.org/10.1071/CH9951665]

[16] Langa F, de la Cruz P, de la Hoz A, Díaz-Ortiz A, Díez-Barra E. Microwave irradiation: more than just a method for accelerating reactions. Contemp Org Synth 1997; 4(5): 373-86.
[http://dx.doi.org/10.1039/CO9970400373]

[17] Pelle Lidstrom BSP, Jacob Westman BSP, Anthony Lewis BSP. Enhancement of combinatorial chemistry by microwave-assisted organic synthesis. Comb Chem High Throughput Screen 2002; 5(6): 441-58.
[http://dx.doi.org/10.2174/1386207023330147] [PMID: 12470274]

[18] Algul O, Kaessler A, Apcin Y, Yilmaz A, Jose J. Comparative studies on conventional and microwave synthesis of some benzimidazole, benzothiazole and indole derivatives and testing on inhibition of hyaluronidase. Molecules 2008; 13(4): 736-48.
[http://dx.doi.org/10.3390/molecules13040736] [PMID: 18463575]

[19] Collins MJ Jr. Future trends in microwave synthesis. Future Med Chem 2010; 2(2): 151-5.
[http://dx.doi.org/10.4155/fmc.09.133] [PMID: 21426181]

[20] Saxena VK, Chandra U. Microwave Synthesis: A Physical Concept. In: Chandra DU, ed. Microwave Heating. New Delhi: D.U. Chandra; 2011. p. 3-22.

[21] Larhed M, Hallberg A. Microwave-assisted high-speed chemistry: a new technique in drug discovery. Drug Discov Today 2001; 6(8): 406-16.
[http://dx.doi.org/10.1016/S1359-6446(01)01735-4] [PMID: 11301285]

[22] Shen Y. Atom Transfer Radical Polymerization and Its Continuous Processes [PhD dissertation]. Ottawa: National Library of Canada Bibliothèque nationale du Canada; 2004.

[23] Lew A, Krutzik PO, Hart ME, Chamberlin AR. Increasing rates of reaction: microwave-assisted organic synthesis for combinatorial chemistry. J Comb Chem 2002; 4(2): 95-105.
[http://dx.doi.org/10.1021/cc010048o] [PMID: 11886281]

[24] Stadler A, Yousefi BH, Dallinger D, et al. Scalability of Microwave-Assisted Organic Synthesis. From Single-Mode to Multimode Parallel Batch Reactors. Org Process Res Dev 2003; 7(5): 707-16.
[http://dx.doi.org/10.1021/op034075+]

[25] Wilson NS, Sarko CR, Roth GP. Development and Applications of a Practical Continuous Flow Microwave Cell. Org Process Res Dev 2004; 8(3): 535-8.

[http://dx.doi.org/10.1021/op034181b]

[26] Ley SV, Baxendale IR. New tools and concepts for modern organic synthesis. Nat Rev Drug Discov 2002; 1(8): 573-86.
[http://dx.doi.org/10.1038/nrd871] [PMID: 12402498]

[27] Gaba M, Dhingra N. Microwave Chemistry: GeneralFeatures and Applications. Indian J PharmEduc Res 2011; 45(2): 175-83.

[28] Morent R, De Geyter N, Desmet T, *et al.* Plasma surface modification of biodegradable polymers: a review. Plasma Process Polym 2011 Mar 22; 8(3): 171–90.
[http://dx.doi.org/10.1002/ppap.201000153]

[29] Montes I, Sanabria D, García M, Castro J, Fajardo J. A Greener Approach to Aspirin Synthesis Using Microwave Irradiation. J Chem Educ 2006; 83(4): 628-31.
[http://dx.doi.org/10.1021/ed083p628]

[30] Panda J, Patro VJ, Sahoo BM, Mishra J. Green Chemistry Approach for Efficient Synthesis of Schiff Bases of Isatin Derivatives and Evaluation of Their Antibacterial Activities. Journal of Nanoparticles 2013; 2013: 1-5.
[http://dx.doi.org/10.1155/2013/549502]

[31] Mordini A, Faigl F. New Methodologies and Techniques for a Sustainable Organic Chemistry. Springer, Netherlands 2008; pp. 193-223.
[http://dx.doi.org/10.1007/978-1-4020-6793-8]

[32] Loupy A, Petit A, Hamelin J, Texier-Boullet F, Jacquault P, Mathé D. New solvent free organicsynthesis using focused microwaves. Synthesis 1998; 1998(9): 1213-34.
[http://dx.doi.org/10.1055/s-1998-6083]

[33] Kerton FM. Applying the principles of green chemistry to achieve a sustainable polymer industry. Curr Opin Green Sustainable Chem. 2024;50:100997.
[http://dx.doi.org/10.1016/j.cogsc.2024.100997]

[34] Baghurst DR, Mingos DMP. Superheating effects associated with microwave dielectric heating. J Chem Soc Chem Commun 1992; (9): 674-7.
[http://dx.doi.org/10.1039/c39920000674]

[35] Surati MA, Jauhari S, Desai KR. A brief review:Microwave-assisted organic reaction. Indian JPharmEducRes 2012; 4(1): 645-61.

[36] Charde MS, Shukla A, Bukhariya V, Chakole RD. Areview on: a significance of microwave assist techniquein green chemistry. Int J Phytopharm 2012; 2(2): 39-50.
[http://dx.doi.org/10.7439/ijpp.v2i2.441]

[37] Hayes BL. Recent Advances in Microwave-Assisted Synthesis. Aldrichim Acta 2004; 37(2): 66-76.

[38] Chen JJ, Deshpande SV. Rapid synthesis of α-ketoamides using microwave irradiation–simultaneous cooling method. Tetrahedron Lett 2003; 44(49): 8873-6.
[http://dx.doi.org/10.1016/j.tetlet.2003.09.180]

[39] Chemat-Djenni Z, Hamada B, Chemat F. Atmospheric pressure microwave assisted heterogeneous catalytic reactions. Molecules 2007; 12(7): 1399-409.
[http://dx.doi.org/10.3390/12071399] [PMID: 17909495]

[40] Liu Y, Lu Y, Liu P, Gao R, Yin Y. Effects of microwaves on selective oxidation of toluene to benzoic acid over a V_2O_5/TiO_2 system. Appl Catal A Gen 1998; 170(2): 207-14.
[http://dx.doi.org/10.1016/S0926-860X(98)00045-3]

[41] Lin Z, Wragg DS, Morris RE. Microwave-assisted synthesis of anionic metal–organic frameworks under ionothermal conditions. Chem Commun (Camb) 2006; (19): 2021-3.
[http://dx.doi.org/10.1039/B600814C] [PMID: 16767262]

[42] Boscencu R. Microwave synthesis under solvent-free conditions and spectral studies of some

mesoporphyrinic complexes. Molecules 2012; 17(5): 5592-603.
[http://dx.doi.org/10.3390/molecules17055592] [PMID: 22576229]

[43] Baghbanzadeh M, Carbone L, Cozzoli PD, Kappe CO. Microwave-assisted synthesis of colloidal inorganic nanocrystals. Angew Chem Int Ed 2011; 50(48): 11312-59.
[http://dx.doi.org/10.1002/anie.201101274] [PMID: 22058070]

[44] Rao KJ, Vaidhyanathan B, Ganguli M, Ramakrishnan PA. Synthesis of inorganic solids using microwaves. Chem Mater 1999; 11(4): 882-95.
[http://dx.doi.org/10.1021/cm9803859]

[45] Sreeram KJ, Nidhin M, Nair BU. Microwave assisted template synthesis of silver nanoparticles. Bull Mater Sci 2008; 31(7): 937-42.
[http://dx.doi.org/10.1007/s12034-008-0149-3]

[46] Roy MD, Herzing AA, De Paoli Lacerda SH, Becker ML. Emission-tunable microwave synthesis of highly luminescent water soluble CdSe/ZnS quantum dots. Chem Commun (Camb) 2008; (18): 2106-8.
[http://dx.doi.org/10.1039/b800060c] [PMID: 18438483]

[47] Polshettiwar V, Nadagouda MN, Varma RS. Microwave-assisted chemistry: a rapid and sustainableroute to synthesis of organics and nanomaterials. Aust J Chem 2009; 62(1): 16-26.
[http://dx.doi.org/10.1071/CH08404]

[48] Jiang ZL, Feng ZW, Shen XC. Microwave Synthesis ofAu Nanoparticles with the System of $AuCl_4^-$–CH_3CH_2OH. Chin Chem Lett 2001; 12(6): 551-4.

[49] Ambrozic G, Orel ZC, Zigon M. Microwave-assistednon-aqueous synthesis of ZnO nanoparticles. Mater Technol 2011; 45(3): 173-7.

[50] Amin RS, Elzatahry AA, El-Khatib KM, Youssef ME. Nanocatalysts Prepared by Microwave and Impregnation Methods for Fuel Cell Application. Int J Electrochem Sci 2011; 6(10): 4572-80.
[http://dx.doi.org/10.1016/S1452-3981(23)18349-0]

[51] Bohnemann J, Libanori R, Moreira ML, Longo E. High-efficient microwave synthesis and characterisation of SrSnO3. Chem Eng J 2009; 155(3): 905-9.
[http://dx.doi.org/10.1016/j.cej.2009.09.004]

[52] Jacob J. Microwave Assisted Reactions in Organic Chemistry: A Review of Recent Advances. Int J Chem 2012; 4(6): 29-43.
[http://dx.doi.org/10.5539/ijc.v4n6p29]

[53] Synthesis and characterization of graft copolymers by photoinduced CuAAC click chemistry,European Polymer Journal, Volume 66, 2015, Pages 282-289, ISSN 0014-3057.

[54] Akyel C, Bilgen E. Microwave and radio-frequency curing of polymers: Energy requirements, cost and market penetration. Energy 1989; 14(12): 839-51.
[http://dx.doi.org/10.1016/0360-5442(89)90038-8]

[55] Pedersen SL, Tofteng AP, Malik L, Jensen KJ. Microwave heating in solid-phase peptide synthesis. Chem Soc Rev 2012; 41(5): 1826-44.
[http://dx.doi.org/10.1039/C1CS15214A] [PMID: 22012213]

[56] Bacsa B, Desai B, Dibó G, Kappe CO. Rapid solid-phase peptide synthesis using thermal and controlled microwave irradiation. J Pept Sci 2006; 12(10): 633-8.
[http://dx.doi.org/10.1002/psc.771] [PMID: 16789045]

[57] Erdmenger T, Guerrero-Sanchez C, Vitz J, Hoogenboom R, Schubert US. Recent developments in the utilization of green solvents in polymer chemistry. Chem Soc Rev 2010; 39(8): 3317-33.
[http://dx.doi.org/10.1039/b909964f] [PMID: 20601997]

[58] Čemažar M, Craik DJ. Microwave-assisted Boc-solid phase peptide synthesis of cyclic cysteine-rich peptides. J Pept Sci 2008; 14(6): 683-9.

[http://dx.doi.org/10.1002/psc.972] [PMID: 18044816]

[59] Wu CS. Grafting of biodegradable polyesters: mechanisms and applications. Polymers (Basel) 2020; 12(3): 612.

[60] Matsushita T, Hinou H, Fumoto M, *et al.* Construction of highly glycosylated mucin-type glycopeptides based on microwave-assisted solid-phase syntheses and enzymatic modifications. J Org Chem 2006; 71(8): 3051-63.
[http://dx.doi.org/10.1021/jo0526643] [PMID: 16599599]

[61] Emran AM. Chemist's Views of Imaging Centers. New York: Springer 1995; pp. 445-54.
[http://dx.doi.org/10.1007/978-1-4757-9670-4]

[62] Willert-Porada M. Advances in Microwave and Radio Frequency Processing. Berlin, Heidelberg: Springer; 2007. p. 359-69.

[63] Hwang DR, Moerlein SM, Lang L, Welch MJ. Application of microwave technology to the synthesis of short-lived radiopharmaceuticals. J Chem Soc Chem Commun 1987; (23): 1799-801.
[http://dx.doi.org/10.1039/c39870001799]

[64] Wild D, Wicki A, Mansi R, *et al.* Exendin-4-based radiopharmaceuticals for glucagonlike peptide-1 receptor PET/CT and SPECT/CT. J Nucl Med 2010; 51(7): 1059-67.
[http://dx.doi.org/10.2967/jnumed.110.074914] [PMID: 20595511]

[65] Bashir A, Warsi MH, Sharma PK. An overview of natural gums as pharmaceutical excipient: Their chemical modification. World J Pharm Pharm Sci 2016; 5(4): 2025-39.

[66] Acta A, Ogaji IJ, Nep EI, Audu-Peter JD. Advances in natural polymers as pharmaceutical excipients. Pharmaceutica 2012; 3(1): 1-16.

[67] Bhattacharya A, Misra BN. Grafting: a versatile means to modify polymersTechniques, factors and applications. Prog Polym Sci 2004; 29(8): 767-814.
[http://dx.doi.org/10.1016/j.progpolymsci.2004.05.002]

[68] Alves NM, Mano JF. Chitosan derivatives obtained by chemical modifications for biomedical and environmental applications. Int J Biol Macromol 2008; 43(5): 401-14.
[http://dx.doi.org/10.1016/j.ijbiomac.2008.09.007] [PMID: 18838086]

[69] Xie W, Xu P, Wang W, Liu Q. Preparation and antibacterial activity of a water-soluble chitosan derivative. Carbohydr Polym 2002; 50(1): 35-40.
[http://dx.doi.org/10.1016/S0144-8617(01)00370-8]

[70] Hosseinzadeh H. Chemical modification of sodium hyaluronate *via* graft copolymerization of acrylic acid using ammonium persulfate. Res J Pharm Biol Chem Sci 2012; 3(1): 756-61.

[71] Yoshikawa S, Takayama T, Tsubokawa N. Grafting reaction of living polymer cations with amino groups on chitosan powder. J Appl Polym Sci 1998; 68(11): 1883-9.

[72] Szwarc M. Living polymers. Their discovery, characterization, and properties. Polym Sci Part A Polym Chem 1997; 36: 9-15.

[73] Lizotte JR, Long TE. Stable free-radical polymerization of styrene in combination with 2-vinylnaphthalene initiation. Macromol Chem Phys 2003; 204(4): 570-6.
[http://dx.doi.org/10.1002/macp.200390020]

[74] Mishra V, Kumar R. Living radical polymerization: A review. J Sci Res 2012; 56: 141-76.

[75] Bulbul Sonmez H, Senkal BF, Sherrington DC, Bıcak N. Atom transfer radical graft polymerization of acrylamide from N-chlorosulfonamidated polystyrene resin, and use of the resin in selective mercury removal. React Funct Polym 2003; 55(1): 1-8.
[http://dx.doi.org/10.1016/S1381-5148(02)00193-1]

[76] Arslan H. Block and graft copolymerization by controlled/living radical polymerization methods. InTechOpen 2012; 13: 279-320.

[http://dx.doi.org/10.5772/45970]

[77] Liu ZM, Xu ZK, Wang J-Q, Wu J, Fu J-J. Surface modification of polypropylene microfiltration membranes by graft polymerization of N-vinyl-2-pyrrolidone. Eur Polym J 2004; 40(9): 2077-87.
[http://dx.doi.org/10.1016/j.eurpolymj.2004.05.020]

[78] Chen T, Kumar G, Harris MT, Smith PJ, Payne GF. Enzymatic grafting of hexyloxyphenol onto chitosan to alter surface and rheological properties. Biotechnol Bioeng 2000; 70(5): 564-73.
[PMID: 11042553]

[79] Yamaguchi T, Yamahara S, Nakao S, Kimura S. Preparation of pervaporation membranes for removal of dissolved organics from water by plasma-graft filling polymerization. J Membr Sci 1994; 95(1): 39-49.
[http://dx.doi.org/10.1016/0376-7388(94)85027-5]

[80] Kim YJ, Kang IK, Huh MW, Yoon SC. Surface characterization and in vitro blood compatibility of poly(ethylene terephthalate) immobilized with insulin and/or heparin using plasma glow discharge. Biomaterials 2000; 21(2): 121-30.
[http://dx.doi.org/10.1016/S0142-9612(99)00137-4] [PMID: 10632394]

[81] Contreras-García A, Bucio E, Concheiro A, et al. Surface functionalization of polypropylene devices with hemocompatible DMAAm and NIPAAm grafts for norfloxacin sustained release. J Bioact Compat Polym 2011; 26(4): 405-19.
[http://dx.doi.org/10.1177/0883911511407788]

[82] Hanh TT, Huy HT, Hien NQ. Pre-irradiation grafting of acrylonitrile onto chitin for adsorption of arsenic in water. Radiat Phys Chem 2015; 106: 235-41.
[http://dx.doi.org/10.1016/j.radphyschem.2014.08.004]

[83] Singh DK, Ray AR. Graft copolymerization of 2-hydroxyethylmethacrylate onto chitosan films and their blood compatibility. J Appl Polym Sci 1994; 53(8): 1115-21.
[http://dx.doi.org/10.1002/app.1994.070530814]

[84] Sah SK, Tiwari AK, Bairwa R, Bishnoi N. Natural gums emphasized grafting technique: Applications and perspectives in floating drug delivery system. Asian J Pharm 2016; 10(2): 72-83.

[85] Dailey LA, Wittmar M, Kissel T. The role of branched polyesters and their modifications in the development of modern drug delivery vehicles. J Control Release 2005; 101(1-3): 137-49.
[http://dx.doi.org/10.1016/j.jconrel.2004.09.003] [PMID: 15588900]

[86] Jung T, Kamm W, Breitenbach A, Hungerer KD, Hundt E, Kissel T. Tetanus toxoid loaded nanoparticles from sulfobutylated poly(vinyl alcohol)-graft-poly(lactide-co-glycolide): evaluation of antibody response after oral and nasal application in mice. Pharm Res 2001; 18(3): 352-60.
[http://dx.doi.org/10.1023/A:1011063232257] [PMID: 11442276]

[87] Ms ME, Am O, Ma W, Tm T, Ms AB, Sa I. Novel smart pH sensitive chitosan grafted alginate hydrogel microcapsules for oral protein delivery: I. Preparation and characterization. Int J Pharm Pharm Sci 2015; 7(10): 320-6.

[88] Cai H, Zhang ZP, Chuan Sun P, Lin He B, Xia Zhu X. Synthesis and characterization of thermo- and pH- sensitive hydrogels based on Chitosan-grafted N-isopropylacrylamide via γ-radiation. Radiat Phys Chem 2005; 74(1): 26-30.
[http://dx.doi.org/10.1016/j.radphyschem.2004.10.007]

[89] Hu Y, Jiang X, Ding Y, Ge H, Yuan Y, Yang C. Synthesis and characterization of chitosan–poly(acrylic acid) nanoparticles. Biomaterials 2002; 23(15): 3193-201.
[http://dx.doi.org/10.1016/S0142-9612(02)00071-6] [PMID: 12102191]

[90] Kumar R, Sharma K. Biodegradable polymethacrylic acid grafted psyllium for controlled drug delivery systems. Front Chem Sci Eng 2013; 7(1): 116-22.
[http://dx.doi.org/10.1007/s11705-013-1310-0]

[91] Vihola H, Laukkanen A, Tenhu H, Hirvonen J. Drug release characteristics of physically cross-linked

thermosensitive poly(N-vinylcaprolactam) hydrogel particles. J Pharm Sci 2008; 97(11): 4783-93.
[http://dx.doi.org/10.1002/jps.21348] [PMID: 18306245]

[92] Frauke Pistel K, Breitenbach A, Zange-Volland R, Kissel T. Brush-like branched biodegradable polyesters, part III. J Control Release 2001; 73(1): 7-20.
[http://dx.doi.org/10.1016/S0168-3659(01)00231-0] [PMID: 11337055]

[93] Kulkarni RV, Sa B. Evaluation of pH-sensitivity and drug release characteristics of (polyacrylamide-grafted-xanthan)-carboxymethyl cellulose-based pH-sensitive interpenetrating network hydrogel beads. Drug Dev Ind Pharm 2008; 34(12): 1406-14.
[http://dx.doi.org/10.1080/03639040802130079] [PMID: 18785037]

[94] Mundargi RC, Patil SA, Agnihotri SA, Aminabhavi TM. Evaluation and controlled release characteristics of modified xanthan films for transdermal delivery of atenolol. Drug Dev Ind Pharm 2007; 33(1): 79-90.
[http://dx.doi.org/10.1080/03639040600975030] [PMID: 17192254]

[95] Siraj S, Sudhakar P, Rao US, Sekharnath KV, Rao KC, Subha MC. Interpenetrating polymer network microspheres of poly (Vinyl alcohol)/methyl cellulose for controlled release studies of 6-thioguanine. Int J Pharm Pharm Sci 2014; 6(9): 101-6.

[96] Kumbar SG, Soppimath KS, Aminabhavi TM. Synthesis and characterization of polyacrylamide-grafted chitosan hydrogel microspheres for the controlled release of indomethacin. J Appl Polym Sci 2003; 87(9): 1525-36.
[http://dx.doi.org/10.1002/app.11552]

[97] Yoshioka H, Nonaka K, Fukuda K, Kazama S. Chitosan-derived polymer-surfactants and their micellar properties. Biosci Biotechnol Biochem 1995; 59(10): 1901-4.
[http://dx.doi.org/10.1271/bbb.59.1901] [PMID: 8534983]

[98] Liu WG, De Yao K, Liu QG. Formation of a DNA/ N-dodecylated chitosan complex and salt-induced gene delivery. J Appl Polym Sci 2001; 82(14): 3391-5.
[http://dx.doi.org/10.1002/app.2199]

[99] Nam JP, Lee KJ, Choi JW, Yun CO, Nah JW. Targeting delivery of tocopherol and doxorubicin grafted-chitosan polymeric micelles for cancer therapy: In vitro and in vivo evaluation. Colloids Surf B Biointerfaces 2015; 133: 254-62.
[http://dx.doi.org/10.1016/j.colsurfb.2015.06.018] [PMID: 26117805]

[100] Sakhare MS, Rajput HH. Polymer grafting and applications in pharmaceutical drug delivery systems - a brief review. Asian J Pharm Clin Res 2017; 10(6): 59.
[http://dx.doi.org/10.22159/ajpcr.2017.v10i6.18072]

[101] Sheldon RA, Kochi JK. Metal-Catalyzed Oxidations of Organic Compounds. New York: Academic Press; 1981.

[102] Hudlicky M. Oxidations in Organic Chemistry. ACS Monogr No 186. Washington (DC): American Chemical Society; 1990.

[103] Huang NM, Lim HN, Radiman S, et al. Sucrose ester micellar-mediated synthesis of Ag nanoparticles and the antibacterial properties. Colloids Surf A Physicochem Eng Asp 2010; 353(1): 69-76.
[http://dx.doi.org/10.1016/j.colsurfa.2009.10.023]

[104] Hebeish AA, El-Rafie MH, Abdel-Mohdy FA, Abdel-Halim ES, Emam HE. Carboxymethyl cellulose for green synthesis and stabilization of silver nanoparticles. Carbohydr Polym 2010; 82(3): 933-41.
[http://dx.doi.org/10.1016/j.carbpol.2010.06.020]

[105] Bar H, Bhui DK, Sahoo GP, Sarkar P, De SP, Misra A. Green synthesis of silver nanoparticles using latex of Jatropha curcas. Colloids Surf A Physicochem Eng Asp 2009; 339(1-3): 134-9.
[http://dx.doi.org/10.1016/j.colsurfa.2009.02.008]

[106] Cravotto G, Cintas P. Power ultrasound in organic synthesis: moving cavitational chemistry from academia to innovative and large-scale applications. Chem Soc Rev 2006; 35(2): 180-96.

[http://dx.doi.org/10.1039/B503848K] [PMID: 16444299]

[107] Saini P, Arora M, Kumar M. Free radical grafting of biodegradable polymers: recent developments. Prog Polym Sci 2023; 138: 101694.

CHAPTER 4

Renewable Resources for Green Grafting: Types, Benefits, and Challenges

Deepali D. Bhandari[1]**, Dattatraya M. Shinkar**[2,*]**, Ramanlalal N. Kachave**[1]**, Unmesh G. Bhamare**[2] **and Sunil V. Amrutkar**[1]

[1] *Department of Pharmaceutical Chemistry, GES's Sir Dr. M. S. Gosavi College of Pharmaceutical Education and Research, Nashik, Maharashtra 422001, India*

[2] *Department of Pharmaceutics, GES's Sir Dr. M. S. Gosavi College of Pharmaceutical Education and Research, Nashik, Maharashtra 422001, India*

Abstract: The pharmaceutical sector has been using renewable energy solutions more in recent years as part of its environmental stewardship and sustainability efforts. This development reflects an increasing understanding of the necessity of lowering carbon emissions, lessening the effects of climate change, and minimizing the environmental impact of pharmaceutical production processes. Globally, hospitals and health systems are making investments in clean, renewable energy to safeguard the health of their patients. Hospitals can continue to operate amid severe weather conditions or other disruptions by combining renewable energy with power storage. The EPA defines renewable energy as those sources that depend on non-depleting fuel sources that replenish themselves over brief periods. In reality, wind and solar energy provide the majority of renewable electricity generated in the United States since they are affordable, easily accessible, and clean. Compared to electricity, thermal energy (such as steam, heat, and hot water) presents greater challenges. Many healthcare facilities employ biomass-powered combined heat and power plants, although concerns exist regarding their impact on local health, carbon emissions, and sustainable forestry practices. Though they will need to grow in size before they can be economically utilized for healthcare, emerging technologies like green hydrogen may be useful in decarbonizing thermal energy. Green electrospinning is a promising field for sustainable nanomaterial production, offering eco-efficient methods and reducing environmental impacts. Biological nanofibers, which control drug administration and environmental cleanup, utilize strategies such as solvent removal, integration of renewable energy sources, and waste utilization. Additional investigation is required in the areas of materials engineering, scaling up procedures, achieving multifunctionality, and evaluating entire life cycle sustainability. Research on natural gums and mucilages could be used for drug delivery systems. This study examines the integration of renewable energy sources in pharmaceutical production processes, focusing on both the challenges and benefits.

---

[*] **Corresponding author Dattatraya M. Shinkar:** Department of Pharmaceutics, GES's Sir Dr. M. S. Gosavi College of Pharmaceutical Education and Research, Nashik, Maharashtra 422001, India; E-mail: dattashinkar@gmail.com

Kuldeep Vinchurkar, Satish Polshettiwar, Nilesh Mahajan &Yogeshwar Bachhav (Eds.)
All rights reserved-© 2026 Bentham Science Publishers

Keywords: Eco-friendly, Green grafting, Green electrospinning, Gums, Nanofibers, Renewable energy, Sustainable.

INTRODUCTION

The surrounding environment has bestowed upon humans a vast array of resources for directly or indirectly balancing the health of all living creatures. The pharmaceutical industry is compliant in using the majority of polymers in different dosage forms. Particularly polymers obtained from natural sources, such as various gums and mucilages, are extensively used as resources from nature for both novel and established drug delivery systems. Short half-lives, low bioavailability, and chemical and physical instability are among the drawbacks of pharmaceutical and biological therapies. The primary cause of physical instability is the modification of highly organized protein structures, which can result in unfavorable processes such as breakdown, aggregation, and precipitation. The drug's chemical instability is a result of various chemical processes, including oxidation, deamination, hydrolysis, and racemization [1]. Due to their numerous pharmaceutical uses, including binders, solvents, disintegrating agents in tablets, stabilizers for colloidal particles in suspension, thickening agents in oral solutions, bases in suppositories, gelling agents in gels, and polymers derived from plant sources, these substances have garnered remarkable attention in recent years. Polymers are also used in colorants, textiles, paper manufacturing, and cosmetics.

It is possible to successfully deliver medications to a targeted location at a particular level for a specific amount of time by using the right polymer or polymers. Easy production, a single dosage for prolonged release of the combined medication, enhanced stability, and maintenance of the intended therapeutic concentration are key advantages of using polymers with the appropriate properties in drug delivery systems. Therefore, selecting the appropriate copolymer system is a crucial step in preparing the drug for dosage form formulation. The drug's mechanism of action, degree of discharge, and stability are primarily dictated by the type of polymers utilized in the formulation. Macromolecules called polymers are made up of monomers bonded together covalently. Jöns Jacob Berzelius was the one who originally used the term "polymer" [2]. Due to their numerous pharmaceutical uses as binding agents, solvents, fillers in tablets, protective colloidal particles in a suspended form, gelling components in gels, thickening agents in oral preparations and bases in suppositories, polymers derived from microbial, marine, plant, and synthetic sources have garnered immense attention in recent years [3]. However, not all of the qualities required must be present in the polymers; formulation scientists may find that certain traits are missing from the polymers that are now on the market. This highlighted the importance of polymer grafting in the formulation development process. A wide range of advantageous qualities can be added to polymers and their undesirable ones reduced by altering their chemical functional groups.

Gums can be altered or substituted with desired chemicals by various methods to address the shortcomings of both natural and synthetic gums. This allows gums to be comparable to their counterparts while also modulating the exact location of drug release and its kinetics—specific requirements designed for the intended uses. Reactive functional compounds, such as carboxylic acids, amino groups, hydroxyl groups, and thiol groups, show potential locations for grafting or chemical alteration in the chemical composition of polymers. In the era of polymers, it is crucial to tailor a polymer's characteristics to meet custom requirements tailored for specific uses. Polymer characteristics can be altered through various methods, including mixing, blending, and curing [4].

Natural polymers, known as biopolymers, are present in living organisms. A biological polymer is a molecule with a lengthy chain composed of monomeric units joined by covalent bonds to form a biodegradable molecule. Biopolymers primarily originate from natural sources, including bacteria, plants, and trees. Synthetic polymers are more random and simpler than biopolymers, which are complex molecules with well-defined three-dimensional geometries [5]. Renewable resources are utilized to produce a range of biopolymer materials with diverse chemical and physical properties. Nature contains lignin, cellulose, starches, hemicelluloses, and a wide variety of other biopolymers [6]. Future production for biopolymers will place high expectations on suppliers of innovative materials. Nonetheless, the material's cost-effectiveness must increase because they are being donated specifically for sustainable development. These polymer properties are utilized in new material applications, and goods are made with those characteristics in mind. Recent research has focused on developing new recyclable polymers with strong mechanical and skeletal properties. Massive amounts of biopolymers derived from living things have been produced. The existence of specific microorganisms and enzymes with unique degradable characteristics has been linked to the biological degradation of biopolymers [7]. Because the skeletal backbone of biopolymers contains nitrogen and oxygen atoms, they biodegrade readily. In the process of biodegradation, a biopolymer breaks down into carbon dioxide, liquid water, biomass, and various other natural elements, including damp water. Numerous bioengineering applications, including the engineering of tissues, systems for drug delivery, and wound dressings, are possible using biodegradable polymers [8]. Biopolymers thicken and solidify more successfully in abrasive pool settings due to their distinctive helical form, stiffness, charge-free chains, and strong resilience to cold and salt [9].

GRAFTING

A graft is an element that is inserted or implanted into another to become a part of it. Chemical alteration of both synthetic and natural macromolecular moieties (such as polysaccharides, polyvinyl alcohol, *etc.*) can be achieved by graft procedures, which is an exciting new field of research. Enhancing a surface polymer's wettability, biocompatibility, and mechanical properties is the primary goal of surface modification.

Grafting can be divided into two main categories:

- Using one monomer for the grafting process
- Utilizing a combination for grafting, comprising two or more monomers

In addition, there are two methods, termed graft from and graft to, that are used to generate grafting copolymers. The steps involved in the grafting procedure and its main advantages are mentioned in Fig. (**1**). Several process-related parameters influence grafting metrics, including grafting productivity and total percentage grafting, as well as process temperature, time, monomer concentration, and initiator type and concentration. The length, quantity, and molecular structure of these side chains have a significant influence on the behaviour of the finished graft copolymers [10, 11]. Graft polymerization is a useful approach for modifying the properties of polymers and graft gums, such as acacia, Arabic, and xanthan gum, among others [12]. The various grafting techniques and methods are summarized in Fig. (**2**).

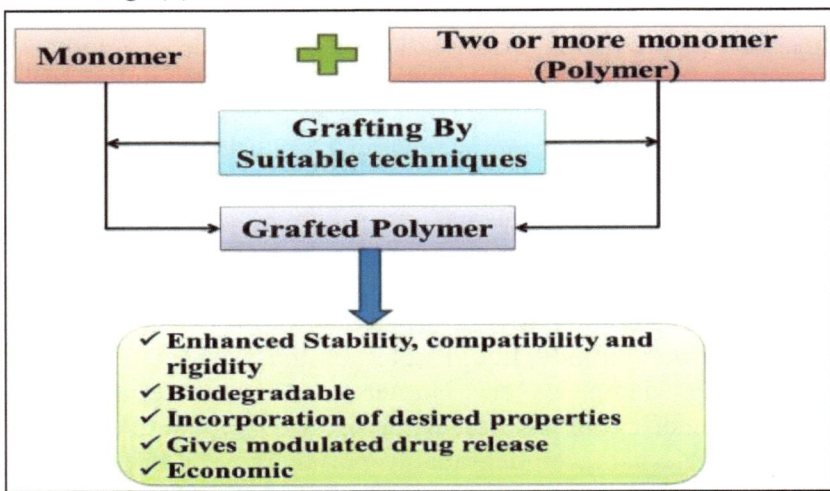

Fig. (**1**). The steps involved in the grafting procedure and its main advantages.

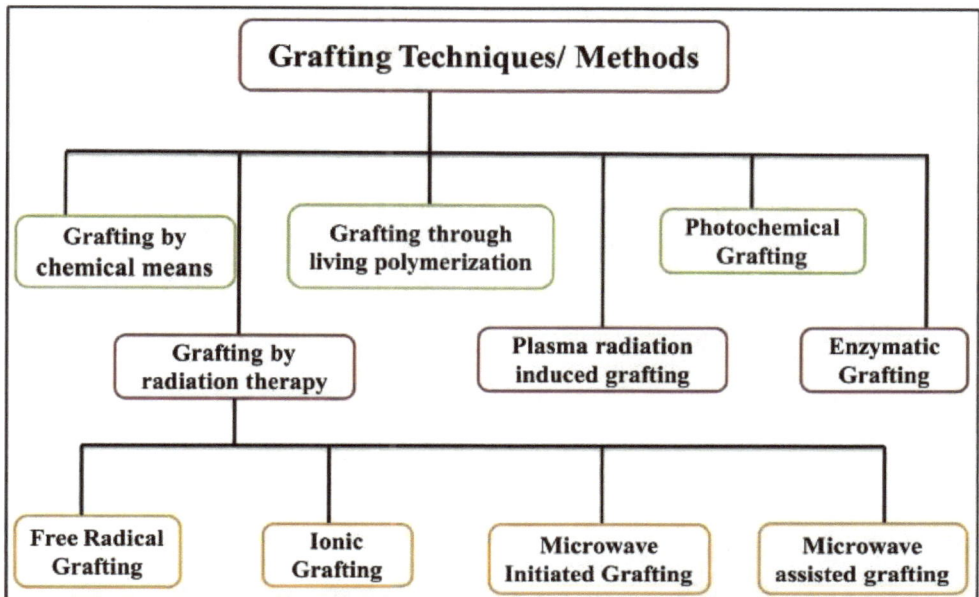

Fig. (2). Different techniques and methods for grafting.

Grafting by Chemical Process

These techniques enable the chemical initiation of grafting through two main pathways: ionic grafting and free radical grafting. Since it controls the rate at which the grafting operation progresses, the initiator is essential to the process. It is additionally divided into ionization and free radical grafting [13].

Grafting with Free Radicals

Free radicals produced by the initiators are transferred to the substrate, where they combine with the monomers to generate graft copolymers. When performing oxygen-based reactive species (which are free radicals) initiated grafting, the following initiator processes are used:

- CAN: Ceric ammonium nitrate
- PPS: Potassium persulfate
- PDC: Potassium diperiodatocuprate (III)
- TCPB: Thiocarbonationpotassium bromate
- FAS: Ferrous ammonium sulfate
- APS: Ammonium persulfate

By using the APS initiator, Xie *et al.* and coworkers [14] prepared hydroxypropyl chitosan-grafted MAA (methylacrylamide), achieving a derivative with good water solubility.

Ionic Grafting

In ionic settings, grafting can be initiated primarily through ionic mode. Grafting can be carried out by a cationic mechanism, which utilizes alkaline metal salts, such as alkyl aluminum (R3Al), or an anionic mechanism, which employs the alkoxide form of alkali metals, like sodium methoxide [4]. Ionic grafting is classified further into two types:

- Anionic grafting
- Cationic grafting

- **Anionic grafting:**

By the copolymerization of grafts of acrylic acid, Hossein Hosseinzadeh altered sodium hyaluronate, a polysaccharide. The method of graft copolymerization was carried out using ammonium persulfate, an anionic initiator. The ultimate goal of grafting was believed to increase the molecular size of sodium hyaluronate after homopolymer is extracted; this also provided the basis for grafting parameter determination. The impact of grafting temperature and the concentrations of acrylic acids, sodium hyaluronate, as well as ammonium persulfate, on the grafting process is included in this study [15].

- **Cationic grafting:**

Grafting of chitosan was performed by Yoshikawa *et al.* using cationic living polymers, such as poly(isobutyl vinyl ether) and poly(2-methyl-2-oxazoline). This work investigated the effect of the molecular weight of the living polymer cation on the molecular weight of the grafted polymer. It was found that the viscosity of the final polymer produced by the grafting process increased with increasing grafting percentage. Additionally, the water-permeable nature of the aforementioned grafted polymer has been found [16].

Living Polymerization-induced Grafting

When a polymer can reach its maximal size and reproduce for an extended period without transferring very little chain, it is said to be alive. If there is not enough absorption of light, photosensitizers such as benzoin ethyl ether can be added to continue the process [17, 18]. Regulated free radical polymerizations combine the best features of both conventional and ionic free radical polymerizations. If a living polymerization occurs, it yields live polymers with regulated molecular weights and low polydispersities [4].

Modifying the microscopic attributes of the polymer is extremely difficult with conventional radical polymerization because it lacks the ability to regulate the molecular weight of the molecules, their molecular weight dispersion, and polymer structure. The production of copolymers with specific features, such as defined copolymer arrangements, branching, end-group operations, regulated molecular weights, and constrained molecular weight distributions, can be achieved; however, this is typically done using living ionic polymerizations.

- The application of an active radical site to the polymerization process, along with the controllable homolytic fragmentation of a dormant chain end to generate a stable free radical, is involved in the process of stable free radical polymers (SFRP) . The literature focused primarily on styrenic monomers, even though SFRP was primarily used as acrylamides, styrenics, and acrylates [19].
- Reversed addition-fragmentation chain transfer, or RAFT, is achieved by performing a free radical polymerization in the presence of dithio compounds, which act as efficient RAFT agents [20].
- ATRP, which is Atom transfer radical addition, involves the reversible transfer of halogen atoms, which cap inactive chains, to complexes of metals in their reduced oxidation state. As a result, greater oxidation state complexes and transiently increasing radicals are produced. The dynamic equilibrium mechanism of activation-deactivation is the central reaction of ATRP [20].

Acrylamide grafting by ATRP was described by Sonmez *et al.* [21]; the initiation appears to be caused by redox reactions on N-chlorosulfonamide groups, using CuBr to produce radicals. All these techniques rely on quickly creating a dynamic balance between a large number of dormant chains, which are more resistant to functional groups and contaminants, and a small concentration of active chains that are propagating or terminating [22].

Photochemical Grafting

Photochemical reactions in the grafting process produce reactive free radicals when light is absorbed by chromophores on macromolecules. Photosensitizers can be used to complete the process if light absorption is insufficient. The grafted copolymer is formed by reacting free radicals with monomer free radicals. Liu and Xu used ultraviolet and γ-radiations to photochemically graft poly (N-vinyl-2-pyrrolidone) onto a polypropylene microfiltration membrane, increasing its hydrophilicity and biocompatibility [23, 24].

Radiation-based Grafting Approach

The process of homolytic fission causes macromolecule irradiation; polymers can be produced in this process. The lifespan of free radicals is dependent on the kind of polymer backbone. Three methods exist for radiation technology grafting:

- **Pre-irradiation:** Free radicals are produced when polymeric substrates are treated with monomers in a liquid or gaseous state or solvent, either in a vacuum or with inert gas.
- **Peroxidation:** High-energy radiation creates hydroperoxides or diperoxides in polymers, which are then grafted by reacting peroxy compounds with monomer at higher temperatures.
- **Mutual irradiation:** In this method, monomers and polymers are exposed to radiation simultaneously in order to produce radicals that are free [25].

Enzymatic Grafting

It is predicated on the idea that the chemical as well as electrochemical grafting process can be started with the aid of an enzyme [27]. Enzyme application can provide a safer, more cost-effective, and environmentally friendly method for grafting processes by eliminating the need for reactive reagents. Moreover, the specificity of enzymes may present a possibility for accurately modifying macromolecular characteristics to desired ones. Tyrosinase was identified as the initiator of peptide grafting onto the amine-containing polysaccharide chitosan [27].

Plasma- Radiation-Induced Grafting

Plasma radiation exposure can be achieved through slow discharge, allowing for similar grafting probabilities as ionizing radiation. High-energy electrons stimulate, ionize, and dissociate polymeric structures, leading to graft copolymerization. Heparin immobilized on grafted polymers is blood-compatible, and acrylic acid-grafted polyethylene terephthalate reduces antithrombogenic properties [26 - 29].

Radiation Grafting

Exposed to radiation with significant energy along the polymeric backbone, free radicals are formed as active sites that serve as grafting sites for side-chain graft propagation. Without the need for chemical initiators, these radicals readily react with suitable functional monomers to generate bonds of covalent attraction and, as a result, the formation of macromolecular chains [30, 31].

Singh and Roy investigated the use of N,N-dimethylaminoethylmethacrylate in radiation grafting of chitosan. The process of transplanting and homopolymerization was found to be influenced by radiation factors, including radiation dose rate and cumulative dose/time. According to this investigation, carefully selecting these radiation parameters under grafting conditions enabled the achievement of the required degree of grafting [32].

- **Grafting with free radicals:** This method is mentioned under point 4 of Grafting initiated by radiation technique.
- **Ionic grafting:** An ionic mode can also be used to carry out radiation grafting. Ionic grafting can produce either anionic or cationic ions, which are produced by high-energy irradiation. The grafted copolymer is created by the polymer reacting with the monomer after being exposed to radiation, producing the polymeric ion. A high response rate is one potential benefit of ionic grafting [4].
- **Microwave-initiated grafting:** This strategy does not employ initiators. A free radical-mediated grafting process wherein the processes of grafting during microwave irradiation are slowed down by the use of hydroquinone as an inhibitor, is depicted in Fig. (3) [33, 34].
- **Microwave Assisted Grafting:** Redox initiators convert microwave energy into thermal energy, aiding in the production of radicals free from initiators through grafting reactions under microwave dielectric heating [33].

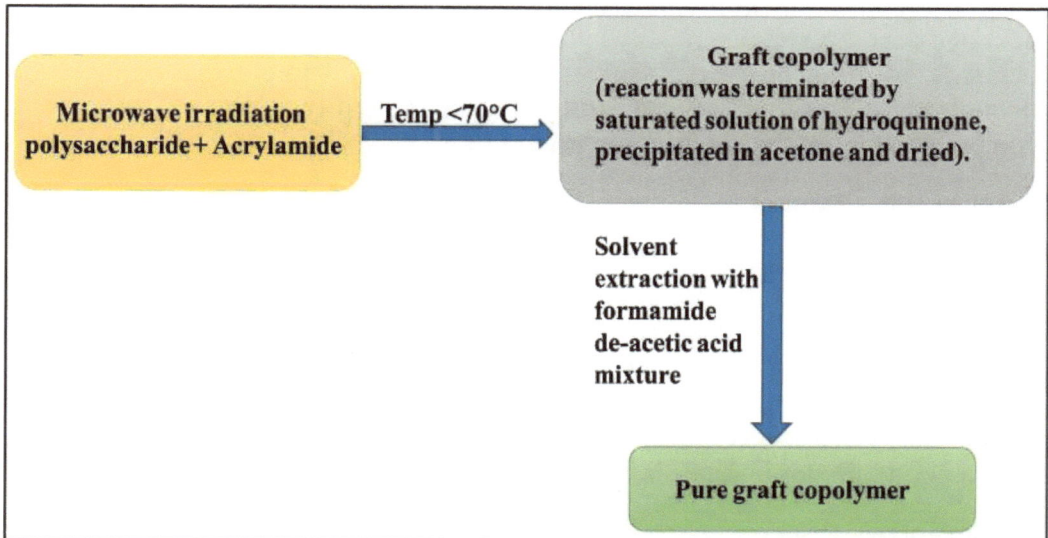

Fig. (3). General method of microwave-initiated technique.

FACTORS AFFECTING GRAFTING

The Polymer Backbone's Nature

The grafting process, which involves a covalent bond between the monomer and a pre-formed polymeric backbone, is significantly influenced by the chemical and physical characteristics of the polymer backbone. According to Ng *et al.*, cellulose's insolubility prevents it from supporting grafting processes in water. Because wool contains a large amount of polymerized chain bonding among amino acids, its features may be determined and set by cysteine linkages and intramolecular hydrogen bonding. UV radiation initiates oxidative processes, leading to the formation of free radicals, and, if monomers are present, grafting occurs [35].

The Impact of the Monomer

In grafting, the monomer's reactivity, such as the polymer backbone's nature, is essential. Monomer concentration, backbone swellability in the condition of monomers, and polar as well as steric nature are some of the variables that affect a monomer's reactivity [35].

Solvent's Effect

The grafting mechanism involves a solvent carrier transporting monomers to the backbone area. The choice of solvent depends on factors like monomer solubility, backbone swelling properties, solvent miscibility, and free radical generation. Alcohols are useful for styrene grafting due to their ability to dissolve styrene and decrease grafting as the reaction progresses [36 - 38].

The Initiator's Impact

The grafting process is significantly impacted by the initiator's character. Chemical grafting techniques, except for radiation, necessitate an initiator, which must be chosen based on its type, concentration, solubility, and function. Azobisobutyronitrile (AIBN) and potassium persulfate ($K_2S_2O_8$) are two examples of initiators. Azobisobutyronitrile (AIBN) reduces grafting when 2-hydroxy methacrylate (HEMA) is grafted on cellulose, and potassium persulfate ($K_2S_2O_8$) is not a good initiator because it breaks down the cellulose chain [39].

Temperature's Effect

The primary regulator of graft copolymerization kinetics is temperature. Grafting yield rises in tandem with temperature before a limit is reached. This could be explained by the greater rate of monomeric diffusion in the backbone, which rises with temperature and smooths the grafting process [40]. According to Gangadharappa HV *et al.*, this behavior generates free radicals in the base polymer as the temperature increases due to an increase in the initiator's thermal breakdown rate and efficiency, which raises the concentration of polymer macro radicals and enhances graft polymerization [41]. Temperature increases grafting yield and peroxide breakdown, but grafting yield drops with rising temperatures when acrylamide is grafted onto cellulose acetate [42]. The breakdown of peroxides created when the base polymer is exposed to air radiation makes the required radicals accessible for grafting, resulting in an initial rise in grafting. The subsequent decrease in grafting is caused by elevated molecular movement with rising temperature, which leads to enhanced radical decay [42].

Additives' Impact on Grafting

The grafting, co-polymerization, and yield are affected by additives like metallic ions, acidic substances, and inorganic salts. Monomer-backbone reactions must counteract additive reactions, even if a few additives enhance the effectiveness of the polymer-backbone reaction [43 - 45]. All the factors have been summarized in Figure **4**.

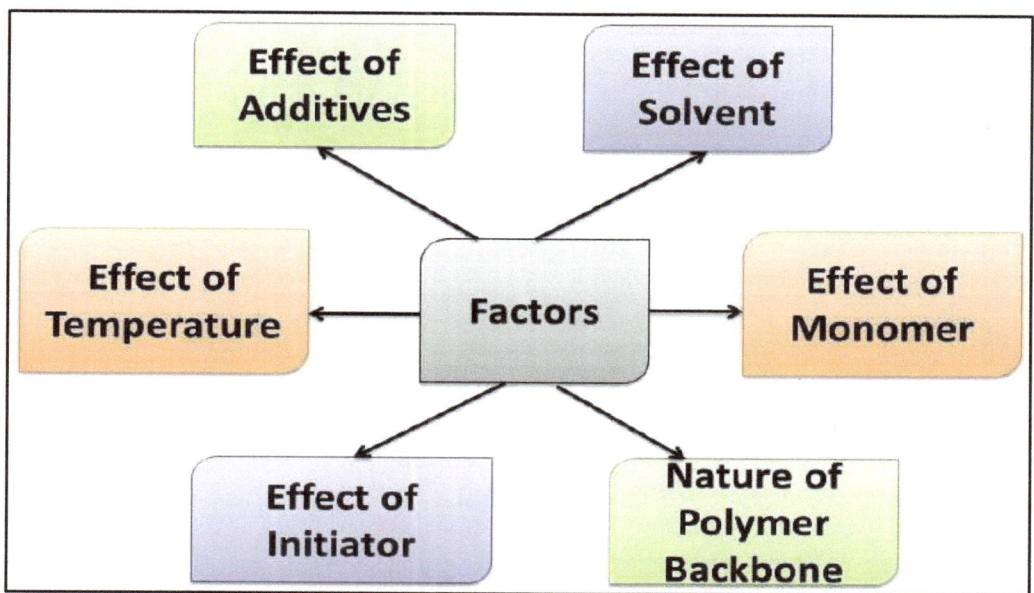

Fig. (4). Summary of factors affecting grafting.

ASSESSMENT OF GRAFTED POLYMERS USING ANALYTICAL TECHNIQUES

The analytical methods and their characteristic of grafted polymer are listed in Table **1** [46]. Various analytical methods are employed to evaluate grafted polymeric materials, including FTIR, NMR, XRD, differential scanning calorimeter, C, H, N analysis, and microwave-assisted analysis. The different analytical methods and the characteristics of grafted polymers are listed in Table **1**.

Table 1. The different analytical methods and the characteristics of the grafted polymer.

S. no.	Method of Analysis	Characterization
1	13C-solid state NMR	For the characterization of the hydrocarbon backbone
2	FT-IR	Particular functional groups that are grafted into blend in
3	DSC	The examination of the melting point and polymer purity
4	XRD	To characterize Polymer polymorphism
5	An analysis of Elements	By using this procedure, elements like C, H, and N content can be identified.
6	Microwave-assisted analysis	An increase in molecular weight upon grafting is described.

(DSC: Differential scanning calorimeter, XRD: X-ray diffractometer, FTIR: Fourier transmission infrared).

THE PHARMACEUTICAL ADVANTAGES OF GRAFTED POLYMERS

To Modify a Drug's Biological Carrying Capacity

Dailey *et al.* enhanced PLGA's ability to transport biomolecules by grafting PVA polymer onto it. It is simple to develop either negatively or positively charged surface properties by securing amines as well as sulfobutyl chemical groups onto the PVA backbone. The act of grafting led to an increase in hydrophilicity, which subsequently enhanced the ability of biomolecules to be carried, particularly those that are sensitive, like DNA, peptides, and proteins [47]. By creating the grafting material sulfobutylated poly(vinyl alcohol)-graft-poly(lactide-co-glycolide) along with cholera toxin, Jung *et al.* made tetanus toxoid available for use as an oral and nasal vaccination [48].

pH-dependent polyelectrolyte complex hydrogel microcapsules were produced and examined by Omer *et al.* for the oral delivery of proteins. In this study, Omer *et al.* used the "grafting to" approach to synthesize chitosan (CS) grafting alginate hydrogel. NH_2 groups were used to click-graft chitosan chains after P-Benzoquinone was employed as the coupling agent [49].

To Use Polymer Grafting to Obtain Customized Physicochemical Attributes

The derivatives with side chains of carboxyl groups exhibited zwitterionic properties, which provided excellent flocculation properties in both acidic and basic media, despite the fact that chitosan is only an efficient flocculating agent in acidic media [10]. Cai *et al.* created hydrogels using irradiation and poly-N-isopropylacrylamide grafted chitosan. The tendency for these chitosan-g-PNIPAAm hydrogels to swell increased as the number of grafted branches, or grafting percentage, was increased [50]. In recent years, a lot of research has been done on the possible applications of chitosan grafted with acrylic acid to produce aqueous and mucoadhesive polymers. Using Silk peptide as a model drug, Hu *et al.* demonstrated that chitosan with poly-crylic acid has a potential grafting delivery strategy for hydrophilic medicines, proteins, and peptides [51].

The pH sensitivity, swellability, sustainability, and drug release mechanism of the psyllium were altered by chemical grafting with N-hydroxymethylacrylamide, acrylamide, and 2-acrylamido-2-methylpropane sulfonic acid combinations, as well as succinic acid treatment [52]. The thermostatic system was created by Vihola *et al.* by grafting poly (ethylene oxide) with thermosensitive polymers, such as poly (N-vinylcaprolactam) (PVCL) to stabilize thermally responsive particles and create PVCL-graft-C11EO42. Drug release can be regulated using these novel polymeric materials, which provide hydrogel particles with a more thermostable system [53].

To Attain the Intended Features of the Dosage Form

Poly lactic co-glycolic acid, which is PLGA, was grafted onto water-soluble polyvinyl alcohol backbones, as discovered by Pistel *et al.* According to the arrangement and molecular composition of the copolymer, biodegradable polyesters produced showed great promise for parenteral delivery methods of medications having controlled release patterns and less burst effects [54].

Kulkarni and Sa created pH-dependent controlled systems for the distribution of ketoprofen using pH-dependent interpenetrating polymeric networks (IPN) beads prepared from NaCMC and PAAm-g-XG. The main structure of xanthan gum (XG) and the IPN of PAAm-g-XG-NaCMC were grafted [55].

Patil S. A. *et al.* developed transdermal films using modified XG, achieved through acrylamide grafting, enabling effective controlled release of atenolol [56].

The polypropylene film was grafted with N, N-dimethylacrylamide and N-isopropylacrylamide, which was produced by Melendez-Ortiz HI *et al.* These functionalized polypropylene films offer excellent bioproperties in addition to a sustained release of norfloxacin [57].

PVAL with Methylcellulose (MC), which is a naturally occurring polymer, and 6-thioguanine was found by Siraj *et al.* In the in vivo study, this mixing resulted in an interpenetrating polymer matrix microsphere that effectively regulated drug release [58].

To Accomplish Site-specific Distribution by Polymer Grafting

Indomethacin is encapsulated in polyacrylamide-grafted chitosan microspheres that have been crosslinked using glutaraldehyde to treat arthritis. At longer durations, indomethacin is primarily released from the completely inflated polymer *via* regulated molecular diffusion; however, initially, it is released from these microspheres *via* a chain of polymer relaxation processes [59].

Polymeric grafted systems like nifidifine and N-lauryl carboxymethyl chitosan [60] have been studied for site-specific drug delivery, including Taxol to cancerous tissues [10].

Liu *et al.* produced N-dodecylated chitosan-based materials for intracellular distribution for gene therapy. It was also discovered that the complex increased activity and shielded DNA from DNA nuclease [61].

Two medications are carried in grafted-chitosan polymeric micelles. The HP-TOC-DOX polymeric micelles were created by conjugating chitosan derivatives with α-tocopherol, linking it to doxorubicin, and targeting an anti-HER2 peptide. The authors' in vivo investigation demonstrated that using the anti-HER targeted peptide improved both therapeutic effectiveness and cellular uptake [62].

APPLICATIONS OF NATURAL POLYMERS

Applications of natural polymers in physics and chemistry have been developed for lignin as well as carbohydrate polymers, which are naturally occurring polymers present in lignocellulosic biomass. This article will delve into the current and significant applications of renewable polymers in the food industry, for medical delivery, corrosion inhibition, and catalysis.

Catalytic Applications

As a fundamental field of modern research, catalysis plays a vital role in promoting economic prosperity and creating materials that support civilization

[63]. In various industries, including the manufacture of bulk and fine chemicals, food and beverage production, crude oil refining, pollution abatement, and many others, catalytic technologies have demonstrated an exceptional ability [64]. Particularly, heterogeneous catalytic assemblies have garnered considerable attention due to their numerous advantages over homogeneous catalysts, which include regeneration, reusability, ease of handling and storage, and facile separation from the reaction fluid [65, 66].

The shift from petrochemical-based raw materials to bio-based feedstocks is a recent development driven by the goal of greener and more sustainable chemistry. Researchers are increasingly focusing on environmentally acceptable catalytic system support materials due to their environmental and economic benefits.67, 68]. Thus, the catalytic system needs to be easy to handle, have long endurance, be moisture and air inert, exhibit high heat and chemical stability, and allow for quick modification. It also needs to be environmentally acceptable.

Due to their high availability, biocompatibility, and renewable nature, natural polymers, particularly polysaccharides, have garnered considerable interest as superior support materials in this regard [69]. Furthermore, they are superior substitutes for petrochemical-based polymers along with other catalytic support components due to their high performance, low cost, greater loading capacity, and chemical inertness [70]. In order to graft organometallic moieties and give them better qualities than other catalytic supports, polysaccharides offer a variety of functions. Additionally, they can stabilize nanoparticles (NPs) through electrostatic forces between their functional structures and the NPs [71]. The utilization of naturally generated polysaccharides as catalytic scaffolds is still in its infancy, despite the fact that their exploitation is rapidly increasing to produce a multitude of industrially significant products. Many biopolymers, such as starch, cellulose, chitosan, lignin, pectin, hemicellulose, and chitin, have recently been investigated for use as ecologically friendly catalytic supports [72 - 77].

This section covers the catalytic applications of metallic nanoparticles based on natural polymers, including lignocellulose, hemicellulose, and nanocellulose.

Lignocellulose

One of the largest lignocellulosic biomass sources, Oil palm fronds (OPF), from the oil palm industry, has recently been described as a renewable form of support for metal nanoparticles (NPs). In addition to stabilizing and preventing NPs from aggregating, this support material improves their characteristics. In this context, extremely scattered cobalt and copper mixed NPs on OPF support were synthesized in an easy, affordable, and environmentally friendly manner by Sohni *et al.* [78]. The catalyst was created using a wet co-impregnation method at room temperature. The hydrophilic property of OPF and its intrinsic porosity facilitated the immobilization of copper and cobalt ions. The artificially produced OPF-derived nanocatalyst (Cu+Co@OPF) was investigated for its morphological structure, thermal behavior, crystalline structure, and functional groups using a range of physico-chemical techniques, such as spectroscopy with energy dispersive (EDS), thermogravimetric measurement (TGA), X-ray diffraction (XRD), the Fourier transform for infrared (FTIR), and scanning electron microscopy (SEM). Due to the porosity and three-dimensional structure of biomass, mixed metal nanoparticles (NPs) can produce the greatest number of catalytic sites possible, thereby increasing the contact between NPs and pollutant molecules during reduction processes. This nanocatalyst thus demonstrated a significant catalytic capacity for the removal of toxic dyes and nitroarenes, including the dye methyl orange (MO), the dye rhodamine B (RB), congo red (CR), ortho, meta, and para-nitrophenol (ONP, MNP, PNP), methylene blue (MB), and 2, 4-dinitrophenol (DNP), using sodium borohydride (NaBH4) as a reducing agent.

The kinetics of catalytic reduction for dyes were found to be MO$\tilde{>}$RB$\tilde{>}$MB$\tilde{>}$CR, while the rate for nitrophenols was reported to be PNP$\tilde{>}$MNP$\tilde{>}$ONP$\tilde{>}$DNP. The use of more environmentally friendly, affordable, and sustainable OPF support, easy room-temperature nanocatalyst preparation, relatively inexpensive Cu NPs, simple catalyst recovery, and rapid reduction of harmful organic substances in wastewater treatment are some of the key features of this work.

Hemicellulose

Hemicellulose is another highly prevalent and plentiful polymer that has not received much attention as a catalytic support. Xylan-type hemicellulose (XH) is one of the main components of hardwood and herbaceous plant cell walls [79]. Because it primarily consists of a β (1→4)-D-xylopyranose structure with replaced compounds located in the 2- or 3-position of the hydroxyl group, including ether groups on the XH chain, it is a suitable polymer substrate for metal nanoparticles [80]. XH is a widely accessible waste product from the paper and pulp industries, as well as biorefineries. Chen *et al.* proposed a simple method for creating PdNPs@XH by anchoring PdNPs on XH due to its high abundance. After that, this was used in the aerobic C–C bond-forming processes and investigated by Thermo Gravimetric Analysis, FT, X-ray photoelectron spectroscopy (XPS), and XRD [81]. The resultant catalyst was reported to exhibit significant catalytic activity in Suzuki, Heck, and Sonogashira cross-coupling reactions, in comparison to earlier carbon-supported and biopolymer-based catalysts. The catalytic system boasts minimal palladium loading, which operates under ligand-free conditions, exhibits substantial activity, demonstrates strong thermal and chemical stability, facilitates ease of recovery, and allows for reuse up to six times without a noticeable decrease in efficiency.

Nanocellulose

Nanocellulose (NC) is a member of the cellulose derivative family that has attracted special attention due to its unique combination of cellulose and nano-scale material characteristics. Usually, cellulose from plants or microbes is hydrolyzed using a strong acid, like sulfuric acid, to produce it. Among many other applications, nanocelluloses have proven to be effective as adaptable catalytic supports in the creation of sustainable catalysts. Their large surface area, thermal stability, chiral characteristics, functionalizable motifs, improved stability, non-toxicity, and affordability are the reasons behind this. Additionally, it is straightforward to graft organometallic species onto NC surfaces [82]. Sodium borohydride and nickel chloride were recently employed by Prathap *et al.* in a combination of TEMPO-oxidized CNF with water (0.01%) to develop a simple and effective protocol for the in-situ reduction of nitroarenes and other aliphatic nitrocompounds to amines in higher yields [83]. This work utilized a minimal nickel catalyst due to the stabilizing effect of the nanocellulosic scaffold and the increase in turnover number. Jabasingh and co-workers developed a composite NC-TiO2 catalyst for cellulosic ethanol production from bagasse. The composite was found to yield more glucose than enzymatic hydrolysis due to its increased surface area. Azad and Mirajalili utilized nanocellulose as a catalyst to prepare magnetic core-shell nanoparticles, which were then treated with titanium tetrachloride. The catalyst demonstrated remarkable efficacy in the production of 4H-pyrimido [2,1-b]benzothiazole derivatives. By combining titanium catalytic properties with nanocellulose fibers and magnetic material nanoparticles, this eco-friendly process produces a bio-based, recyclable magnetic nanocatalyst that exhibits excellent yields, a simple setup, fast reaction times, and recyclability [84].

Corrosion Inhibitor's Application

Green Obstructers (inhibitors)

Corrosion is the simple method by which metals, as well as alloys, deteriorate, destroy, or "eat away" as a result of an electrochemical process involving a corrosive environment that contains either moisture (H_2O), air (O_2), or both. Regarding industry, infrastructure, and technological innovation for sustainable growth, it is a major issue. As a result, resources are depleted, maintenance expenses increase, and structural materials fail prematurely [85]. Mineral acids, such as HCl and H_2SO_4, are frequently employed in various industries for metallurgical processes, including descaling, chemical cleaning, and acid pickling

[86]. A corrosion inhibitor can inhibit corrosion in two ways: (i) green corrosion inhibitors adhere to metal surfaces to provide an inhibitory coating that shields the metals, or (ii) by converting the corrosive substance into a less corrosive or non-corrosive environment with the use of environment modifiers.

The ability of numerous synthetic organic compounds to limit corrosion in various acidic conditions has been studied, but the potentially harmful effects of these inhibitors on marine life and animal life have not received as much attention as they should have [87]. Researchers have been exploring non-toxic, biodegradable, environmentally friendly, and financially feasible methods to suppress corrosion and its obstructers for several decades [88]. A review of the literature on green corrosion inhibitors indicates that the choice of natural products must meet certain criteria, such as the presence of phytochemicals or isolated organic compounds with heteroatoms, that is, P, O, S, and N, which have a π-electrons system [89]. The inhibitory efficiency of these compounds can be quantified by examining the density of electrons at heteroatoms, the type of orbital holding free electrons, and the number of mobile electrons present [90].

Electrochemical techniques like potentiodynamic polarizing and impedance spectroscopy can evaluate the corrosion inhibition efficacy of organic compounds. Density Function Theory (DFT) can establish the relationship between inhibitor structures and efficacy. Investigations have shown the highest occupied molecular orbital (HOMO), the lowest unoccupied molecular orbital (LUMO) energies, optimal quantum chemical structure, and structural characteristics. This research establishes the association between HOMO and inhibitory qualities [91 - 93].

Experimental and Computational Analysis

Recent studies have reviewed and investigated various carbohydrate polymers as green inhibitors for metallic corrosion under various corrosive conditions [91 - 95].

Chitosan (CH) and its 5-chloromethyl-8-hydroxyquinoline derivative (CH-HQ) were recently demonstrated to exhibit inhibitory efficiencies of 78% and 93% for the acidic corrosion of mild steel, respectively, with a relatively small amount of 10-2 g/L of the derivative [51]. The high inhibitory activity of CH-HQ is believed to be caused by its large molecular size and the presence of functional groups with polarity, which enhance its ability to dissolve in polar electrolytes, including acidic minerals HCl and H_2SO_4. A DFT study was performed to investigate the mechanistic aspects of metal protection in corrosive environments by directly correlating the macromolecular measurements, biochemical compositions, and electronic configurations of CH and CH-HQ [91]. It is worth noting that water, a green solvent, was utilized in the synthesis of CH-HQ [89].

Studies have shown that substituted/modified CH chitosans, exudate gums, cellulose, starch, pectin, and pectates can effectively prevent metal corrosion in acidic environments. These compounds contain free amine and hydroxyl groups, which can chelate metal ions and form protective inhibitor films over metal surfaces. The efficiency of carbohydrate biopolymers as environmentally friendly inhibitors of metallic corrosion has been examined in recent research [89, 92 - 94].

Natural Polymer Applications in the Food Industry

Reddy and Yang discovered typical cellulose fibers that were [96, 97] extracted from cornhusks. Cornhusk fibers are not only essential for horticulture but also for meeting individual demands for food, fiber, and energy [96]. They are also environmentally beneficial. The pieces, characteristics, and structure of cellulose strands obtained from horticultural materials make them suitable for uses such as composite material, mash, and paper manufacture. These strands are also significantly less expensive because they are derived from horticultural outcomes. Agro-based biofibers have numerous important sources, encompassing the production of sugarcane, pineapple, banana, coconut, rice, wheat, sorghum, and other grains [97].

While nanocomposite films have been produced utilizing kappa-carrageenan and nanofibrillated cellulose, the latter was effectively blended from short, stable cotton strands *via* a chemo-mechanical cycle. The bionanocomposites of NFC were added to a KCRG grid at varying weight percentages (between 0.1 and 1%) using an answer projecting technique. The bionanocomposites' humid characteristics, as determined by DSC, morphological concepts, fluid emissions propagation rate, oxygen transmission rate (OTR), X-ray diffractograms (XRD), and versatile properties, were displayed [98].

Since additional compounds have been extracted from mulberry branch barks using a process involving solvents at 130 °C, followed by sulfuric acid hydrolysis. It is envisaged that cellulose stubble will find usage in the medical field, optical industries, and as a stabilizing stage in composite materials. Based on the compound organization inspection, AFM picture, FT-IR, and XRD data, the stubble composition is 20–40 nm broad. The fiber hairs exhibited a high degree of lignin and hemicellulose extraction, with a crystallization of 73.4%. This has expanded the material's potential usage in pharmaceuticals and optical organization as additional substances, as well as in composites as a fortified stage [99].

As a versatile microbial polysaccharide, bacterial cellulose (BC) has numerous applications in the food industry, including strengthening and clumping, settling, and water filtration, as well as pressing materials. Given its great purity, it has significant promise as a food additive. In addition, as a nanofiber, it can form a three-dimensional network structure [100].

Whey protein detaches and cellulose filaments are produced when cellulose strands are combined with the corrosive-induced gelatinous protein from whey in a concurrent cycle. Flower-like miniatures or nanomaterials have been found surrounding the regenerated cellulose threads made from ammonia copper sulphate hexahydrate. These cellulose/WPI filaments are useful for discharge materials in dynamic repairs, tissue engineering, and medicine [101].

Microcrystalline cellulose (MCC) is becoming more and more popular due to the increased demand for substitutes for finite and scarce fossil fuels. It is used in food, medicine, and polymer composites. Haafiz *et al.* compiled a review of techniques for extracting and representing microcrystalline cellulose (MCC). This review integrates new sources, novel division techniques, and emerging medicines that stimulate the development of new types of MCC substances for industrialization. Prospects for MCC-based composite polymers are outlined and examined [102].

Not many foods are offered unpackaged, and food bundling is at the core of the modern food market. Food bundling has been taken into consideration in the context of continuous historical innovation in the display of bioplastic materials. Certain materials used for bundling, such as specific types of plastic, polythene, and Styrofoam, have the potential to release toxins when heated, posing a risk to buyers. A diagram of the basic ingredients used to supply biobased films, their arrangements, obstacles, and possible uses is included in the report [103]. It also a best-in-class analysis of bioplastics that are successfully used as food packaging materials.

Guar gum, a natural product derived from the endosperm of the bunch bean, is used as a laxative, treating conditions such as diabetes, heart disease, and obesity, and as a thickener and stabilizer in tablets, ointments, and creams [104]. Nanostructured materials are gaining popularity in the food industry for their applications in food additives, packaging materials, preparation aids, and food quality and safety sensors. Research is underway to create nanofiber-based non-woven mats for food applications [105].

The planned applications of wood gum polysaccharides and their nanostructures in the food, beverage, healthcare, biotechnology, and pharmaceutical industries have garnered considerable interest lately. Despite the widespread use of tree gums in food, these commercial gums also provide a wealth of non-food applications. Their accessibility, additional variety, and remarkable qualities as "green" bio-based sustainable materials have made them increasingly desirable. Tree gums can be obtained as common polysaccharides from several tree genera that include unique characteristics, such as being infinite, biocompatible, biodegradable, and non-toxic, as well as having the ability to undergo minor chemical changes [106].

The development of cellulose specialized strands from soybean straw, with properties similar to those of common cellulose filaments, could provide a cost-effective and abundant source of cellulose filaments, addressing concerns about future expenses and the availability of current filaments, and increasing the value of food crops [107]. Nano- and miniature particles can enhance the multifunctional properties of cellulose fiber networks, offering potential applications in food packaging, record storage, electrofunctional materials, electromagnetic devices, and antimicrobial injury repair [108].

Biomedical and Biotechnological Applications

Regeneration and Tissue Engineering

- **Engineering Scaffolds for Tissue**

Because green electrospun nanofibers may mirror the structural characteristics of natural extracellular matrices, they have considerable potential for use in tissue engineering and the development of medical devices. These nanofibers are remarkable in their mechanical strength, tunable biodegradability, high porosity, and nanoscale fiber diameters. They enhance tissue regeneration, aid in cell infiltration, and promote nutrient transport, making them perfect for scaffold construction. They are applied in various fields, including the restoration of cartilage, skin, and bones. Utilizing renewable materials supports ecologically friendly practices and results in sustainable healing. To enhance regeneration results, future studies will focus on optimizing the microarchitecture of the scaffold, degrading kinetics, and functionalization [109 - 113].

- **Venous Grafts**

When it comes to small-diameter vasculature grafts for blood artery regeneration, green electrospun nanofibers show promise. The strength degradation balance can be adjusted over time with these materials, which replicate the mechanical characteristics of native blood vessels. Additionally, they create nonthrombogenic fiber surfaces, which encourage the invasion of endothelial cells and the formation of new blood vessels. Green electrospun structures, with enhanced biological degradation kinetics and functionalization techniques, offer environmentally friendly alternatives to petroleum-derived vascular conduits. Its potential to facilitate cell penetration and remodeling, as well as its capacity to maintain adjustable strength-degrading qualities over time, are its major features [114 - 118].

Modified Drug Delivery

• Supply of Antibiotics and Anticancer Drugs

Green electrospinning methods have been employed to develop biopolymer carriers that can sustain the release of medication for more effective treatments. For three weeks, gelatin fibers release norfloxacin, while chitosan nanofibers sustain doxorubicin for 4 weeks. These biodegrading platforms can target specific sites, reducing side effects and maintaining potency. These sustainable platforms offer green alternatives to permanent drug depots and can be optimized for personalized therapies [119 - 123].

• Gene Delivery

There is promise for creating nonviral gene transfer vectors using green electrospinning. The positively charged backbones of chitosan, gelatin, and cellulose nanofibers allow for complexation by therapeutic nucleic acids. This preserves genetic payloads until they are absorbed by cells. For instance, plasmid DNA is condensed by gelatin/chitosan fibers and gradually released from a scaffold for seven days. Nontoxic carriers circumvent the risks associated with viral vectors. Applications for tissue regeneration include scaffolds with growth factor genes integrated to promote angiogenesis or late wound healing [124]. Viral transfection efficiencies are modeled by the sustained delivery of nucleic acids. Scaffold-based distribution is made easier locally by fiber nanoarchitecture and biopolymer biocompatibility. For example, targeted gene therapy with ligand targeting may allow for the regeneration of particular organ systems [124 - 126]. All things considered, such environmentally friendly nanocarriers hold the potential as safe, nonviral gene delivery tools for sophisticated tissue engineering uses that call for DNA/RNA payloads. They meet the requirements for sustainable biomaterial design since they are renewable and biodegradable materials.

• Hemostatic Dressing

Hemostatic wound dressings are made from green electrospun fibers, which help limit bleeding and prevent infection. Through platelet interactions and the mechanical absorption of blood, collagen, and cellulose fibers, clotting is accelerated. Hemostasis agents are localized at wound beds by their porous features. Gelatin nanofibers coated with silver nanoparticles or chlorhexidine release antimicrobials in regulated bursts to reduce the risk of infection as wounds heal. When compared to gauze, the effectiveness of hemostats reduces the amount of time needed to stop bleeding. Materials escape the biological incompatibility of synthetics because they are biopolymers. Future research aimed at maximizing loading levels and customizing release profiles is expected to result in the least amount of bacteria during the suturing and healing phases [127, 128]. Sophisticated 3D fiber architectures have the potential to replicate intricate tissue geometries, allowing for uniform interaction with uneven wound surfaces and promoting rapid, infection-free healing. With their safe combination of hemostatic and antibacterial properties achieved through the use of renewable biomaterials, these enhanced green wound dressings demonstrate promise.

- **Burn Wound Dressing**

Green electrospun fiber wound dressings efficiently collect exudate while maintaining a moist environment. Nanoscale cellulose and gelatin dressings are porous and quickly absorb fluids through a large surface area, thereby halting further damage to tissues from contact with dry surfaces. Re-epithelialization is accelerated by moisture-managing skills as opposed to air-drying techniques. For the patient's comfort, hydrophilic fibers absorb excess wound fluid. When it comes to open burn injuries, biocompatible materials eliminate the potential for sensitivity to synthetic polymers. Because biodegradable fibers do not need to be removed, unpleasant alterations are avoided. Healing trajectories may be optimized through future research into enhancing mechanical properties and developing intelligent dressings that respond to wound conditions. Regrowth factor-infused advanced formulations more closely mimic the extracellular matrix (ECM) milieu, strengthening the potential for tissue regeneration in green wound care solutions [129 - 132]. These green fiber dressings promote the healing of burn wounds by efficiently managing exudate and by providing a moist, safe surface made of non-irritating, renewable materials.

- **Sutures/Medical Textiles**

Green electrospinning creates soft tissue-repairable fibrous meshes and structures. Meshes made of collagen, silk, and polyester function as renewable surgical sutures that maintain their strength over time in a controlled manner. Gelatin/PLGA nanofibrous meshes are utilized as scaffolds for skin or heart tissue regeneration. Cell penetration is encouraged by porous structures, which prevent negative polymer reactions. Non-absorbable mesh and sutures made from petroleum-based materials are being replaced by these renewable textile structures. Foreign body reactions are eliminated *via* biocompatibility. Anatomical replacements, such as ligaments or tendons, may be developed in the future that optimizes mechanical characteristics and fiber alignment. For quick tissue regeneration, endothelial cells on fiber surfaces that vascularize efficiently can be included in cardiovascular patches [133 - 136]. Overall, due to their timeline-appropriate characteristics and ability to eliminate sensitivities to synthetic polymers, sustainable electrospun medical textiles offer safe and sustainable alternatives to artificial substances in soft tissue applications.

- **Implants**

Green electrospun implants that degrade slowly while maintaining mechanical integrity are biodegradable. Composite fibers that blend collagen and polycaprolactone exhibit porosities that promote osteoblast ingrowth and customize the stability of implants over months as new bone grows. As the tissue ages, the calcium/phosphate reinforcement in gelatin/hydroxyapatite scaffolds, which facilitates the development of new cartilage, breaks down [137 - 139]. To expedite healing, future guidelines recommend developing intelligent implants that incorporate cellular connectivity and enhancing fiber architectures for anatomical congruency. Certain tissue lineages may be stimulated by printed conductive networks that combine biosignaling with synthetic resorption [140 - 142]. Permanent metal or plastic implants can be replaced with renewable fiber composites, which provide uniform resorption appropriate for tissue healing in bone/cartilage regeneration. Green electrospun implants are bioresorbable alternatives that utilize sustainable materials to restore function without causing long-term foreign body reactions.

Due to their large surface area and volume, electrospun nanofibers have an advantage over large-scale fibers, such as those used in spinning or drawing. This is because, in comparison to microscale fibers, their nanoscale diameters enable higher loading with drugs and more efficient release. Additionally, reduced diameters and substantial porosity, as well as improved cell connections, are achieved by more closely resembling the extracellular matrix of tissue than larger fibers. Manufacturing can be simplified by producing fibers directly from polymers during electrospinning, eliminating the need for downstream operations, such as spinning. A wide range of synthetic and natural polymers allows for the tuning of specific qualities for various uses. Nanoparticles in electrospun fibers enable controlled release, providing sustained discharge and protection against colloids, compared to diffusion micelles/nanoparticles.

Environmental Remediation

Water Purification

Through adsorption and filtration, green water is produced by electrospun nanofiber membranes, facilitating the treatment of contaminated water. Due to their large surface area and reactive hydroxyl groups, cellulose nanofiber mats are an effective method for removing contaminants from wastewater, including dyes and heavy metals. Composite chitosan-gellan gum nanowebs, functioning with metal-chelating groups, capture hazardous ions during filtration. High porosity offers an environmentally responsible alternative to petroleum-based purification techniques, preserving throughput while utilizing renewable biomass [143 - 146]. Subsequent efforts to optimize the thickness of membranes and multilayer designs may lead to practical applications that address real-world challenges, such as industrial wastewater cleanup and desalination. A further benefit of parallel sensor integration would be automated filtration monitoring [147 - 149]. Overall, these environmentally friendly nanofibrous systems, with nanomaterial-enabled adsorptive and sieving mechanisms offer a green alternative for water purification applications.

Air Filters

Green nanofibers that have been electrospun are a useful medium for air filtration in medical and industrial settings. Antibacterial silver nanoparticle-coated nonwoven gelatin/PVA sheets can be used by hospitals and HVAC systems to capture airborne pathogens. Through winding diffusion routes, loose nanofiber networks catch particles and microorganisms larger than 100 nm, allowing for high flow rates. Heavy metals and volatile organic compounds are among the industrial contaminants that chitosan/cellulose composites chemisorb from industrial emissions [150 - 153]. These filters offer sustainable alternatives to synthetic solutions, such as HEPA filters, as they are made from renewable materials. Sterilized lab and industrial stacked gas disinfection in real-world conditions can be achieved by optimizing fiber charges and creating self-cleaning properties [154 - 157]. In conclusion, green electrospun filters for air have the potential to filter chemical and microbiological air contaminants with high efficiency, utilizing renewable energy, and making them a promising option for both environmental and industrial applications.

Pesticides and Fertilizers with Controlled Release

Pesticide carriers that provide long-term crop/nutrient protection are created using green electrospinning. Fertilizer salts are encapsulated by gelatin/clay composite fibers, which release them gradually as the gelatin breaks down. This matches plant uptake during growth cycles. Insecticide levels are maintained for weeks using biopesticide-loaded cellulose nanofibers as mesh barriers to keep pests away. Accurately measuring pesticides maximizes harvests while reducing runoff pollution. Nonbiodegradable formulations are replaced with renewable polymers [158 - 161]. Additionally, controlled release lowers the frequency and quantity of applications. Different soil and climate conditions are addressed in future formulations that customize release profiles. In conclusion, by using degradable fiber networks to transport nutrients and other substances in a personalized manner, these green platforms strike a compromise between environmental sustainability and agricultural output.

Applications in Food Packaging

Food packaging with antibacterial properties and active ingredients is made possible by green electrospun nanofibers. Polyvinyl alcohol coatings coated with zinc oxide nanoparticles on cellulose-derived substrates prevent the growth of bacteria on perishable goods. Incorporating grapefruit extract into electrospun gelatin fibers kills microorganisms and inhibits their resuspension by creating a physical network of protection. When compared to passive storage containers, these attributes extend the shelf lifespan through multifunctional antioxidant and antibacterial effects [162 - 165]. Paper packaging is being replaced with recyclable materials. Degradability ensures that, after disposal, materials do not persist in the environment. Optimizing microbial inhibitory effects can be achieved by customizing critical oils designed for specific produce varieties [166 - 169]. In conclusion, green active nanofiber coatings, which utilize natural, biodegradable preservation technologies, provide a sustainable approach to reducing foodborne illness.

Cell Encapsulation

Platforms for encapsulating cells for medicinal applications are created *via* green electrospinning. Alginate microcapsules embedded with chitosan-gelatin nanofiber coatings codeliver islet cells. The viability of encapsulated hepatocytes is preserved by embedding them within collagen-HA fibers as cell-interactive hydrogels respond to intrinsic signaling pathways. Soft biomaterials help cell treatments by shielding the immune system from harm without the need for pharmaceuticals. Future research utilizing sophisticated bioprinting to optimize the mechanical and mass transport characteristics of capsules may scale up production for the treatment of liver disease or diabetes [170 - 172]. Through adaptable biomaterial cell interactions, these renewable encapsulation technologies provide biocompatible, localized cell immunoprotection.

Prodrug Activation

Synthetic biology is integrated onto green electrospun scaffolds to provide regulated multidrug release. Enzyme-expressing *E. coli* and passive prodrug conjugates are coencapsulated in polyester fibers. Bacteria metabolize conjugates at infection sites, causing medications to activate in short bursts. Drug-resistant bacteria and anodes are localized by conductive mixed-ligand hydrogels, which release medications and kill cells by producing reactive oxygen species [173 - 177]. By combining synergistic therapy with enzymatic activation, these living materials circumvent numerous issues related to drug resistance. The implantation of these modified bacterial systems is made easier by biocompatible materials. Advanced in vivo diagnostics and treatments could be realized through future work, improving signals and population dynamics. In conclusion, these platforms show promise for creating tailored prodrug delivery systems at the intersection of synthetic biology and green materials.

Electrospinning Fiber Diameter

Because it is influenced by the voltage, rate of flow, and polymer characteristics, fiber diameter is important in electrospinning. PLA can reach dimensions of 200 nm to 5 μm, whereas electrospun fibers can have a range of 50 to 500 nm. This sets electrospun fibers apart from traditional microfiber manufacturing techniques. Thin fibers are produced, *via* charge-induced bending, whereby fiber diameter is a direct result of the electrospinning process. The multifaceted application landscape and distinct character of electrospinning render it a unique technology.

Environmental Impact and Sustainability

Examining Green Electrospun Nanofiber Materials' Environmental Impact

The use of hazardous chemicals and energy-intensive procedures in conventional electrospinning results in a significant carbon footprint and harm to the environment. However, by employing a range of strategies, green electrospinning approaches aim to mitigate this environmental impact [110, 178, 179]. Biopolymers, including chitosan, gelatin, and cellulose, are readily obtained from marine and agricultural waste. This offers a steady and plentiful supply of raw resources. By eliminating organic solvents from the manufacturing process, toxicity issues associated with their development, application, and disposal are avoided.

Additionally, it lowers application-related regulatory barriers. Electricity consumption is significantly reduced by technologies including near-field electrospinning, centrifugal, and solar. This reduces carbon emissions compared to the high power requirements of earlier technology. Some methods combine the production of nanofibers with CO2 sequestration or wastewater treatment. By doing this, waste is converted into a resource, increasing eco-efficiency. Closing the circle on material fluxes, biodegradable nanofibers produced through environmentally friendly production may return to the biosphere after use. In general, green electrospinning strategies reduce pollution, utilize industrial and agricultural waste streams, encourage the development of more circular material economies, and decrease reliance on nonrenewable resources. This reduces the environmental harm caused by nanomanufacturing.

Green electrospinning technologies focus on resource efficiency, cleaner production, energy efficiency, environmental performance, social responsibility, and carbon mitigation. They use renewable biomass sources, recover materials from waste streams, implement closed-loop manufacturing, eliminate toxic organic solvent emissions, and utilize aqueous, solvent-free fabrication routes. Green electrospinning also reduces electrical energy usage and can be powered by renewable energy sources. It creates green jobs, supports a circular bioeconomy, and upcycles agricultural residues, achieving sustainability across environmental, economic, and social dimensions.

NEWLY SUBMITTED PATENTS CONCERNING POLYMER GRAFTING

- **Procedure for forming solid dose forms (Patent number: US 7419685 B2)**

In order to create solid dosage forms, the water-swellable graft copolymer or a combination of graft copolymers is used as a polymeric binder [180].

- **Films that dissolve quickly for oral medication delivery (Patent number: US 20040208931 A1)**

A polyvinyl alcohol-polyethylene glycol graft copolymer (PVA-PEG) is the second polymer used in the preparation of the dosage form, which consists of a first polymer, a deposit containing an active substance, and a cover layer [180].

- **In graft copolymers, solid dispersion of poorly soluble medicines (Patent number: WO 2007115381 A2)**

It specifically concerned a system and technique for improving the solubility and dissolution of drugs that are physiologically active but have low water solubility and dissolution rates, scuh as BCS Class II or Class IV pharmacological molecules. It entailed employing a PVA-PEG copolymer, such as Kollicoat IR, to create solid dispersions with poor water solubility and bioactive compounds that dissolve at a faster rate [180].

- **Films that dissolve quickly for medication delivery orally (Patent number: WO 2004060298 A2)**

A dosage unit is composed of a variety of polymers that are involved in many layers, such as

PVA-PEG and PVA-PEG graft polymers [180].

FUTURE PERSPECTIVES AND CHALLENGES

The topic of the environmental impact of sustainable electrospun nanofiber materials highlights the sharp differences between more environmentally friendly methods and traditional electrospinning techniques. Traditional techniques mostly rely on energy-intensive procedures and hazardous chemicals, which could result in a large carbon footprint and harm to the environment. Green electrospinning approaches, on the other hand, are designed to reduce these environmental effects using a variety of strategies. On the other hand, green electrospinning approaches are designed to mitigate these environmental effects in various ways. First and foremost, green electrospinning utilizes biopolymers such as cellulose, gelatin, and chitosan, all of which are easily obtained from marine and agricultural waste, making them a priority when it comes to using renewable resources. This reduces the demand on natural resources while also ensuring a steady and environmentally friendly supply of basic materials. Second, a crucial component of green electrospinning is the removal of dangerous chemicals from the manufacturing process. It avoids the toxicity issues associated with the production, use, and disposal of organic solvents, making regulatory compliance easier for a wider range of applications. Third, the amount of energy required is significantly reduced through green electrospinning. Compared to the high-voltage requirements of conventional processes, innovative technologies, including solar energy, centrifugal, and with near-field electrospinning, significantly cut electricity consumption, lowering carbon emissions. Fourth, some eco-friendly electrospinning techniques combine the production of nanofibers with the treatment of wastewater or the sequestration of carbon dioxide, so transforming waste into an advantageous resource.

This method improves eco-efficiency overall. Last but not least, green electrospinning helps create biodegradable nanofibers that can rejoin the biosphere once they have served their intended purpose [181 - 183]. Such closed-loop applications demonstrate a dedication to improving material utilization through an approach that prioritizes material lifecycle management. All things considered, green electrospinning techniques reduce pollution, utilize industrial and agricultural waste, and foster the development of further circular material economies. They also lessen dependency on nonrenewable resources. By taking a holistic approach, the environmental effects of nanomanufacturing are successfully mitigated, aligning with the need to advance cutting-edge technology while minimizing environmental impact.

The sustainability features of green electrospinning technologies encompass a wide range of factors. These elements demonstrate how committed green electrospinning is to social responsibility and environmental accountability, as demonstrated by research findings [58 - 60]. A key component of green electrospinning is resource efficiency, which is achieved by utilizing sustainable biomass sources that do not compromise food production. Green electrospinning utilizes waste streams, such as lignin and food processing waste, promoting disassembly and reprocessing designs that minimize waste and optimize resource consumption through closed-loop manufacturing. Another important factor is cleaner production, since wastewater effluents and hazardous organic solvent emissions are eliminated by green electrospinning. The company utilizes water-based, solvent-free, and mechanistic production techniques for wastewater treatment and the removal of volatile substances, maximizing energy efficiency through the use of solar energy, centrifugal techniques, and near-field methods. To reduce dependency on nonrenewable energy sources, green electrospinning also explores the incorporation of energy from renewable sources, such as the sun and wind. This aligns with applications for sustainable energy generation. Green electrospinning, a green manufacturing process, produces biodegradable nanofibers that reduce greenhouse gas emissions and sequester carbon through the use of bacterial and algal cultures. Comprehensive lifetime impact assessments offer a comprehensive picture of their environmental contributions, especially when considering the potential for global warming [184 - 186]. The development of green jobs resulting from the use of renewable technology in the environmentally friendly electrospinning industry underscores social responsibility. This strategy contributes to mitigating climate change by upcycling agricultural leftovers that would otherwise release greenhouse gases, promoting a circular bioeconomy. All things considered, green electrospinning is a comprehensive strategy that encompasses carbon reduction, clean manufacturing techniques, resource conservation, adoption of renewable energy sources, and socioeconomic advantages. Together, these elements support sustainability on the social, economic, and environmental fronts, indicating a thoughtful step toward a future where social justice and environmental responsibility are prioritized. Even though the study of green electrospinning has advanced significantly in recent years, there is still much potential for this technology to be advanced through focused research projects. Future research has indicated several interesting directions [184 - 186]. First, a potential path is provided by the creation of biopolymers from waste materials. By modifying the structures and characteristics of these biopolymers, sustainable materials of the future generation may be produced. Furthermore, the emulation of multipurpose composite materials would increase their broader range of applications. Another crucial area of attention is advanced process optimization. To fully realize the potential of techniques such as aqueous electrospinning, extensive parametric investigations, modeling, and scale-up experiments are necessary. Efficiency and scalability could be improved by integrating these manufacturing processes. In particular, the creation of stimuli-responsive fibers for applications such as controlled theranostics and the incorporation of electronic elements into composites necessitates further investigation into multifunctional materials engineering. Additionally, green electrospinning can have a big impact on environmental and sustainable energy applications. Combining nanofiber technologies with desalination, energy storage, and water treatment could result in significant and sustainable solutions. Researchers will be able to benchmark progress successfully by using meaningful sustainability metrics that are provided through thorough life cycle evaluations.

Industry collaboration is crucial for the widespread adoption of green nanofibers, which can be advanced through biopolymer substrate design, process development, renewable applications, environmental effects, and commercialization routes, supporting sustainable nanomanufacturing [187 - 189]. Researchers are working to overcome challenges in biomaterials science, including precise control over biopolymer characteristics, sourcing renewable inputs, scaling up green approaches, and optimizing sustainability metrics. They are also collaborating to strengthen biomass supply chains and use modeling and machine learning techniques for materials-by-design. This will lead to the realization of green electrospinning's benefits, sustainable nanomanufacturing, and a future that goes beyond fundamental studies. Polymers and their derivatives are increasingly used in drug delivery systems.

CONCLUSION

This makes polymer grafting an emerging method in dosage form design, utilizing the grafting technique to modify properties acquired by the polymer. Polymers are superior to their synthesized and natural counterparts due to polymer grafting. To give the polymer the appropriate features, a variety of methods are employed, such as the recently invented plasma radiation grafting, radiation-induced, enzyme-induced grafting, live polymerization, and conventional one-chemical grafting. These grafted polymers were effectively described and examined using FTIR, NMR, and XRD, three contemporary analytical techniques. In the field of polymer grafting, several new patent applications have been made recently. The industrialist should be required to demonstrate a strong interest in the field of polymer grafting. Polymer grafting will be used extensively in medication delivery systems in the future.

Green electrospinning is a promising field for sustainable nanomaterial production, offering eco-efficient methods and reducing environmental impacts. Techniques such as solvent elimination, renewable energy integration, and waste utilization have been developed, enhancing eco-efficiency. Applications for these technologies include controlled medication delivery, environmental cleanup, and biomedical nanofibers. Nonetheless, further research is required in the areas of life cycle sustainability assessments, materials engineering, scale-up procedures, and the realization of multifunctionality. Through the substitution of plentiful waste resources for nonrenewable inputs, the sector can support the development of sustainable nanomaterials and the circular bioeconomy. Early commercial achievements help to accelerate the shift to a green bioeconomy paradigm by indicating that renewable nanotechnologies are commercially viable. The remarkable properties of recently isolated and identified gums have led to their widespread use and modification, even though the use of conventional gums has persisted. Research on more recent natural gums and mucilages, as well as how to modify them through grafting, is incredibly broad. It has the potential to be further utilized in the near and distant future as innovatively altered natural polymers for the creation of diverse systems for drug delivery in the pharmaceutical industry.

REFERENCES

[1]   Priya James H, John R, Alex A, Anoop KR. Smart polymers for the controlled delivery of drugs – a concise overview. Acta Pharm Sin B 2014; 4(2): 120-7.
[http://dx.doi.org/10.1016/j.apsb.2014.02.005] [PMID: 26579373]

[2]   Bashir A, Warsi MH, Sharma PK. An overview of natural gums as pharmaceutical excipient: their chemical modification. World J Pharm Pharm Sci 2016; 5(4): 2025-39.

[3]   Ogaji IJ, Nep EI, Audu-Peter JD. Advances in natural polymers as pharmaceutical excipients.
[http://dx.doi.org/10.4172/2153-2435.1000146]

[4]   Bhattacharya A, Misra BN. Grafting: a versatile means to modify polymersTechniques, factors and applications. Prog Polym Sci 2004; 29(8): 767-814.
[http://dx.doi.org/10.1016/j.progpolymsci.2004.05.002]

[5]   Mohan S, Oluwafemi OS, Kalarikkal N, Thomas S, Songca SP. Biopolymers–application in nanoscience and nanotechnology. Recent advances in biopolymers. 2016 Mar 9; 1(1): 47-66.
[http://dx.doi.org/10.5772/62225]

[6]   Hernández N, Williams RC, Cochran EW. The battle for the "green" polymer. Different approaches for biopolymer synthesis: bioadvantaged vs. bioreplacement. Org Biomol Chem 2014; 12(18): 2834-49.
[http://dx.doi.org/10.1039/C3OB42339E] [PMID: 24687118]

[7]   Rao MG, Bharathi P, Akila RM. A comprehensive review on biopolymers. Sci Revs Chem Commun 2014; 4(2): 61-8.

[8]   Yadav P, Yadav H, Shah VG, Shah G, Dhaka G. Biomedical biopolymers, their origin and evolution in biomedical sciences: a systematic review. J Clin Diagn Res 2015; 9(9): ZE21-5.
[http://dx.doi.org/10.7860/JCDR/2015/13907.6565] [PMID: 26501034]

[9]   Pu W, Shen C, Wei B, Yang Y, Li Y. A comprehensive review of polysaccharide biopolymers for enhanced oil recovery (EOR) from flask to field. J Ind Eng Chem 2018; 61: 1-11.
[http://dx.doi.org/10.1016/j.jiec.2017.12.034]

[10] Alves NM, Mano JF. Chitosan derivatives obtained by chemical modifications for biomedical and environmental applications. Int J Biol Macromol 2008; 43(5): 401-14.
[http://dx.doi.org/10.1016/j.ijbiomac.2008.09.007] [PMID: 18838086]

[11] Bhosale R, Gangadharappa HV, Moin A, Gowda DV, Osmani R. Gangadharappa HV, Moin A, Gowda DV, Osmani AM. Grafting technique with special emphasis on natural gums: applications and perspectives in drug delivery. Nat Prod J 2015; 5(2): 124-39.
[http://dx.doi.org/10.2174/2210315505021507021422228]

[12] Vendruscolo C, Andreazza I, Ganter J, Ferrero C, Bresolin T. Xanthan and galactomannan (from M. scabrella) matrix tablets for oral controlled delivery of theophylline. Int J Pharm 2005; 296(1-2): 1-11.
[http://dx.doi.org/10.1016/j.ijpharm.2005.02.007] [PMID: 15885450]

[13] Russell KE. Free radical graft polymerization and copolymerization at higher temperatures. Prog Polym Sci 2002; 27(6): 1007-38.
[http://dx.doi.org/10.1016/S0079-6700(02)00007-2]

[14] Xie W, Xu P, Wang W, Liu Q. Preparation and antibacterial activity of a water-soluble chitosan derivative. Carbohydr Polym 2002; 50(1): 35-40.
[http://dx.doi.org/10.1016/S0144-8617(01)00370-8]

[15] Hosseinzadeh H. Chemical modification of sodium hyaluronate *via* graft copolymerization of acrylic acid using ammonium persulfate. Res J Pharm Biol Chem Sci 2012; 3(1): 756-61.

[16] Yoshikawa S, Takayama T, Tsubokawa N. Grafting reaction of living polymer cations with amino groups on chitosan powder. J Appl Polym Sci 1998; 68(11): 1883-9.
[http://dx.doi.org/10.1002/(SICI)1097-4628(19980613)68:11<1883::AID-APP21>3.0.CO;2-U]

[17] Coessens V, Pintauer T, Matyjaszewski K. Functional polymers by atom transfer radical polymerization. Prog Polym Sci 2001; 26(3): 337-77.
[http://dx.doi.org/10.1016/S0079-6700(01)00003-X]

[18] Szwarc M. Living polymers. Their discovery, characterization, and properties. J Polym Sci A Polym Chem 1998; 36(1): ix-xv.

[19] Lizotte JR, Long TE. Stable Free-Radical Polymerization of Styrene in Combination with 2-Vinylnaphthalene Initiation. Macromol Chem Phys 2003; 204(4): 570-6.
[http://dx.doi.org/10.1002/macp.200390020]

[20] Mishra V, Kumar R. Living radical polymerization: A review. J Sci Res 2012; 56: 141-76.

[21] Bulbul Sonmez H, Senkal BF, Sherrington DC, Bıcak N. Atom transfer radical graft polymerization of acrylamide from N-chlorosulfonamidated polystyrene resin, and use of the resin in selective mercury removal. React Funct Polym 2003; 55(1): 1-8.
[http://dx.doi.org/10.1016/S1381-5148(02)00193-1]

[22] Arslan H. Block and graft copolymerization by controlled/living radical polymerization methods. Polymerization. InTechOpen. 2012 Sep 12: 279-320.

[23] Kubota H, Suka IG, Kuroda S, Kondo T. Introduction of stimuli-responsive polymers into regenerated cellulose film by means of photografting. Eur Polym J 2001; 37(7): 1367-72.
[http://dx.doi.org/10.1016/S0014-3057(00)00257-3]

[24] Liu ZM, Xu ZK, Wang JQ, Wu J, Fu JJ. Surface modification of polypropylene microfiltration membranes by graft polymerization of N-vinyl-2-pyrrolidone. Eur Polym J 2004; 40(9): 2077-87.

[http://dx.doi.org/10.1016/j.eurpolymj.2004.05.020]

[25] Coviello T, Dentini M, Rambone G, *et al.* A novel co-crosslinked polysaccharide: studies for a controlled delivery matrix. J Control Release 1998; 55(1): 57-66.
[http://dx.doi.org/10.1016/S0168-3659(98)00028-5] [PMID: 9795015]

[26] Uchida E, Uyama Y, Ikada Y. A novel method for graft polymerization onto poly(ethylene terephthalate) film surface by UV irradiation without degassing. J Appl Polym Sci 1990; 41(3-4): 677-87.
[http://dx.doi.org/10.1002/app.1990.070410317]

[27] Chen T, Kumar G, Harris MT, Smith PJ, Payne GF. Enzymatic grafting of hexyloxyphenol onto chitosan to alter surface and rheological properties. Biotechnol Bioeng 2000; 70(5): 564-73.
[PMID: 11042553]

[28] Yamaguchi T, Yamahara S, Nakao S, Kimura S. Preparation of pervaporation membranes for removal of dissolved organics from water by plasma-graft filling polymerization. J Membr Sci 1994; 95(1): 39-49.
[http://dx.doi.org/10.1016/0376-7388(94)85027-5]

[29] Kim YJ, Kang IK, Huh MW, Yoon SC. Surface characterization and in vitro blood compatibility of poly(ethylene terephthalate) immobilized with insulin and/or heparin using plasma glow discharge. Biomaterials 2000; 21(2): 121-30.
[http://dx.doi.org/10.1016/S0142-9612(99)00137-4] [PMID: 10632394]

[30] Contreras-García A, Bucio E, Concheiro A, Alvarez-Lorenzo C. Surface functionalization of polypropylene devices with hemocompatible DMAAm and NIPAAm grafts for norfloxacin sustained release. J Bioact Compat Polym 2011; 26(4): 405-19.
[http://dx.doi.org/10.1177/0883911511407788]

[31] Hanh TT, Huy HT, Hien NQ. Pre-irradiation grafting of acrylonitrile onto chitin for adsorption of arsenic in water. Radiat Phys Chem 2015; 106: 235-41.
[http://dx.doi.org/10.1016/j.radphyschem.2014.08.004]

[32] Singh DK, Ray AR. Graft copolymerization of 2-hydroxyethylmethacrylate onto chitosan films and their blood compatibility. J Appl Polym Sci 1994; 53(8): 1115-21.
[http://dx.doi.org/10.1002/app.1994.070530814]

[33] Pal S, Ghorai S, Dash MK, Ghosh S, Udayabhanu G. Flocculation properties of polyacrylamide grafted carboxymethyl guar gum (CMG-g-PAM) synthesised by conventional and microwave assisted method. J Hazard Mater 2011; 192(3): 1580-8.
[http://dx.doi.org/10.1016/j.jhazmat.2011.06.083] [PMID: 21802849]

[34] Kaity S, Isaac J, Kumar PM, Bose A, Wong TW, Ghosh A. Microwave assisted synthesis of acrylamide grafted locust bean gum and its application in drug delivery. Carbohydr Polym 2013; 98(1): 1083-94.
[http://dx.doi.org/10.1016/j.carbpol.2013.07.037] [PMID: 23987450]

[35] Ng LT, Garnett JL, Zilic E, Nguyen D. Effect of monomer structure on radiation grafting of charge transfer complexes to synthetic and naturally occurring polymers. Radiat Phys Chem 2001; 62(1): 89-98.
[http://dx.doi.org/10.1016/S0969-806X(01)00425-X]

[36] Dilli S, Garnett JL. Radiation-induced reactions with cellulose. III. Kinetics of styrene copolymerization in methanol. J Appl Polym Sci 1967; 11(6): 859-70.
[http://dx.doi.org/10.1002/app.1967.070110608]

[37] Yasukawa T, Sasaki Y, Murakami K. Kinetics of radiation-induced grafting reactions. II. Cellulose acetate–styrene systems. J Polym Sci Polym Chem Ed 1973; 11(10): 2547-56.
[http://dx.doi.org/10.1002/pol.1973.170111010]

[38] Bhattacharyya SN, Maldas D. Radiation-induced graft copolymerization of mixtures of styrene and

acrylamide onto cellulose acetate. I. Effect of solvents. J Polym Sci Polym Chem Ed 1982; 20(4): 939-50.
[http://dx.doi.org/10.1002/pol.1982.170200404]

[39] Nishioka N, Kosai K. Homogeneous graft copolymerization of vinyl monomers onto cellulose in a dimethyl sulfoxide–paraformaldehyde solvent system. I. Acrylonitrile and methyl methacrylate. Polym J 1981; 13(12): 1125-33.
[http://dx.doi.org/10.1295/polymj.13.1125]

[40] Sun T, Xu P, Liu Q, Xue J, Xie W. Graft copolymerization of methacrylic acid onto carboxymethyl chitosan. Eur Polym J 2003; 39(1): 189-92.
[http://dx.doi.org/10.1016/S0014-3057(02)00174-X]

[41] Chandakavathe BN, Kulkarni RG, Dhadde SB. Grafting of natural polymers and gums for drug delivery applications: A perspective review. Critical Review in Therapeutic Drug Carrier Systems. 2022;39(6).
[http://dx.doi.org/10.1615/critrevtherdrugcarriersyst.2022035905]

[42] Maldas D. Radiation induced grafting of styrene and acrylamide on cellulose acetate from a binary mixture. Doctoral dissertation. University of Calcutta; 1983.

[43] Chappas WJ, Silverman J. The effect of acid on the radiation-induced grafting of styrene to polyethylene. Radiat Phys Chem 1979; 14(3-6): 847-52.

[44] El-Assy NB. Effect of mineral and organic acids on radiation grafting of styrene onto polyethylene. J Appl Polym Sci 1991; 42(4): 885-9.
[http://dx.doi.org/10.1002/app.1991.070420402]

[45] Hoffman AS, Ratner BD. The radiation grafting of acrylamide to polymer substrates in the presence of cupric ion—I. A preliminary study. Radiat Phys Chem 1979; 14(3-6): 831-40.

[46] Sah SK. Natural gums emphasized grafting technique: Applications and perspectives in floating drug delivery system. Asian J Pharm 2016; 10(2) [AJP].

[47] Dailey LA, Wittmar M, Kissel T. The role of branched polyesters and their modifications in the development of modern drug delivery vehicles. J Control Release 2005; 101(1-3): 137-49.
[http://dx.doi.org/10.1016/j.jconrel.2004.09.003] [PMID: 15588900]

[48] Jung T, Kamm W, Breitenbach A, Hungerer KD, Hundt E, Kissel T. Tetanus toxoid loaded nanoparticles from sulfobutylated poly(vinyl alcohol)-graft-poly(lactide-co-glycolide): evaluation of antibody response after oral and nasal application in mice. Pharm Res 2001; 18(3): 352-60.
[http://dx.doi.org/10.1023/A:1011063232257] [PMID: 11442276]

[49] MA W, TM T, SA I. Novel smart pH sensitive chitosan grafted alginate hydrogel microcapsules for oral protein delivery: I. Preparation and characterization. Int J Pharm Pharm Sci 2015; 7: 320-6.

[50] Cai H, Zhang ZP, Chuan Sun P, Lin He B, Xia Zhu X. Synthesis and characterization of thermo- and pH- sensitive hydrogels based on Chitosan-grafted N-isopropylacrylamide *via* γ-radiation. Radiat Phys Chem 2005; 74(1): 26-30.
[http://dx.doi.org/10.1016/j.radphyschem.2004.10.007]

[51] Hu Y, Jiang X, Ding Y, Ge H, Yuan Y, Yang C. Synthesis and characterization of chitosan–poly(acrylic acid) nanoparticles. Biomaterials 2002; 23(15): 3193-201.
[http://dx.doi.org/10.1016/S0142-9612(02)00071-6] [PMID: 12102191]

[52] Kumar R, Sharma K. Biodegradable polymethacrylic acid grafted psyllium for controlled drug delivery systems. Front Chem Sci Eng 2013; 7(1): 116-22.
[http://dx.doi.org/10.1007/s11705-013-1310-0]

[53] Vihola H, Laukkanen A, Tenhu H, Hirvonen J. Drug release characteristics of physically cross-linked thermosensitive poly(N-vinylcaprolactam) hydrogel particles. J Pharm Sci 2008; 97(11): 4783-93.
[http://dx.doi.org/10.1002/jps.21348] [PMID: 18306245]

[54]　Frauke Pistel K, Breitenbach A, Zange-Volland R, Kissel T. Brush-like branched biodegradable polyesters, part III. J Control Release 2001; 73(1): 7-20.
[http://dx.doi.org/10.1016/S0168-3659(01)00231-0] [PMID: 11337055]

[55]　Kulkarni RV, Sa B. Evaluation of pH-sensitivity and drug release characteristics of (polyacrylamide-grafted-xanthan)-carboxymethyl cellulose-based pH-sensitive interpenetrating network hydrogel beads. Drug Dev Ind Pharm 2008; 34(12): 1406-14.
[http://dx.doi.org/10.1080/03639040802130079] [PMID: 18785037]

[56]　Mundargi RC, Patil SA, Agnihotri SA, Aminabhavi TM. Evaluation and controlled release characteristics of modified xanthan films for transdermal delivery of atenolol. Drug Dev Ind Pharm 2007; 33(1): 79-90.
[http://dx.doi.org/10.1080/03639040600975030] [PMID: 17192254]

[57]　Melendez-Ortiz HI, Díaz-Rodríguez P, Alvarez-Lorenzo C, Concheiro A, Bucio E. Binary graft modification of polypropylene for anti-inflammatory drug–device combo products. J Pharma Sci 2014 Apr 1;103(4):1269-77.
[http://dx.doi.org/10.1002/jps.23903]

[58]　Siraj S, Sudhakar P, Rao US, Sekharnath KV, Rao KC, Subha MC. Interpenetrating polymer network microspheres of poly (vinyl alcohol)/methyl cellulose for controlled release studies of 6-thioguanine. Int J Pharm Pharm Sci 2014; 6(9): 101-6.

[59]　Kumbar SG, Soppimath KS, Aminabhavi TM. Synthesis and characterization of polyacrylamide-grafted chitosan hydrogel microspheres for the controlled release of indomethacin. J Appl Polym Sci 2003; 87(9): 1525-36.
[http://dx.doi.org/10.1002/app.11552]

[60]　Yoshioka H, Nonaka K, Fukuda K, Kazama S. Chitosan-derived polymer-surfactants and their micellar properties. Biosci Biotechnol Biochem 1995; 59(10): 1901-4.
[http://dx.doi.org/10.1271/bbb.59.1901] [PMID: 8534983]

[61]　Liu WG, De Yao K, Liu QG. Formation of a DNA/*N*-dodecylated chitosan complex and salt-induced gene delivery. J Appl Polym Sci 2001; 82(14): 3391-5.
[http://dx.doi.org/10.1002/app.2199]

[62]　Nam JP, Lee KJ, Choi JW, Yun CO, Nah JW. Targeting delivery of tocopherol and doxorubicin grafted-chitosan polymeric micelles for cancer therapy: In vitro and in vivo evaluation. Colloids Surf B Biointerfaces 2015; 133: 254-62.
[http://dx.doi.org/10.1016/j.colsurfb.2015.06.018] [PMID: 26117805]

[63]　Catlow CR, Davidson M, Hardacre C, Hutchings GJ. Catalysis making the world a better place. Philos Trans- Royal Soc, Math Phys Eng Sci 2016; 374(2061): 20150089.
[http://dx.doi.org/10.1098/rsta.2015.0089] [PMID: 26755766]

[64]　Li D, Qu J. The progress of catalytic technologies in water purification: A review. J Environ Sci (China) 2009; 21(6): 713-9.
[http://dx.doi.org/10.1016/S1001-0742(08)62329-3] [PMID: 19803071]

[65]　Vásquez-Céspedes S, Betori RC, Cismesia MA, Kirsch JK, Yang Q. Heterogeneous catalysis for cross-coupling reactions: an underutilized powerful and sustainable tool in the fine chemical industry? Org Process Res Dev 2021; 25(4): 740-53.
[http://dx.doi.org/10.1021/acs.oprd.1c00041]

[66]　Yadav M, Sharma RK. Heterogenized nickel catalysts for various organic transformations. Curr Opin Green Sustain Chem 2019; 15: 47-59.
[http://dx.doi.org/10.1016/j.cogsc.2018.08.010]

[67]　Sin E, Yi SS, Lee YS. Chitosan-g-mPEG-supported palladium (0) catalyst for Suzuki cross-coupling reaction in water. J Mol Catal Chem 2010; 315(1): 99-104.
[http://dx.doi.org/10.1016/j.molcata.2009.09.007]

[68] Nasrollahzadeh M, Rostami-Vartooni A, Ehsani A, Moghadam M. Fabrication, characterization and application of nanopolymer supported copper (II) complex as an effective and reusable catalyst for the CN bond cross-coupling reaction of sulfonamides with arylboronic acids in water under aerobic conditions. J Mol Catal Chem 2014; 387: 123-9.
[http://dx.doi.org/10.1016/j.molcata.2014.02.017]

[69] Chtchigrovsky M, Primo A, Gonzalez P, et al. Functionalized chitosan as a green, recyclable, biopolymer-supported catalyst for the [3+2] Huisgen cycloaddition. Angew Chem Int Ed 2009; 48(32): 5916-20.
[http://dx.doi.org/10.1002/anie.200901309] [PMID: 19575432]

[70] Campelo JM, Luna D, Luque R, Marinas JM, Romero AA. Sustainable preparation of supported metal nanoparticles and their applications in catalysis. ChemSusChem 2009; 2(1): 18-45.
[http://dx.doi.org/10.1002/cssc.200800227] [PMID: 19142903]

[71] Sureshkumar M, Siswanto DY, Lee CK. Magnetic antimicrobial nanocomposite based on bacterial cellulose and silver nanoparticles. J Mater Chem 2010; 20(33): 6948-55.
[http://dx.doi.org/10.1039/c0jm00565g]

[72] Baruah D, Konwar D. Cellulose supported copper nanoparticles as a versatile and efficient catalyst for the protodecarboxylation and oxidative decarboxylation of aromatic acids under microwave heating. Catal Commun 2015; 69: 68-71.
[http://dx.doi.org/10.1016/j.catcom.2015.05.029]

[73] Chen M, Kang H, Gong Y, Guo J, Zhang H, Liu R. Bacterial cellulose supported gold nanoparticles with excellent catalytic properties. ACS Appl Mater Interfaces 2015; 7(39): 21717-26.
[http://dx.doi.org/10.1021/acsami.5b07150] [PMID: 26357993]

[74] Budarin VL, Clark JH, Luque R, Macquarrie DJ, White RJ. Palladium nanoparticles on polysaccharide-derived mesoporous materials and their catalytic performance in C–C coupling reactions. Green Chem 2008; 10(4): 382-7.
[http://dx.doi.org/10.1039/B715508E]

[75] Quignard F, Choplin A, Domard A. Chitosan: A natural polymeric support of catalysts for the synthesis of fine chemicals. Langmuir 2000; 16(24): 9106-8.
[http://dx.doi.org/10.1021/la000937d]

[76] Zhou J, Dong Z, Yang H, Shi Z, Zhou X, Li R. Pd immobilized on magnetic chitosan as a heterogeneous catalyst for acetalization and hydrogenation reactions. Appl Surf Sci 2013; 279: 360-6.
[http://dx.doi.org/10.1016/j.apsusc.2013.04.113]

[77] Gopiraman M, Bang H, Yuan G, et al. Noble metal/functionalized cellulose nanofiber composites for catalytic applications. Carbohydr Polym 2015; 132: 554-64.
[http://dx.doi.org/10.1016/j.carbpol.2015.06.051] [PMID: 26256382]

[78] Sohni S, Khan SA, Akhtar K, et al. Room temperature preparation of lignocellulosic biomass supported heterostructure (Cu+Co@OPF) as highly efficient multifunctional nanocatalyst using wetness co-impregnation. Colloids Surf A Physicochem Eng Asp 2018; 549: 184-95.
[http://dx.doi.org/10.1016/j.colsurfa.2018.04.015]

[79] Mosier N, Wyman C, Dale B, et al. Features of promising technologies for pretreatment of lignocellulosic biomass. Bioresour Technol 2005; 96(6): 673-86.
[http://dx.doi.org/10.1016/j.biortech.2004.06.025] [PMID: 15588770]

[80] Habibi Y, Dufresne A. Highly filled bionanocomposites from functionalized polysaccharide nanocrystals. Biomacromolecules 2008; 9(7): 1974-80.
[http://dx.doi.org/10.1021/bm8001717] [PMID: 18510360]

[81] Chen W, Zhong L, Peng X, Wang K, Chen Z, Sun R. Xylan-type hemicellulose supported palladium nanoparticles: a highly efficient and reusable catalyst for the carbon–carbon coupling reactions. Catal Sci Technol 2014; 4(5): 1426-35.

[http://dx.doi.org/10.1039/C3CY00933E]

[82]   Kaushik M, Moores A. Review: nanocelluloses as versatile supports for metal nanoparticles and their applications in catalysis. Green Chem 2016; 18(3): 622-37.
[http://dx.doi.org/10.1039/C5GC02500A]

[83]   Prathap KJ, Wu Q, Olsson RT, Dinér P. Catalytic reductions and tandem reactions of nitro compounds using in situ prepared nickel boride catalyst in nanocellulose solution. Org Lett 2017; 19(18): 4746-9.
[http://dx.doi.org/10.1021/acs.orglett.7b02090] [PMID: 28858520]

[84]   Jabasingh SA, Lalith D, Prabhu MA, Yimam A, Zewdu T. Catalytic conversion of sugarcane bagasse to cellulosic ethanol: TiO2 coupled nanocellulose as an effective hydrolysis enhancer. Carbohydr Polym 2016; 136: 700-9.
[http://dx.doi.org/10.1016/j.carbpol.2015.09.098] [PMID: 26572403]

[85]   Mourya P, Banerjee S, Singh MM. Corrosion inhibition of mild steel in acidic solution by Tagetes erecta (Marigold flower) extract as a green inhibitor. Corros Sci 2014; 85: 352-63.
[http://dx.doi.org/10.1016/j.corsci.2014.04.036]

[86]   Mobin M, Rizvi M. Adsorption and corrosion inhibition behavior of hydroxyethyl cellulose and synergistic surfactants additives for carbon steel in 1 M HCl. Carbohydr Polym 2017; 156: 202-14.
[http://dx.doi.org/10.1016/j.carbpol.2016.08.066] [PMID: 27842815]

[87]   Issaadi S, Douadi T, Zouaoui A, Chafaa S, Khan MA, Bouet G. Novel thiophene symmetrical Schiff base compounds as corrosion inhibitor for mild steel in acidic media. Corros Sci 2011; 53(4): 1484-8.
[http://dx.doi.org/10.1016/j.corsci.2011.01.022]

[88]   Rajan JP, Shrivastava R, Mishra RK. Corrosion inhibition effect of clerodendron colebrookianum Walp leaves (Phuinam) extract on the acid corrosion of mild steel. Prot Met Phys Chem Surf 2017; 53(6): 1161-72.
[http://dx.doi.org/10.1134/S2070205118010264]

[89]   Hashim NM, Rahim AA, Osman H, Raja PB. Quinazolinone compounds as corrosion inhibitors for mild steel in sulfuric acid medium. Chem Eng Commun 2012; 199(6): 751-66.
[http://dx.doi.org/10.1080/00986445.2011.617801]

[90]   Rbaa M, Lgaz H, El Kacimi Y, Lakhrissi B, Bentiss F, Zarrouk A. Synthesis, characterization and corrosion inhibition studies of novel 8-hydroxyquinoline derivatives on the acidic corrosion of mild steel: experimental and computational studies. Materials discovery. 2018 Jun 1; 12: 43-54.

[91]   Efil K, Bekdemir Y. Theoretical study on corrosion inhibitory action of some aromatic imines with sulphanilic acid: a DFT study. Can Chem Trans 2015; 3(1): 85-93.

[92]   Rbaa M, Fardioui M, Verma C, et al. 8-Hydroxyquinoline based chitosan derived carbohydrate polymer as biodegradable and sustainable acid corrosion inhibitor for mild steel: Experimental and computational analyses. Int J Biol Macromol 2020; 155: 645-55.
[http://dx.doi.org/10.1016/j.ijbiomac.2020.03.200] [PMID: 32224172]

[93]   Verma C, Olasunkanmi LO, Ebenso EE, Quraishi MA, Obot IB. Adsorption behavior of glucosamine-based, pyrimidine-fused heterocycles as green corrosion inhibitors for mild steel: experimental and theoretical studies. J Phys Chem C 2016; 120(21): 11598-611.
[http://dx.doi.org/10.1021/acs.jpcc.6b04429]

[94]   Umoren SA, Eduok UM. Application of carbohydrate polymers as corrosion inhibitors for metal substrates in different media: A review. Carbohydr Polym 2016; 140: 314-41.
[http://dx.doi.org/10.1016/j.carbpol.2015.12.038] [PMID: 26876859]

[95]   Umoren SA, Obot IB, Madhankumar A, Gasem ZM. Performance evaluation of pectin as ecofriendly corrosion inhibitor for X60 pipeline steel in acid medium: Experimental and theoretical approaches. Carbohydr Polym 2015; 124: 280-91.
[http://dx.doi.org/10.1016/j.carbpol.2015.02.036] [PMID: 25839822]

[96]   Reddy N, Yang Y. Properties and potential applications of natural cellulose fibers from cornhusks.

Green Chem 2005; 7(4): 190-5.
[http://dx.doi.org/10.1039/b415102j]

[97] Reddy N, Yang Y. Biofibers from agricultural byproducts for industrial applications. Trends Biotechnol 2005; 23(1): 22-7.
[http://dx.doi.org/10.1016/j.tibtech.2004.11.002] [PMID: 15629854]

[98] Savadekar NR, Karande VS, Vigneshwaran N, Bharimalla AK, Mhaske ST. Preparation of nano cellulose fibers and its application in kappa-carrageenan based film. Int J Biol Macromol 2012; 51(5): 1008-13.
[http://dx.doi.org/10.1016/j.ijbiomac.2012.08.014] [PMID: 22940239]

[99] Li R, Fei J, Cai Y, Li Y, Feng J, Yao J. Cellulose whiskers extracted from mulberry: A novel biomass production. Carbohydr Polym 2009; 76(1): 94-9.
[http://dx.doi.org/10.1016/j.carbpol.2008.09.034]

[100] Shi Z, Zhang Y, Phillips GO, Yang G. Utilization of bacterial cellulose in food. Food Hydrocoll 2014; 35: 539-45.
[http://dx.doi.org/10.1016/j.foodhyd.2013.07.012]

[101] Tomczyńska-Mleko M, Terpiłowski K, Mleko S. Physicochemical properties of cellulose/whey protein fibers as a potential material for active ingredients release. Food Hydrocoll 2015; 49: 232-9.
[http://dx.doi.org/10.1016/j.foodhyd.2015.03.027]

[102] Trache D, Hussin MH, Hui Chuin CT, *et al.* Microcrystalline cellulose: Isolation, characterization and bio-composites application—A review. Int J Biol Macromol 2016; 93(Pt A): 789-804.
[http://dx.doi.org/10.1016/j.ijbiomac.2016.09.056] [PMID: 27645920]

[103] Peelman N, Ragaert P, De Meulenaer B, *et al.* Application of bioplastics for food packaging. Trends Food Sci Technol 2013; 32(2): 128-41.
[http://dx.doi.org/10.1016/j.tifs.2013.06.003]

[104] Mudgil D, Barak S, Khatkar BS. Guar gum: processing, properties and food applications—A Review. J Food Sci Technol 2014; 51(3): 409-18.
[http://dx.doi.org/10.1007/s13197-011-0522-x] [PMID: 24587515]

[105] Kriegel C, Arrechi A, Kit K, McClements DJ, Weiss J. Fabrication, functionalization, and application of electrospun biopolymer nanofibers. Crit Rev Food Sci Nutr 2008; 48(8): 775-97.
[http://dx.doi.org/10.1080/10408390802241325] [PMID: 18756399]

[106] Padil VVT, Wacławek S, Černík M, Varma RS. Tree gum-based renewable materials: Sustainable applications in nanotechnology, biomedical and environmental fields. Biotechnol Adv 2018; 36(7): 1984-2016.
[http://dx.doi.org/10.1016/j.biotechadv.2018.08.008] [PMID: 30165173]

[107] Reddy N, Yang Y. Natural cellulose fibers from soybean straw. Bioresour Technol 2009; 100(14): 3593-8.
[http://dx.doi.org/10.1016/j.biortech.2008.09.063] [PMID: 19345577]

[108] Bayer IS, Fragouli D, Attanasio A, *et al.* Water-repellent cellulose fiber networks with multifunctional properties. ACS Appl Mater Interfaces 2011; 3(10): 4024-31.
[http://dx.doi.org/10.1021/am200891f] [PMID: 21902239]

[109] Kanu NJ, Gupta E, Vates UK, Singh GK. Electrospinning process parameters optimization for biofunctional curcumin/gelatin nanofibers. Mater Res Express 2020; 7(3): 035022.
[http://dx.doi.org/10.1088/2053-1591/ab7f60]

[110] Moon JY, Lee J, Hwang TI, Park CH, Kim CS. A multifunctional, one-step gas foaming strategy for antimicrobial silver nanoparticle-decorated 3D cellulose nanofiber scaffolds. Carbohydr Polym 2021; 273: 118603.
[http://dx.doi.org/10.1016/j.carbpol.2021.118603] [PMID: 34561003]

[111] Chen Y, Dong X, Shafiq M, Myles G, Radacsi N, Mo X. Recent advancements on three-dimensional

electrospun nanofiber scaffolds for tissue engineering. Adv Fiber Mater 2022; 4(5): 959-86.
[http://dx.doi.org/10.1007/s42765-022-00170-7]

[112] Imani F, Karimi-Soflou R, Shabani I, Karkhaneh A. PLA electrospun nanofibers modified with polypyrrole-grafted gelatin as bioactive electroconductive scaffold. Polymer. 2021;18(218):123487.
[http://dx.doi.org/10.1016/j.polymer.2021.123487]

[113] Wang B, Lu G, Song K, Chen A, Xing H, Wu J, Sun Q, Li G, Cai M. PLGA-based electrospun nanofibers loaded with dual bioactive agent loaded scaffold as a potential wound dressing material. Colloids and Surfaces B: Biointerfaces. 2023 Nov 1;231:113570.
[http://dx.doi.org/10.1016/j.colsurfb.2023.113570]

[114] Zhu G, Wu M, Ding Z, Zou T, Wang L. Biological properties of polyurethane: Issues and potential for application in vascular medicine. Euro Poly J 2023 Dec 11;201:112536.
[http://dx.doi.org/10.1016/j.eurpolymj.2023.112536]

[115] Kausar A, Ahmad I, Zhao T, Aldaghri O, Ibnaouf KH, Eisa MH. Nanocomposite Nanofibers of Graphene—Fundamentals and Systematic Developments. J Compos Sci 2023; 7(8): 323.
[http://dx.doi.org/10.3390/jcs7080323]

[116] Hayat U, Raza A, Bilal M, Iqbal HMN, Wang JY. Biodegradable polymeric conduits: Platform materials for guided nerve regeneration and vascular tissue engineering. J Drug Deliv Sci Technol 2022; 67: 103014.
[http://dx.doi.org/10.1016/j.jddst.2021.103014]

[117] Wu J, Wang S, Zheng Z, Li J. Fabrication of Biologically Inspired Electrospun Collagen/Silk fibroin/bioactive glass composited nanofibrous scaffold to accelerate the treatment efficiency of bone repair. Regen Ther 2022; 21: 122-38.
[http://dx.doi.org/10.1016/j.reth.2022.05.006] [PMID: 35844293]

[118] Li G, Yang T, Liu Y, et al. The proteins derived from platelet-rich plasma improve the endothelialization and vascularization of small diameter vascular grafts. Int J Biol Macromol 2023; 225: 574-87.
[http://dx.doi.org/10.1016/j.ijbiomac.2022.11.116] [PMID: 36395946]

[119] Contreras-Cáceres R, Cabeza L, Perazzoli G, et al. Electrospun nanofibers: Recent applications in drug delivery and cancer therapy. Nanomaterials (Basel) 2019; 9(4): 656.
[http://dx.doi.org/10.3390/nano9040656] [PMID: 31022935]

[120] Shahriar SMS, Mondal J, Hasan MN, Revuri V, Lee DY, Lee YK. Electrospinning nanofibers for therapeutics delivery. Nanomaterials (Basel) 2019; 9(4): 532.
[http://dx.doi.org/10.3390/nano9040532] [PMID: 30987129]

[121] Berdimurodov E, Dagdag O, Berdimuradov K, et al. Green Electrospun Nanofibers for Biomedicine and Biotechnology. Technologies (Basel) 2023; 11(5): 150.
[http://dx.doi.org/10.3390/technologies11050150]

[122] Torres-Martínez EJ, Pérez-González GL, Serrano-Medina A, et al. Drugs loaded into electrospun polymeric nanofibers for delivery. J Pharm Pharm Sci 2019; 22(1): 313-31.
[http://dx.doi.org/10.18433/jpps29674] [PMID: 31329535]

[123] Bhattarai RS, Bachu RD, Boddu SHS, Bhaduri S. Biomedical applications of electrospun nanofibers: Drug and nanoparticle delivery. Pharmaceutics 2018; 11(1): 5.
[http://dx.doi.org/10.3390/pharmaceutics11010005] [PMID: 30586852]

[124] Jiang J, Ceylan M, Zheng Y, Yao L, Asmatulu R, Yang SY. Poly-ε-caprolactone electrospun nanofiber mesh as a gene delivery tool. AIMS Bioeng 2016; 3(4): 528-37.
[http://dx.doi.org/10.3934/bioeng.2016.4.528]

[125] Rao GSNK, Kurakula M, Yadav KS. Application of Electrospun Materials in Gene Delivery. In: Inamuddin, Boddula R, Ahamed MI, Asiri AM, eds. Electrospun Materials and Their Allied Applications. Singapore: Springer Nature Singapore; 2020. p. 265-306.

[http://dx.doi.org/10.1002/9781119655039.ch10]

[126] Zhang J, Duan Y, Wei D, *et al.* Co-electrospun fibrous scaffold–adsorbed DNA for substrate-mediated gene delivery. J Biomed Mater Res A 2011; 96A(1): 212-20.
[http://dx.doi.org/10.1002/jbm.a.32962] [PMID: 21105170]

[127] Ranjbar-Mohammadi M, Rabbani S, Bahrami SH, Joghataei MT, Moayer F. Antibacterial performance and in vivo diabetic wound healing of curcumin loaded gum tragacanth/poly(ε-caprolactone) electrospun nanofibers. Mater Sci Eng C 2016; 69: 1183-91.
[http://dx.doi.org/10.1016/j.msec.2016.08.032] [PMID: 27612816]

[128] Abrigo M, McArthur SL, Kingshott P. Electrospun nanofibers as dressings for chronic wound care: advances, challenges, and future prospects. Macromol Biosci 2014; 14(6): 772-92.
[http://dx.doi.org/10.1002/mabi.201300561] [PMID: 24678050]

[129] Gao C, Zhang L, Wang J, *et al.* Electrospun nanofibers promote wound healing: theories, techniques, and perspectives. J Mater Chem B Mater Biol Med 2021; 9(14): 3106-30.
[http://dx.doi.org/10.1039/D1TB00067E] [PMID: 33885618]

[130] Chen S, Liu B, Carlson MA, Gombart AF, Reilly DA, Xie J. Recent advances in electrospun nanofibers for wound healing. Nanomedicine (Lond) 2017; 12(11): 1335-52.
[http://dx.doi.org/10.2217/nnm-2017-0017] [PMID: 28520509]

[131] Memic A, Abdullah T, Mohammed HS, Joshi Navare K, Colombani T, Bencherif SA. Latest progress in electrospun nanofibers for wound healing applications. ACS Appl Bio Mater 2019; 2(3): 952-69.
[http://dx.doi.org/10.1021/acsabm.8b00637] [PMID: 35021385]

[132] Liu M, Duan XP, Li YM, Yang DP, Long YZ. Electrospun nanofibers for wound healing. Mater Sci Eng C 2017; 76: 1413-23.
[http://dx.doi.org/10.1016/j.msec.2017.03.034] [PMID: 28482508]

[133] Özen İ, Wang X. Biomedicine: electrospun nanofibrous hormonal therapies through skin/tissue—a review. Int J Polym Mater 2023; 72(1): 21-39.
[http://dx.doi.org/10.1080/00914037.2021.1985493]

[134] Bakhori NM, Ismail Z, Hassan MZ, Dolah R. Emerging trends in nanotechnology: Aerogel-based materials for biomedical applications. Nanomaterials (Basel) 2023; 13(6): 1063.
[http://dx.doi.org/10.3390/nano13061063] [PMID: 36985957]

[135] Xiao Z, Liu H, Zhao Q, Niu Y, Chen Z, Zhao D. Application of microencapsulation technology in silk fibers. J Appl Polym Sci 2022; 139(25): e52351.
[http://dx.doi.org/10.1002/app.52351]

[136] Samatya Yılmaz S, Aytac A. The highly absorbent polyurethane/polylactic acid blend electrospun tissue scaffold for dermal wound dressing. Polym Bull 2023; 80(12): 12787-813.
[http://dx.doi.org/10.1007/s00289-022-04633-0]

[137] Wang J, Zhan L, Zhang X, Wu R, Liao L, Wei J. Silver nanoparticles coated poly (L-lactide) electrospun membrane for implant associated infections prevention. Front Pharmacol 2020; 11: 431.
[http://dx.doi.org/10.3389/fphar.2020.00431] [PMID: 32322206]

[138] Chen Z, Benecke L, Kempert P, *et al.* Simulation and Development of Biomimetic Electrospun PCL Nanofibrous Tympanic Membrane Implants. Proc Appl Math Mech 2021; 20(1): e202000100.
[http://dx.doi.org/10.1002/pamm.202000100]

[139] Jia W, Cui D, Liu Y, *et al.* Polyether-ether-ketone/poly(methyl methacrylate)/carbon fiber ternary composites prepared by electrospinning and hot pressing for bone implant applications. Mater Des 2021; 209: 109893.
[http://dx.doi.org/10.1016/j.matdes.2021.109893]

[140] Hernandez JL, Park J, Yao S, *et al.* Effect of tissue microenvironment on fibrous capsule formation to biomaterial-coated implants. Biomaterials 2021; 273: 120806.
[http://dx.doi.org/10.1016/j.biomaterials.2021.120806] [PMID: 33905960]

[141] Morais M, Coimbra P, Pina ME. Comparative analysis of morphological and release profiles in ocular implants of acetazolamide prepared by electrospinning. Pharmaceutics 2021; 13(2): 260.
[http://dx.doi.org/10.3390/pharmaceutics13020260] [PMID: 33671936]

[142] Ji H, Wang Y, Liu H, et al. Programmed core-shell electrospun nanofibers to sequentially regulate osteogenesis-osteoclastogenesis balance for promoting immediate implant osseointegration. Acta Biomater 2021; 135 274-88.
[http://dx.doi.org/10.1016/j.actbio.2021.08.050] [PMID: 34492371]

[143] Cui J, Li F, Wang Y, Zhang Q, Ma W, Huang C. Electrospun nanofiber membranes for wastewater treatment applications. Separ Purif Tech 2020; 250: 117116.
[http://dx.doi.org/10.1016/j.seppur.2020.117116]

[144] Blanco M, Monteserín C, Angulo A, et al. TiO2-doped electrospun nanofibrous membrane for photocatalytic water treatment. Polymers (Basel) 2019; 11(5): 747.
[http://dx.doi.org/10.3390/polym11050747] [PMID: 31027371]

[145] Fahimirad S, Fahimirad Z, Sillanpää M. Efficient removal of water bacteria and viruses using electrospun nanofibers. Sci Total Environ 2021; 751: 141673.
[http://dx.doi.org/10.1016/j.scitotenv.2020.141673] [PMID: 32866832]

[146] Cárdenas Bates II, Loranger É, Mathew AP, Chabot B. Cellulose reinforced electrospun chitosan nanofibers bio-based composite sorbent for water treatment applications. Cellulose 2021; 28(8): 4865-85.
[http://dx.doi.org/10.1007/s10570-021-03828-4]

[147] Yanar N, Liang Y, Yang E, et al. Robust and fouling-resistant ultrathin membranes for water purification tailored via semi-dissolved electrospun nanofibers. J Clean Prod 2023; 418: 138056.
[http://dx.doi.org/10.1016/j.jclepro.2023.138056]

[148] Abd Halim NS, Wirzal MDH, Hizam SM, et al. Recent development on electrospun nanofiber membrane for produced water treatment: a review. J Environ Chem Eng 2021; 9(1): 104613.
[http://dx.doi.org/10.1016/j.jece.2020.104613]

[149] Kugarajah V, Ojha AK, Ranjan S, et al. Future applications of electrospun nanofibers in pressure driven water treatment: A brief review and research update. J Environ Chem Eng 2021; 9(2): 105107.
[http://dx.doi.org/10.1016/j.jece.2021.105107]

[150] Orlando R, Polat M, Afshari A, Johnson MS, Fojan P. Electrospun Nanofibre Air Filters for Particles and Gaseous Pollutants. Sustainability. 2021; 13(12): 6553.
[http://dx.doi.org/10.3390/su13126553]

[151] Demirel O, Kolgesiz S, Yuce S, Hayat Soytaş S, Koseoglu-Imer DY, Unal H. Photothermal electrospun nanofibers containing polydopamine-coated halloysite nanotubes as antibacterial air filters. ACS Appl Nano Mater 2022; 5(12): 18127-37.
[http://dx.doi.org/10.1021/acsanm.2c04026]

[152] He W, Guo Y, Zhao YB, et al. Self-supporting smart air filters based on PZT/PVDF electrospun nanofiber composite membrane. Chem Eng J 2021; 423: 130247.
[http://dx.doi.org/10.1016/j.cej.2021.130247]

[153] Mamun A, Blachowicz T, Sabantina L. Electrospun nanofiber mats for filtering applications—Technology, structure and materials. Polymers (Basel) 2021; 13(9): 1368.
[http://dx.doi.org/10.3390/polym13091368] [PMID: 33922156]

[154] Zhou Y, Liu Y, Zhang M, Feng Z, Yu DG, Wang K. Electrospun nanofiber membranes for air filtration: A review. Nanomaterials (Basel) 2022; 12(7): 1077.
[http://dx.doi.org/10.3390/nano12071077] [PMID: 35407195]

[155] Niu Z, Bian Y, Xia T, Zhang L, Chen C. An optimization approach for fabricating electrospun nanofiber air filters with minimized pressure drop for indoor $PM_{2.5}$ control. Build Environ 2021; 188: 107449.

[http://dx.doi.org/10.1016/j.buildenv.2020.107449]

[156] Bian Y, Wang S, Zhang L, Chen C. Influence of fiber diameter, filter thickness, and packing density on $PM_{2.5}$ removal efficiency of electrospun nanofiber air filters for indoor applications. Build Environ 2020; 170: 106628.
[http://dx.doi.org/10.1016/j.buildenv.2019.106628]

[157] Rajak A, Hapidin DA, Iskandar F, Munir MM, Khairurrijal K. Electrospun nanofiber from various source of expanded polystyrene (EPS) waste and their characterization as potential air filter media. Waste Manag 2020; 103: 76-86.
[http://dx.doi.org/10.1016/j.wasman.2019.12.017] [PMID: 31865038]

[158] Naz M, Jabeen S, Gull N, et al. Novel silane crosslinked chitosan based electrospun nanofiber for controlled release of benzocaine. Front Mater 2022; 9: 826251.
[http://dx.doi.org/10.3389/fmats.2022.826251]

[159] Chadha S. Recent advances in nano-encapsulation technologies for controlled release of biostimulants and antimicrobial agents. Advances in nano-fertilizers and nano-pesticides in agriculture. 2021 Jan 1:29-55.
[http://dx.doi.org/10.1016/B978-0-12-820092-6.00002-1]

[160] Zaim NSHBH, Tan HL, Rahman SMA, et al. Recent advances in seed coating treatment using nanoparticles and nanofibers for enhanced seed germination and protection. J Plant Growth Regul 2023; 42(12): 7374-402.
[http://dx.doi.org/10.1007/s00344-023-11038-4]

[161] Das KP, Sharma D, Satapathy BK. Electrospun fibrous constructs towards clean and sustainable agricultural prospects: SWOT analysis and TOWS based strategy assessment. J Clean Prod 2022; 368: 133137.
[http://dx.doi.org/10.1016/j.jclepro.2022.133137]

[162] Rashidi M, Seyyedi Mansour S, Mostashari P, Ramezani S, Mohammadi M, Ghorbani M. Electrospun nanofiber based on Ethyl cellulose/Soy protein isolated integrated with bitter orange peel extract for antimicrobial and antioxidant active food packaging. Int J Biol Macromol 2021; 193(Pt B): 1313-23.
[http://dx.doi.org/10.1016/j.ijbiomac.2021.10.182] [PMID: 34728303]

[163] Bodbodak S, Shahabi N, Mohammadi M, Ghorbani M, Pezeshki A. Development of a novel antimicrobial electrospun nanofiber based on polylactic acid/hydroxypropyl methylcellulose containing pomegranate peel extract for active food packaging. Food Bioprocess Technol 2021; 14(12): 2260-72.
[http://dx.doi.org/10.1007/s11947-021-02722-y]

[164] Zhang C, Li Y, Wang P, Zhang H. Electrospinning of nanofibers: Potentials and perspectives for active food packaging. Compr Rev Food Sci Food Saf 2020; 19(2): 479-502.
[http://dx.doi.org/10.1111/1541-4337.12536] [PMID: 33325166]

[165] Kowsalya E, Christas KM, Balashanmugam P, Tamil Selvi A, Jaquline Chinna Rani I. Biocompatible silver nanoparticles/poly(vinyl alcohol) electrospun nanofibers for potential antimicrobial food packaging applications. Food Packag Shelf Life 2019; 21: 100379.
[http://dx.doi.org/10.1016/j.fpsl.2019.100379]

[166] Shi C, Zhou A, Fang D, et al. Oregano essential oil/β-cyclodextrin inclusion compound polylactic acid/polycaprolactone electrospun nanofibers for active food packaging. Chem Eng J 2022; 445: 136746.
[http://dx.doi.org/10.1016/j.cej.2022.136746]

[167] Duan M, Yu S, Sun J, et al. Development and characterization of electrospun nanofibers based on pullulan/chitin nanofibers containing curcumin and anthocyanins for active-intelligent food packaging. Int J Biol Macromol 2021; 187: 332-40.
[http://dx.doi.org/10.1016/j.ijbiomac.2021.07.140] [PMID: 34303741]

[168] Cristofoli NL, Lima AR, Tchonkouang RDN, Quintino AC, Vieira MC. Advances in the food

packaging production from agri-food waste and by-products: market trends for a sustainable development. Sustainability (Basel) 2023; 15(7): 6153.
[http://dx.doi.org/10.3390/su15076153]

[169] Topuz F, Uyar T. Antioxidant, antibacterial and antifungal electrospun nanofibers for food packaging applications. Food Res Int 2020; 130: 108927.
[http://dx.doi.org/10.1016/j.foodres.2019.108927] [PMID: 32156376]

[170] Zhu C, Cao R, Zhang Y, Chen R. RETRACTED: Metallic Ions Encapsulated in Electrospun Nanofiber for Antibacterial and Angiogenesis Function to Promote Wound Repair. Front Cell Dev Biol 2021; 9: 660571.
[http://dx.doi.org/10.3389/fcell.2021.660571] [PMID: 33842486]

[171] Miller B, Hansrisuk A, Highley CB, Caliari SR. Guest–host supramolecular assembly of injectable hydrogel nanofibers for cell encapsulation. ACS Biomater Sci Eng 2021; 7(9): 4164-74.
[http://dx.doi.org/10.1021/acsbiomaterials.1c00275] [PMID: 33891397]

[172] Diep E, Schiffman JD. Encapsulating bacteria in alginate-based electrospun nanofibers. Biomater Sci 2021; 9(12): 4364-73.
[http://dx.doi.org/10.1039/D0BM02205E] [PMID: 34128000]

[173] Eom S, Park SM, Hong H, et al. Hydrogel-assisted electrospinning for fabrication of a 3D complex tailored nanofiber macrostructure. ACS Appl Mater Interfaces 2020; 12(46): 51212-24.
[http://dx.doi.org/10.1021/acsami.0c14438] [PMID: 33153261]

[174] Duan F, Sun T, Zhang J, Wang K, Wen Y, Lu L. Recent innovations in immobilization of β-galactosidases for industrial and therapeutic applications. Biotechnol Adv 2022; 61: 108053.
[http://dx.doi.org/10.1016/j.biotechadv.2022.108053] [PMID: 36309245]

[175] Law KCL, Mahmoudi N, Zadeh ZE, et al. A selective, hydrogel-based prodrug delivery system efficiently activates a suicide gene to remove undifferentiated human stem cells within neural grafts. Adv Funct Mater 2023; 33(43): 2305771.
[http://dx.doi.org/10.1002/adfm.202305771]

[176] Dart A, Roy D, Vlaskin V, et al. A nanofiber based antiviral (TAF) prodrug delivery system. Biomater Adv 2022; 133: 112626.
[http://dx.doi.org/10.1016/j.msec.2021.112626] [PMID: 35039198]

[177] Ye J, Gong M, Song J, et al. Integrating inflammation-responsive prodrug with electrospun nanofibers for anti-inflammation application. Pharmaceutics 2022; 14(6): 1273.
[http://dx.doi.org/10.3390/pharmaceutics14061273] [PMID: 35745845]

[178] Chen J, Zhang T, Hua W, Li P, Wang X. 3D Porous poly(lactic acid)/regenerated cellulose composite scaffolds based on electrospun nanofibers for biomineralization. Colloids Surf A Physicochem Eng Asp 2020; 585: 124048.
[http://dx.doi.org/10.1016/j.colsurfa.2019.124048]

[179] Zhijiang C, Ping X, Shiqi H, Cong Z. Soy protein nanoparticles modified bacterial cellulose electrospun nanofiber membrane scaffold by ultrasound-induced self-assembly technique: characterization and cytocompatibility. Cellulose 2019; 26(10): 6133-50.
[http://dx.doi.org/10.1007/s10570-019-02513-x]

[180] Sakhare MS, Rajput HH. Polymer grafting and applications in pharmaceutical drug delivery systems-a brief review. Asian J Pharm Clin Res 2017; 10(6): 59.
[http://dx.doi.org/10.22159/ajpcr.2017.v10i6.18072]

[181] Zhang Y, Wang T, Li J, et al. Bilayer membrane composed of mineralized collagen and chitosan cast film coated with berberine-loaded PCL/PVP electrospun nanofiber promotes bone regeneration. Front Bioeng Biotechnol 2021; 9: 684335.
[http://dx.doi.org/10.3389/fbioe.2021.684335] [PMID: 34350160]

[182] Huo P, Han X, Zhang W, Zhang J, Kumar P, Liu B. Electrospun nanofibers of

polycaprolactone/collagen as a sustained-release drug delivery system for artemisinin. Pharmaceutics 2021; 13(8): 1228.
[http://dx.doi.org/10.3390/pharmaceutics13081228] [PMID: 34452189]

[183] Xie X, Chen Y, Wang X, *et al.* Electrospinning nanofiber scaffolds for soft and hard tissue regeneration. J Mater Sci Technol 2020; 59: 243-61.
[http://dx.doi.org/10.1016/j.jmst.2020.04.037]

[184] Zarei M, Samimi A, Khorram M, Abdi MM, Golestaneh SI. Fabrication and characterization of conductive polypyrrole/chitosan/collagen electrospun nanofiber scaffold for tissue engineering application. Int J Biol Macromol 2021; 168: 175-86.
[http://dx.doi.org/10.1016/j.ijbiomac.2020.12.031] [PMID: 33309657]

[185] Xu F, Wang H, Zhang J, Jiang L, Zhang W, Hu Y. A facile design of EGF conjugated PLA/gelatin electrospun nanofibers for nursing care of *in vivo* wound healing applications. J Ind Text 2022; 51(1_suppl) (Suppl.): 420S-40S.
[http://dx.doi.org/10.1177/1528083720976348]

[186] Ajmal G, Bonde GV, Mittal P, Pandey VK, Yadav N, Mishra B. PLGA/Gelatin-based electrospun nanofiber scaffold encapsulating antibacterial and antioxidant molecules for accelerated tissue regeneration. Mater Today Commun 2023; 35: 105633.
[http://dx.doi.org/10.1016/j.mtcomm.2023.105633]

[187] Fathi-Karkan S, Banimohamad-Shotorbani B, Saghati S, Rahbarghazi R, Davaran S. A critical review of fibrous polyurethane-based vascular tissue engineering scaffolds. J Biol Eng 2022; 16(1): 6.
[http://dx.doi.org/10.1186/s13036-022-00286-9] [PMID: 35331305]

[188] Ibrahim HM, Klingner A. A review on electrospun polymeric nanofibers: Production parameters and potential applications. Polym Test 2020; 90: 106647.
[http://dx.doi.org/10.1016/j.polymertesting.2020.106647]

[189] Das A, Shetty S, KN C, Shetty R, Suranjan Salins S. Electrospun nanofibers: transformative innovations in biomedical applications and Future prospects in healthcare advancement. Cogent Engineering. 2024 Dec 31;11(1):2433147.
[http://dx.doi.org/10.1080/23311916.2024.2433147]

CHAPTER 5

Types of Green Solvents in Grafting Processes

Shashikant V. Bhandari[1,*], Shambhavi S. Sing[1], Neha A. Raut[1], Sagar R. Ghanwa[1], Abhishek V. Shitol[1] and Sharayu P. Ninaew[1]

[1] *Department of Pharmaceutical Chemistry, AISSMS College of Pharmacy, Pune, Maharashtra 411001, India*

Abstract: The desire for more ecologically friendly and sustainable synthetic techniques is driving an increasing amount of interest in the use of green solvents in grafting operations. Green solvents are safer and more environmentally friendly than standard solvents like toluene, since they are renewable, biodegradable, and non-toxic.

The creation of organic-inorganic hybrid materials, which have a wide range of uses in many industries, requires specialized grafting procedures. In these reactions, the solvent selection is crucial since it affects the process's efficiency, safety, and environmental sustainability. Although they are often utilized, traditional solvents like toluene carry serious dangers to human health and the environment. Investigating green solvents that are non-toxic, renewable, and biodegradable is therefore becoming more and more important.

The several kinds of green solvents that can be utilized in grafting procedures are highlighted in this abstract, including:

Bio-sourced Solvents: A sustainable substitute for conventional solvents, these solvents come from renewable biomass sources. Dimethyl carbonate, (+)-limonene, (−)-β-pinene, (+)-α-pinene, and 2-methyltetrahydrofuran (MeTHF) are a few examples.

Ionic Liquids: Ionic liquids are a family of low-volatility solvents made completely of ions. They may be employed in a wide variety of pressures and temperatures and are quite successful in grafting reactions.

Deep Eutectic Solvents: These solvents are created by combining two or more low-melting-point substances. They are a sustainable substitute for conventional solvents and are very successful in grafting reactions.

---

[*] **Corresponding author Shashikant V. Bhandari:** Department of Pharmaceutical Chemistry, AISSMS College of Pharmacy, Pune, Maharashtra 411001, India;
E-mail: shashikantbhandari2011@gmail.com

Kuldeep Vinchurkar, Satish Polshettiwar, Nilesh Mahajan &Yogeshwar Bachhav (Eds.)
All rights reserved-© 2026 Bentham Science Publishers

Supercritical Fluids: Gases that are above their critical temperature and pressure are known as supercritical fluids. They provide a sustainable substitute for conventional solvents and can be utilized as solvents in grafting reactions.

Green Synthetic Organic Solvents: These biodegradable solvents come from sustainable resources. Solvents such as 2-methyltetrahydrofuran (MeTHF) and dimethyl carbonate are ex:amples. These green solvents can be used in various grafting techniques, including:

Free Radical Grafting: In order to start the grafting reaction, free radicals are used in this procedure. It has been demonstrated that green solvents such as MeTHF and (+)-α-pinene work well with this method.

Ionic Grafting: This method transfers organic functionalities onto inorganic substrates by using ionic liquids as solvents.

"Click" Chemistry Grafting: This method transfers organic functionalities onto inorganic substrates by use of click chemistry processes. It has been demonstrated that green solvents such as dimethyl carbonate and MeTHF work well in this method.

In addition, green solvents are being used in solvothermal synthesis for the surface organosilylation of hierarchical nanozeolites. This approach offers a low-hazard and effective method for the synthesis of these materials, which have potential applications in various fields.

Overall, the use of green solvents in grafting processes is an emerging area of research, with significant potential for improving the sustainability and environmental impact of these reactions.

Keywords: Eco-friendly, Green solvents, Grafting, Organic solvents, Solvothermal synthesis.

TYPES OF GREEN SOLVENTS IN GRAFTING PROCESSES

In order to solubilize the organic reactant(s), the support particles are dispersed, the reaction's viscosity is reduced, and any exothermic reactions that may occur are controlled. The grafting reaction is often conducted with the use of a solvent. The solvent that is used is very important since it needs to be inert and considerate of both the organic molecules (which are typically very reactive) and the inorganic support's structural integrity. Examples of such compounds are APTS [1].

Furthermore, the solvent makes up a greater percentage of the reaction mass and has a significant effect on both the process's environmental impact and safety. Toluene is typically used in grafting techniques for these reasons [2, 3].

Because toluene has a manageable boiling point, the reaction mixture can be heated to 111 °C if needed, but it can then be readily extracted from the material. Toluene is also inexpensive and commonly accessible in the market. But toluene, which comes from the catalytic reforming of light petroleum fractions, is a nonrenewable solvent that poses a serious risk to human health [4, 5].

These days, finding more environmentally friendly solvents for the grafting process is a challenge toward more sustainable synthetic processes. In this paper, we examine the use of environmentally friendly and sustainable solvents for the grafting reaction as an alternative to toluene, examining the impact on the hybrid materials that are produced [6]. As shown in Fig. (**1**), there are 11 types of green solvents used for the purpose of green grafting techniques. Furthermore, green solvents are categorized into two broad classes, renewable and non-renewable solvents, as shown in Fig. (**2**).

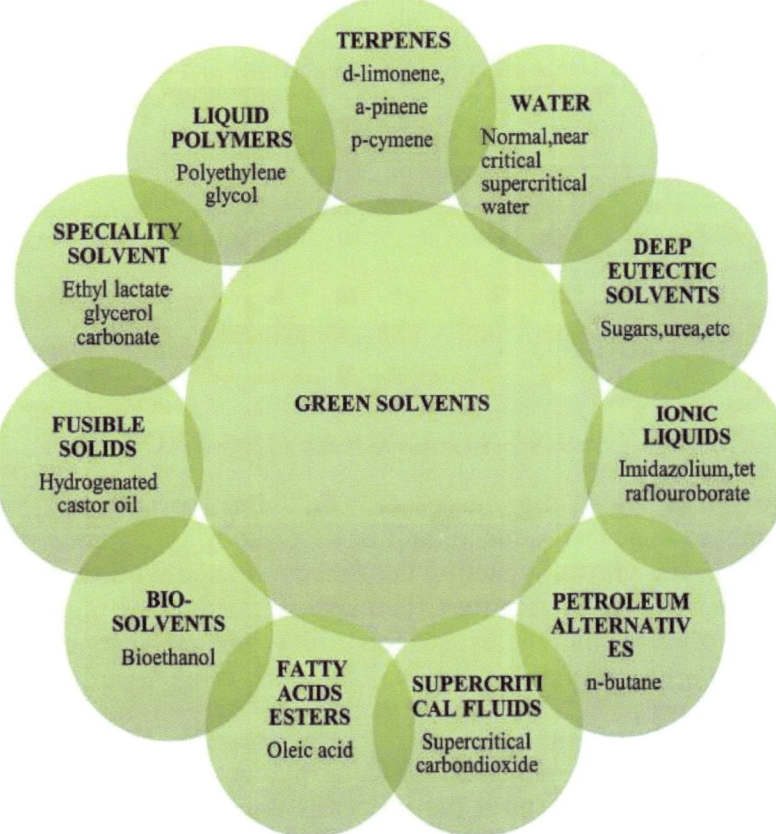

Fig. (1). Types of green solvents

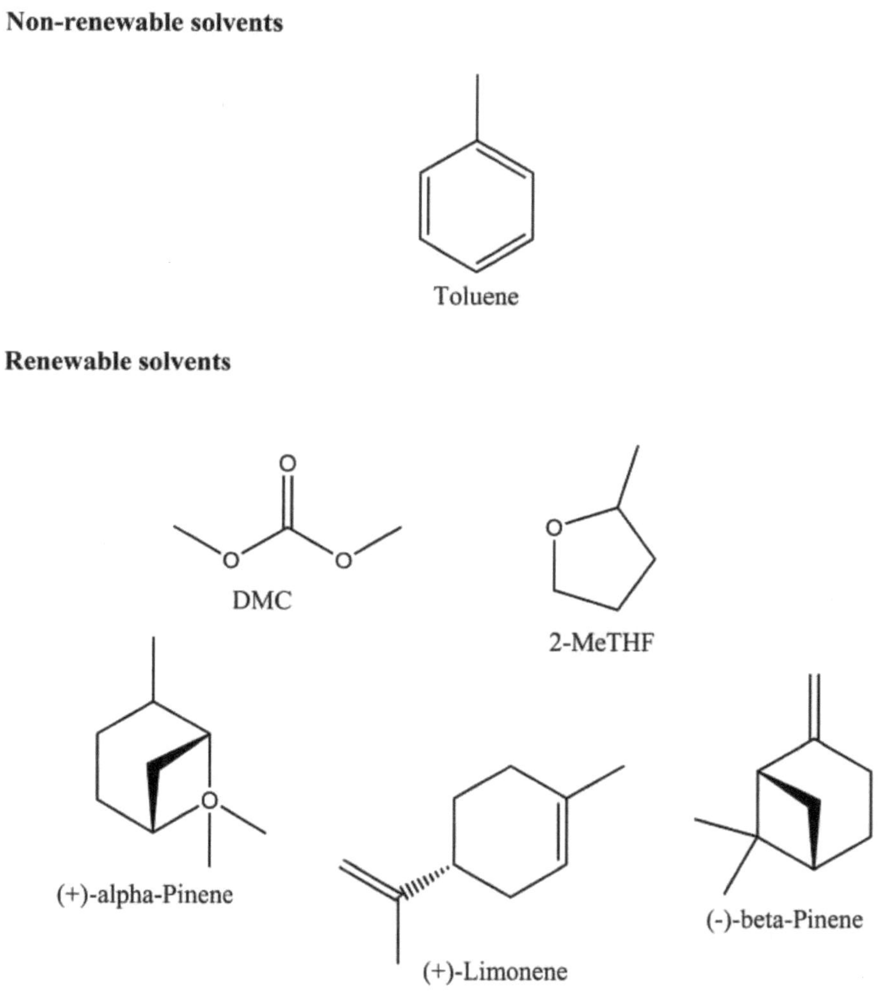

Non-renewable solvents

Toluene

Renewable solvents

DMC

2-MeTHF

(+)-alpha-Pinene

(+)-Limonene

(-)-beta-Pinene

Fig. (2). Renewable and non-renewable solvents.

The Table 2 Describes the Fig. (2) Renewable and Non-renewable solvents.

Table 2. Renewable and non-renewable solvents.

SOLVENT	B. P	D.C	F. P	H.M
toluene	111	2.38	4	Hyb_1
MeTHFa	80	6.97	-10	Hyb_2
(+)-limonene	176	2.37	45	Hyb_3

Types of Green Solvents Green Grafting: Innovations in Polymer Functionalization (Part 1) 173

(Table 2) cont.....

SOLVENT	B. P	D.C	F. P	H.M
DMC*b*	90	3.09	17	Hyb_4
(−)-β-pinene	165	2.50	36	Hyb_5
(+)-α-pinene	157	2.18	33	Hyb_6

The polymer grafting procedures and the reactions they entail are influenced by various parameters, including the chemical composition of the system's constituents (solvent, initiator, monomer, and backbone) and their interactions. Considerations for polymer grafting include temperature and the use of additives, among many other considerations [7]. The synthetic routes of graft polymerization are given below in Fig. (3), with the activators including a variety of exciting and versatile polymer modification routes.

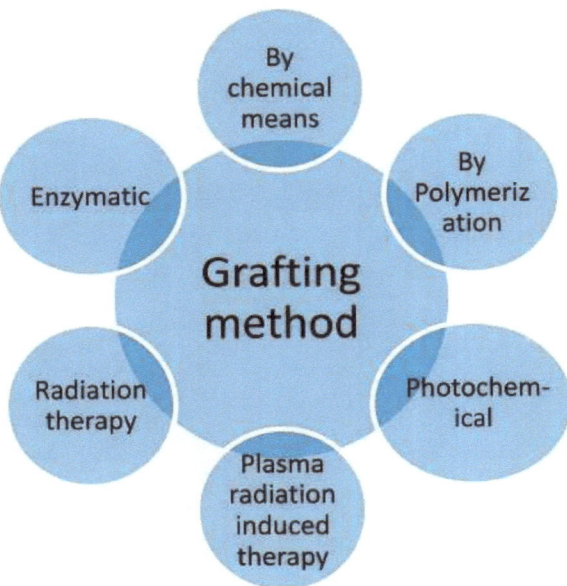

Fig. (3). Synthetic routes of graft polymerization.

GRAFTING TECHNIQUES

Grafting Through Chemical

There are two primary methods of chemical grafting: ionic initiator grafting and free radical grafting. Because it outlines the path of the grafting method, the initiator's role is crucial.

Free-Radical Grafting

These chemical interactions result in the production of free radicals, which are then transmitted to the substrate where they combine with monomers to form graft copolymers. While numerous polymerization methods have been utilized for polymer grafting, the most efficient ones are the free radical processes (FRP, RDRP, and REX), since they can be adjusted to operate with various chemical groups and can withstand impurities. By exposing macromolecules to radiation, homolytic fission can be produced, yielding the polymer. The lifetime of free radicals is determined by the activity of the polymer backbone, which is an essential element in the creation of stable graft polymers. The characteristics that influence polymer grafting by FRP and other reaction methods include the backbone, monomer, initiator, and the chemical composition of the solvent as well as their interactions [7].

$Fe^{2+} + H_2O_2 \rightarrow Fe^{3+} + OH^- + O$

$Fe^{2+} + ^-O_3S - OO - SO$

While many polymerization techniques have been used to achieve polymer grafting, free radical techniques (e.g., FRP, RDRP, and REX) are typically the most effective method because of their versatility in working with different chemical groups and their tolerance to impurities. This process still yields around 60% of all available polymers [8].

Free radicals produced by the initiators, also referred to as polymerization initiators, are conveyed to the substrate and react with the monomer to form graft copolymers. Free radicals can start a polymerization chain reaction by reacting with a monomer to create a new radical [9].

Benzoyl Peroxide (BPO) and 2,2'-azo-bis-isobutyryl nitriles are two common initiators (AIBN) for the free radical-based polymerization. While initiators help to initiate the reaction, they are occasionally mislabeled as catalysts because catalysts continue to function even after the reaction has ended. Any of the following substituents, C_6H_5, Cl, Br, $OCOCH_3$, COOR, or H, can be X in this instance. The workings also use di-substituted monomers, such as vinylidene chloride and methyl methacrylate.

Click Chemistry in Polymerization and Polymer Modification

This section lists the many types of click reactions along with their benefits and drawbacks to help the readers choose the best ones. SuFEx, Diels-Alder, thiol-X reactions, carbonyl-based addition, and Azide-Alkyne Cycloaddition (AAC) are the most often used click reactions for polymerization and modification. More specifically, advancements in click chemistry enable the synthesis of copolymers with distinctive topologies, such as heterolayer dendrimers, dendron-polymer conjugates, and nanoconstructs, as well as conjugation with a greater range of biopolymers and living systems in complex conditions. It has also been found that linear polymers can graft functional cyclic backbone polymers and cyclic graft polymers [10].

Three main parts comprise dendrimer 3D globular topological polymers: a) an exterior group; b) branched cells; and c) a focal point [11]. They have many internal voids, a branching structure, and a higher degree of geometrical symmetry, which can be achieved through careful processing. A few examples of architecture-induced events in dendrimers include branched cell-symmetrical impact, architectural amplification, and nanoscale features [12, 13].

The five often utilized dendrimer groups are poly (lysine), poly(amidoamine), poly(propylenimine), phosphorus, and carbosilane. The dendrimers also include a range of structural components [14].

Dendrimer synthesis employing click chemistry has been adopted with success when convergent and divergent growth techniques are employed. AB2-monomers, with acetylenes making up the B2-functionality and a chloromethyl as the A-functionality, were subjected to the first convergent approach. With the click-reaction enabled by the replacement of the chloride atom and the subsequent coupling step, the azoide moiety was inserted with ease. In the end, the fourth-generation dendrons were connected in 1,2, and 3 triazole and 1-4 di to construct a variety of core molecules that enhanced the parent polymer's characteristics [15].

The primary contribution of this work was demonstrating the use of click chemistry and the degree of efficacy that could be achieved in a synthetic process that is usually difficult. Joralemon *et al.* [16] discussed the divergent method of creating azide- or acetylene-terminated dendrimers using click chemistry for Frechet-type dendrimers. Previously, the divergent method of poly(thioether) dendrimers was explored by the use of thiol-ene click chemistry. Even better for the environment, the synthesis was finished at mild reaction conditions and without the use of a metal catalyst [17].

Other Methods

Bromination. This process was used to make sure that replacement occurred preferentially in the unsaturated region of the molecule, since polystyrene bromination was produced by thermally polymerizing styrene at 70°C and was thought to have little to no branching. The method's disadvantage is that, if growing chains were stopped, cross-linking and the formation of an unmanageable network would result during polymerization; primary disproportionation is necessary to prevent this from happening. The rate of reaction and other kinetic factors, which are often present in the polymerization of styrene, regulate the chain length and branching [18, 19].

Esterification

In the presence of acid, the nucleophilic alcohol and the electron-deficient carbonyl combine to create an ester bond during the esterification process. These bonds provide the polymer with a new property without affecting its inherent characteristics. According to reports, cellulose may be esterified with nitric acid in the presence of acetic, phosphoric, or sulfuric acids to generate cellulose nitrate, which is advantageous. Furthermore, it has been discovered that cellulose esters of economic significance include cellulose acetate, cellulose acetate propionate, and cellulose acetate butyrate [20].

Chain Transfer

Radicals are formed from the cellulose backbone when the polymerization process in cellulose chain transfer is halted by the removal of hydrogen atoms from the cellulose molecule. In cellulose grafting, where the chain transfer reaction is often used, this reaction termination is a concern. It is common practice to include thiol or xanthate ester groups in radical transfer graft copolymerization processes, as this approach does not yield greater graft yields. But before grafting, materials with stronger chain transfer activity, such as thiol groups, can be added to the cellulose molecules. The activating species are produced on the swollen cellulose substrate backbone during the initiation of potassium persulfate. The cotton cellulose was modified by Ghosh and Das through the grafting of acrylic acid, whereas potassium persulfate [21].

Polymer chains are attached to a surface or backbone polymer during the grafting process. The qualities of the grafted material, the process's environmental effect, and the grafting efficiency can all be strongly influenced by the type of solvent utilized. These are a few typical solvents that are employed in the grafting procedure: Green solvents are solvents that are safe for human health and the environment. Their application in the grafting procedure is consistent with green chemistry principles, which minimize or do away with dangerous materials. The following is a list of frequently used green solvents and their importance in grafting procedures:

COMMON GREEN SOLVENTS

- Water
- Supercritical CO_2
- Ethanol and other biobased alcohols
- Ionic liquids
- Deep Eutectic Solvents (DES)

Significance and Key Principles

Water

Significance: Water is the ultimate green solvent due to its non-toxic and non-flammable nature, as well as its abundance. It is widely used in grafting hydrophilic polymers.

Principles:

Safety: Non-toxic to humans and the environment.

Availability: Easily accessible and cost-effective [22].

Supercritical CO_2.

Significance: Supercritical CO_2 is a sustainable option due to its low toxicity and tunable properties (by adjusting pressure and temperature).

Principles:

Reduction of hazardous chemicals: Reduces the need for organic solvents.

Energy efficiency: Low critical temperature and pressure [22].

Ethanol and Other Biobased Alcohols

Significance: These solvents are renewable and biodegradable, making them a preferable alternative to petroleum-based solvents.

Principles

Renewability: Derived from renewable resources like corn or sugarcane.

Biodegradability: Easily broken down in the environment [22].

Ionic Liquids

Significance: Ionic liquids have unique solvating properties and negligible vapor pressure, reducing VOC emissions.

Principles

Safety: Non-volatile and less flammable.

Efficiency: Can dissolve a wide range of compounds, facilitating reactions that are difficult in traditional solvents [23].

Deep Eutectic Solvents (DES)

Significance: DES are formed from natural compounds like choline chloride and urea, offering biodegradable and non-toxic alternatives.

Principles

Design for degradation: Biodegradable and environmentally benign.

Reduction of hazardous substances: Formed from non-toxic, natural components [24].

Key Principles of Green Chemistry in Solvent Selection

- Prevent Waste: Choose solvents that minimize waste production.
- Design Safer Chemicals: Opt for non-toxic, non-hazardous solvents.
- Use Renewable Feedstocks: Prefer solvents derived from renewable resources.
- Increase Energy Efficiency: Select solvents that enable reactions under mild conditions, reducing energy consumption.
- Reduce Derivatives: Avoid unnecessary use of derivatives and protecting groups that require additional solvents.
- Design for Degradation: Choose solvents that break down into innocuous products.
- Inherently Safer Chemistry: Select solvents that reduce the risk of accidents, explosions, and exposure to hazardous chemicals [22, 25].

Applications and Benefits

Improved Safety: Green solvents reduce health risks for workers and environmental contamination.

Compliance: Helps meet regulatory standards and reduce the need for costly disposal processes.

Sustainability: Supports sustainable industrial practices and reduces the carbon footprint.

Economic Advantages: Often lower in cost due to reduced need for handling and disposal of hazardous waste.

Using green solvents in the grafting process is a step towards sustainable and responsible research and industrial practices, emphasizing safety, efficiency, and environmental stewardship [25, 26].

APPLICATIONS OF THE GRAFTING PROCESS IN POLYMER SCIENCE

Surface Modification

- **Biocompatibility**: Grafting polymers onto surfaces like medical implants can improve biocompatibility by reducing the immune response and enhancing integration with biological tissues [27].
- **Anti-fouling Surfaces**: Grafted polymer brushes are used to create surfaces resistant to fouling by proteins, bacteria, and other bioorganisms, which are essential for medical devices, marine applications, and biosensors [28].

Material Strengthening

- **Mechanical Properties**: Grafting polymer chains onto a backbone can enhance the mechanical strength, elasticity, and toughness of materials, making them more suitable for industrial applications like packaging and automotive parts [29].
- **Thermal Stability**: Grafted polymers can provide improved thermal stability, allowing materials to withstand higher temperatures without degradation [30].

Functionalization

- **Chemical Functionalization**: Grafting allows for the introduction of functional groups onto a polymer surface, enabling specific interactions with other chemicals, useful in sensors, catalysis, and drug delivery systems.
- **Responsive Materials**: Grafted polymers can be designed to respond to external stimuli like pH, temperature, or light, leading to applications in smart coatings, drug delivery, and tissue engineering [31].

Adhesion Improvement

- **Adhesives and Coatings**: Grafting can improve the adhesion properties of materials, making them more suitable for use in adhesives, coatings, and composite materials where strong bonding is required [32].

Nanotechnology

- **Nanocomposites**: Grafting polymers onto nanoparticles can enhance the dispersion and interaction of nanoparticles within a polymer matrix, leading to advanced nanocomposites with unique electrical, thermal, and mechanical properties [33].
- **Drug Delivery**: Grafted polymers are used to create nanocarriers for targeted drug delivery, thereby improving the efficacy and minimizing the side effects of treatments [34].

CHALLENGES IN THE GRAFTING PROCESS

Control of Grafting Density

- Achieving uniform and controlled grafting density is challenging, as it can affect the material properties significantly. High density can lead to steric hindrance, while low density may result in inadequate functionalization.

Reproducibility

- Ensuring consistent grafting across different batches can be difficult, especially on industrial scales, leading to variability in material performance.

Complexity of Synthesis

- The grafting process often requires complex chemical reactions, which can be time-consuming and expensive. The need for specific catalysts, solvents, and reaction conditions adds to the complexity of the process.

Surface Characterization

- Characterizing the grafted polymers on a surface or backbone is challenging due to the thinness of the grafted layer and the need for sophisticated analytical techniques like X-ray Photoelectron Spectroscopy (XPS) or Atomic Force Microscopy (AFM).

Environmental and Health Concerns

- The use of certain chemicals and solvents in grafting processes can pose environmental and health risks. Finding safer alternatives or more sustainable methods is an ongoing challenge.

Scalability

- Scaling up the grafting process from laboratory to industrial production can be challenging, particularly in maintaining the desired properties and performance of the grafted materials.

Cost

- The grafting process can be expensive, especially when high-purity materials, specialized equipment, or complex procedures are required, limiting its use in cost-sensitive applications.

Degradation and Stability

- The long-term stability of grafted polymers can be an issue, especially under harsh environmental conditions. Grafted chains may degrade over time, leading to a loss of functionality.

Compatibility with Base Polymer

- Ensuring that the grafted polymer is compatible with the base polymer is crucial to avoid phase separation or poor interfacial adhesion, which can weaken the material.

These applications and challenges highlight the importance of the grafting process in advancing materials science, while also pointing to areas where further research and innovation are needed [35].

CASE STUDY: TYPES OF GREEN SOLVENTS IN THE GRAFTING PROCESS

Introduction

The grafting process, essential in modifying polymers' surface properties, traditionally employs organic solvents, which often have adverse environmental impacts. This case study focuses on the use of green solvents in the grafting process, highlighting their types, benefits, and applications.

Case Study: Grafting of Polypropylene Using Green Solvents

Background

Polypropylene (PP) is widely used in various industries due to its excellent mechanical properties and chemical resistance. However, its hydrophobic nature limits certain applications. Grafting hydrophilic monomers onto PP can enhance its surface properties. The challenge lies in finding suitable solvents that are environmentally friendly and effective.

Objective

To evaluate the performance and environmental impact of different green solvents in the grafting process of polypropylene with Acrylic Acid (AA).

Materials and Methods

1. **Materials**
   - Polypropylene (PP)
   - Acrylic acid (AA)
   - Green solvents: Ionic liquids (ILs), Supercritical carbon dioxide (scCO2), Deep Eutectic Solvents (DES), and Water
   - Traditional solvent for comparison: Toluene
2. **Methods.**
   - Grafting Procedure: The PP samples were pretreated and immersed in a solution containing AA and a green solvent. The mixture was then subjected to irradiation to initiate the grafting process.
   - Characterization: The grafted PP was characterized using Fourier Transform Infrared Spectroscopy (FTIR) to confirm the presence of grafted AA. Contact angle measurements were used to assess changes in hydrophilicity.
   - Environmental Impact Assessment: The solvents were evaluated based on their toxicity, biodegradability, and renewability [36].

Results.

- **Ionic Liquids (ILs)**

Performance: ILs showed high grafting efficiency and uniform graft distribution.

Environmental Impact: ILs are non-volatile and can be recycled, reducing environmental pollution. However, their synthesis can be complex and costly.

- **Supercritical Carbon Dioxide (scCO2)**

Performance: scCO2 achieved moderate grafting efficiency. The process conditions (high pressure) required specialized equipment.

Environmental Impact: scCO2 is non-toxic, non-flammable, and leaves no solvent residue. It is considered one of the most environmentally friendly solvents.

- **Deep Eutectic Solvents (DES)**

Performance: DES provided good grafting efficiency and was easy to handle.

Environmental Impact: DES are biodegradable, non-toxic, and can be prepared from natural compounds, making them highly sustainable.

- **Water**

Performance: Water-based grafting showed lower efficiency compared to ILs and DES, but was straightforward and cost-effective.

Environmental Impact: Water is the most benign solvent, with no toxicity or environmental hazards.

- **Comparison with Toluene**

Performance: Toluene achieved high grafting efficiency but posed significant health and environmental risks.

Environmental Impact: Toluene is toxic, volatile, and contributes to air and water pollution.

DISCUSSION

Efficiency vs. Environmental Impact: While ILs and scCO2 provided high grafting efficiency, their use is limited by cost and the need for specialized equipment. DES and water offer a balance between efficiency and environmental benefits.

Industrial Applicability: DES and water-based processes are more scalable and cost-effective for industrial applications, despite slightly lower grafting efficiencies.

CONCLUSION

The study demonstrates that green solvents like DES and water can be effectively used in the grafting process of polypropylene, providing a more sustainable alternative to traditional organic solvents. By adopting these green solvents, the environmental footprint of the grafting process can be significantly reduced without compromising the quality of the modified polymers.

RECOMMENDATIONS

- Further research should focus on optimizing grafting conditions using DES and water to enhance efficiency.
- Industries should consider investing in green solvent technologies to promote sustainable practices in polymer modification.

This case study illustrates the potential of green solvents in improving the sustainability of grafting processes and encourages their adoption in industrial applications.

REFERENCES

[1] Nicole L, Boissière C, Grosso D, Quach A, Sanchez C. Mesostructured hybrid organic–inorganic thin films. J Mater Chem 2005; 15(35-36): 3598-627.
[http://dx.doi.org/10.1039/b506072a]

[2] Enache DF, Vasile E, Simonescu CM, et al. Schiff base-functionalized mesoporous silicas (MCM-41, HMS) as Pb(II) adsorbents. RSC Advances 2018; 8(1): 176-89.
[http://dx.doi.org/10.1039/C7RA12310H]

[3] Lin YW, Lee WH, Lin KL, Kuo BY. Synthesis and grafted NH2-Al/MCM-41 with amine functional groups as humidity control material from silicon carbide sludge and granite sludge. Processes (Basel) 2021; 9(12): 2107.
[http://dx.doi.org/10.3390/pr9122107]

[4] Grandjean P, Landrigan PJ. Neurobehavioural effects of developmental toxicity. Lancet Neurol 2014; 13(3): 330-8.
[http://dx.doi.org/10.1016/S1474-4422(13)70278-3] [PMID: 24556010]

[5] Williams M. The Merck Index: An Encyclopedia of Chemicals, Drugs, and Biologicals, 15th Edition. Edited by O'Neil MJ. Cambridge (UK): Royal Society of Chemistry; 2013. Reviewed in: Drug Dev Res. 2013; 74(5): 339.

[6] O'Neil MJ, Ed. The Merck index: an encyclopedia of chemicals, drugs, and biologicals. RSC Publishing 2013.

[7] Sosnik A, Gotelli G, Abraham GA. Microwave-assisted polymer synthesis (MAPS) as a tool in biomaterials science: How new and how powerful. Prog Polym Sci 2011; 36(8): 1050-78.
[http://dx.doi.org/10.1016/j.progpolymsci.2010.12.001]

[8] Šebenik A. Living free-radical block copolymerization using thio-iniferters. Prog Polym Sci 1998; 23(5): 875-917.
[http://dx.doi.org/10.1016/S0079-6700(98)00001-X]

[9] Sumerlin BS, Tsarevsky NV, Louche G, Lee RY, Matyjaszewski K. Highly efficient "click" functionalization of poly (3-azidopropyl methacrylate) prepared by ATRP. Macromolecules 2005;

38(18): 7540-5.
[http://dx.doi.org/10.1021/ma0511245]

[10] Opsteen JA, van Hest JCM. Modular synthesis of block copolymers via cycloaddition of terminal azide and alkyne functionalized polymers. Chem Commun (Camb) 2005; 1(1): 57-9.
[http://dx.doi.org/10.1039/b412930j] [PMID: 15614371]

[11] Li N, Cai H, Jiang L, *et al.* Enzyme-sensitive and amphiphilic PEGylated dendrimer-paclitaxel prodrug-based nanoparticles for enhanced stability and anticancer efficacy. ACS Appl Mater Interfaces 2017; 9(8): 6865-77.
[http://dx.doi.org/10.1021/acsami.6b15505] [PMID: 28112512]

[12] Zeng Z, Qi D, Yang L, *et al.* Stimuli-responsive self-assembled dendrimers for oral protein delivery. J Control Release 2019; 315: 206-13.
[http://dx.doi.org/10.1016/j.jconrel.2019.10.049] [PMID: 31672623]

[13] Mignani S, Shi X, Rodrigues J, *et al.* Dendrimers toward translational nanotherapeutics: Concise key step analysis. Bioconjugate Chem 2020 Sep; 31(9): 2060-71.
[http://dx.doi.org/10.1021/acs.bioconjchem.0c00395]

[14] Dzmitruk V, Apartsin E, Ihnatsyeu-Kachan A, Abashkin V, Shcharbin D, Bryszewska M. Dendrimers show promise for siRNA and microRNA therapeutics. Pharmaceutics 2018; 10(3): 126.
[http://dx.doi.org/10.3390/pharmaceutics10030126] [PMID: 30096839]

[15] Wu P, Feldman AK, Nugent AK, *et al.* Titelbild: Efficiency and fidelity in a click-chemistry route to triazole dendrimers by the copper (i)-Catalyzed ligation of azides and alkynes (angew. Chem. 30/2004). Angew Chem 2004; 116(30): 3951.
[http://dx.doi.org/10.1002/ange.200490100]

[16] Joralemon MJ, O'Reilly RK, Matson JB, Nugent AK, Hawker CJ, Wooley KL. Dendrimers clicked together divergently. Macromolecules 2005; 38(13): 5436-43.
[http://dx.doi.org/10.1021/ma050302r]

[17] Killops KL, Campos LM, Hawker CJ. Robust, efficient, and orthogonal synthesis of dendrimers via thiol-ene "click" chemistry. J Am Chem Soc 2008; 130(15): 5062-4.
[http://dx.doi.org/10.1021/ja8006325] [PMID: 18355008]

[18] Carlin RB, Shakespeare NE. The polymerization of p-chlorostyrene in the presence of polymethylacrylate. J Am Chem Soc 1946; 68(5): 876-8.
[http://dx.doi.org/10.1021/ja01209a052] [PMID: 21024896]

[19] Jones MH. Graft copolymers of styrene and methyl methacrylate: Part I: Synthesis. Can J Chem 1956; 34(7): 948-56.
[http://dx.doi.org/10.1139/v56-126]

[20] Klemm D, Heinze T, Philipp B, Wagenknecht W. New approaches to advanced polymers by selective cellulose functionalization. Acta Polym 1997; 48(8): 277-97.
[http://dx.doi.org/10.1002/actp.1997.010480801]

[21] Kumar D, Pandey J, Raj V, Kumar P. A Review on the Modification of Polysaccharide Through Graft Copolymerization for Various Potential Applications. Open Med Chem J 2017; 11(1): 109-26.
[http://dx.doi.org/10.2174/1874104501711010109] [PMID: 29151987]

[22] Cherrafi A, Garza-Reyes JA, Belhadi A, Kamble SS, Elbaz J. A readiness self-assessment model for implementing green lean initiatives. J Clean Prod 2021; 309: 127401.
[http://dx.doi.org/10.1016/j.jclepro.2021.127401]

[23] Plechkova NV, Seddon KR. Applications of ionic liquids in the chemical industry. Chem Soc Rev 2008; 37(1): 123-50.
[http://dx.doi.org/10.1039/B006677J] [PMID: 18197338]

[24] Smith EL, Abbott AP, Ryder KS. Deep eutectic solvents (DESs) and their applications. Chem Rev 2014; 114(21): 11060-82.

[http://dx.doi.org/10.1021/cr300162p] [PMID: 25300631]

[25] Anastas PT, Warner JC. Green chemistry: theory and practice. Oxford university press; 2000 May 25.
[http://dx.doi.org/10.1093/oso/9780198506980.001.0001]

[26] Jessop PG. Searching for green solvents. Green Chem 2011; 13(6): 1391-8.
[http://dx.doi.org/10.1039/c0gc00797h]

[27] Ratner BD. The biocompatibility of implant materials. InHost response to biomaterials 2015 Jan 1 (pp. 37-51). Academic Press.
[http://dx.doi.org/10.1016/B978-0-12-800196-7.00003-7]

[28] Genzer J, Efimenko K. Recent developments in superhydrophobic surfaces and their relevance to marine fouling: a review. Biofouling 2006; 22(5): 339-60.
[http://dx.doi.org/10.1080/08927010600980223] [PMID: 17110357]

[29] Naghashpour A, Van Hoa S. Motion of carbon nanotubes based polymer nanocomposites subjected to multi-directional deformation. Polym Test 2016; 55: 109-14.
[http://dx.doi.org/10.1016/j.polymertesting.2016.08.017]

[30] Kumar A, Sharma K, Dixit AR. A review of the mechanical and thermal properties of graphene and its hybrid polymer nanocomposites for structural applications. J Mater Sci 2019; 54(8): 5992-6026.
[http://dx.doi.org/10.1007/s10853-018-03244-3]

[31] Hua Ti. Polyvinylpyrrolidone, graphene oxide and their composites as potential fluorescence sensing materials for nitrate and nitrite ions [Phd dissertation]. Johor Bahru (MY): Universiti Teknologi Malaysia; 2017.

[32] Park SJ, Lee SY, Jin FL. Surface modification of carbon nanotubes for high-performance polymer composites. In: Lee S, Kim Y, Park SJ, Lee Y, eds. Handbook of Polymer Nanocomposites. Processing, Performance and Application: Volume B: Carbon Nanotube Based Polymer Composites. Cham: Springer International Publishing; 2015. p. 13-59.
[http://dx.doi.org/10.1007/978-3-642-45229-1_34]

[33] Koo JH. Polymer Nanocomposites: Processing, Characterization, and Applications. New York: McGraw-Hill; 2006.

[34] Yang T, Hu Y, Wang C, Binks BP. Fabrication of hierarchical macroporous biocompatible scaffolds by combining pickering high internal phase emulsion templates with three-dimensional printing. ACS Appl Mater Interfaces 2017; 9(27): 22950-8.
[http://dx.doi.org/10.1021/acsami.7b05012] [PMID: 28636315]

[35] Weerasinghe UA, Wu T, Chee PL, *et al.* Deep eutectic solvents towards green polymeric materials. Green Chem 2024; 26(15): 8497-527.
[http://dx.doi.org/10.1039/D4GC00532E]

[36] Wang D, Wang J, He S, Yan Y, Zhang J, Dong J. Efficient approach to produce functional polypropylene via solvent assisted solid-phase free radical grafting of multi-monomers. Appl Petrochem Res 2021; 11(1): 99-111.
[http://dx.doi.org/10.1007/s13203-020-00261-9]

CHAPTER 6

Radiation-induced Green Grafting: Mechanism, Applications and Benefits

Shweta Bhandari[1,*], **Sumeet Dwivedi**[2], **Vishal Garg**[3] and **Idress Hamad Attitalla**[4]

[1] *Department of Pharmacy, Sumandeep Vidyapeeth (Deemed to be University), Vadodara, Gujarat 391760, India*

[2] *Department of Pharmacognosy, Acropolis Institute of Pharmaceutical Education & Research, Indore, Madhya Pradesh 453771, India*

[3] *Department of Pharmaceutics, Jaipur School of Pharmacy, Maharaj Vinayak Global University, Jaipur, Rajasthan 302028, India*

[4] *Department of Microbiology, Faculty of Science, Omar Al-Mukhtar University, Al-Bayda, Libya*

Abstract: The method for the creation of surface-grafting polymeric materials is gaining recognition because it makes it possible to create novel materials from well-known, commercially available polymers with desirable bulk properties like elasticity, permeability, and thermal stability, combined with advantageous, newly tailored surface properties like adhesion, biomimicry, and biocompatibility. Since it produces radicals on most substrates, ionizing radiation is one of the most effective techniques for creating graft copolymers. This process involves the use of radiation, such as UV, plasma, Electron Beam (EB), and γ-rays, to modify polymer substrates. The development of RIGG in pharmaceuticals focuses on the covalent immobilization of biocides to various polymer surfaces. Grafting can now be done under control to produce surfaces with specific and well-defined features. The application of radiation has entered a new era of grafting with the development of living free-radical polymerization techniques. The technique is applied to drug delivery systems, where grafted polymers provide controlled release profiles and targeted delivery, improving therapeutic efficacy and patient compliance. In biomedical applications, grafted polymers are utilized to create biocompatible surfaces for medical implants, ensuring reduced risk of infection and improved integration with biological tissues. Radiation-grafted wound dressings are developed for their enhanced antimicrobial activity and accelerated healing properties. The chapter delves into the scientific principles underlying RIGG, detailing the mechanisms by which radiation induces grafting and the factors influencing the process. The chapter then delves into the specific applications of radiation-induced green grafting in the pharmaceutical and healthcare industries.

---

* **Corresponding author Shweta Bhandari:** Department of Pharmacy, Sumandeep Vidyapeeth (Deemed to be University), Vadodara, Gujarat 391760, India; E-mail: bhandarishweta257@gmail.com

Kuldeep Vinchurkar, Satish Polshettiwar, Nilesh Mahajan &Yogeshwar Bachhav (Eds.)
All rights reserved-© 2026 Bentham Science Publishers

Keywords: Antimicrobial surfaces, Biomedical device, Green grafting, Healthcare products, Radiation-induced green grafting.

INTRODUCTION

Copolymers are polymeric materials that include two or more units of two distinct polymeric species bonded together in a specific configuration. Random, block, and graft copolymers are the three different kinds of copolymers. The two polymeric species are arranged differently in them. When the small structural units of two polymeric species appear in a random linear sequence, the copolymer is called a random copolymer. A linear copolymer including one or more continuous, lengthy sequences of each polymeric species is called a block copolymer. Graft copolymers are branching copolymers that have one or more side chains of one polymeric species linked to the "backbone" of another polymeric species. The surfaces of polymer forms can be functionalized in such a way that makes them suitable for a range of uses in the industrial, environmental, and biological domains through radiation-induced grafting [1]. Compared to other grafting processes, radiation-induced radical polymerization is widely applicable to a variety of polymeric material components and forms.

Graft polymerization with radiation-induced was thought of as a prospective method of modifying the surface of polymeric materials without changing the bulk material by chemically joining polar/nonpolar monomers with functional groups, such as H, -OH, -COOH, -NH2, -SO3, -OR, -R, and their derivatives. It preserves the physical characteristics and shape of the trunk polymer while allowing for the creation and development of new functionalities for polymer materials [2]. In addition, no unnecessary chemicals or catalysts are used when using RIG polymerization to add functional groups to polymers. Since this procedure does not require the use of a catalyst or other dangerous chemicals that could damage human skin or the environment, it can be considered both economical and environmentally friendly [3]. Graft polymerization can be started using four main methods: ultraviolet treatment, plasma treatment, electron beams, and gamma radiation.

IMPORTANCE OF RADIATION-INDUCED GREEN GRAFTING IN PHARMACEUTICALS

Sustainability of the Environment

The use of radiation, like gamma rays or electron beams, for the grafting process limited the use of chemical initiators, which led to the production of less hazardous waste. Chemicals and solvents used in traditional grafting methods have been proven to be harmful to the health of people and the environment.

Thus, grafting using the radiation technique aligns with the principle of green chemistry and proves to be advantageous.

Accuracy and Regulation

Accuracy is achieved through the radiation grafting polymerization technique as it controls variable factors and grafting parameters, such as dose, time of exposure, and rate, which precisely define grafting. Precise grafting is an important aspect in terms of pharmaceutical preparations, as it directly determines the characteristics of the formulation where targeted drug delivery must be achieved.

Improved Properties of the Material

Grafting with radiation improves the properties of materials, such as mechanical strength, hydrophilicity, biocompatibility, drug loading capacity, and swelling behavior for the preparation of different pharmaceutical formulations. The incorporation of functional groups aids in achieving improved characteristics, which makes it more feasible for use in pharmaceutical applications [4, 5].

Flexibility and Wide Range of Use

Radiation grafting methods have multiple applications and adaptability. A variety of polymers, ranging from natural to synthetic, like chitosan, cellulose, polyethylene, and polypropylene, can be modified using the radiation-induced grafting method, which increases their usage to a wide range of pharmaceutical applications, like the development of improved drug delivery systems, wound dressing, and scaffolds for tissue engineering.

Better Systems for Delivering Drugs

Targeted release and controlled release of drugs are the major concerns of pharmaceutical preparations, where safety and efficacy with effective drug therapy shall be delivered, such as in cancer therapy. With radiation grafting, cutting-edge medication can be developed as drug carriers that react to specific physiological variables, like temperature and pH, through the incorporation of specific functional groups.

Less Immunogenicity and Biocompatibility

To design an implant or medical device, the issue of biocompatibility should be considered. Radiation grafting is capable of enhancing biocompatibility by allowing for the incorporation of suitable polymer surfaces grafted until they integrate with surrounding tissues, and an immune system response is lower.

Antimicrobial Characteristics

Grafting antimicrobial compounds on the surface of polymers *via* RIGG can substantially reduce the risk of infections arising from medical devices and implants. Grafting polymers with, for example, silver nanoparticles or other antimicrobial materials allows for the increase in the safety and longevity of medical devices by creating surfaces resistant to bacterial colonization and biofilm formation [6, 7].

TYPES OF RADIATION SOURCES

RIG approaches can be employed in four standard ways for transforming the polymer. There are basically four different types of radiation:

- Gamma
- Electron beam
- Plasma
- UV

Gamma

The most efficient method of polymer grafting, which is often utilized in the functionalization process of solid waste, relies on a gamma radiation-induced technique and is preferable. It uses cobalt-60 as the radiation source. Consequently, gamma radiation-induced grafting could have the ability to produce homogenous and immediate active sites for radical generation *in situ*; thus, leading to an efficient graft yield with no deficiency of catalyst contamination or hazardous initiator. Conversely, in the biomedical domain, gamma is chosen. But because gamma radiation uses a radioactive isotope, it can be very dangerous to one's health. In addition, the radioactive emissions from these types of ores can be harmful to both the environment and human health, making them a highly challenging waste type.

Electron Beam Radiation

The second-greatest method to modify polymers *via* radiation is electron beam radiation. Free radicals or active sites on a polymeric surface can be generated by means of electron beam radiation, and the subsequently irradiated material could react with monomers, forming a graft copolymer in terms of branching type in the main chain. Due to the high grafting efficiency resulting from the copolymerization of polystyrene with natural rubber, radiation by gamma rays has been considered better than using an electron beam, even though both produce more stable results [8].

Plasma Irradiation

It has become an interesting surface modification strategy to provide the material of interest with desired functional groups in solventless, eco-friendly, and rapid processes. It has been reported that Carbon Nanotubes (CNTs) were grafted *via* a plasma-induced grafting process. The reactive sidewalls of the CNTs were observed, and other approaches for incorporating defects into the host CNT framework without compromising them were straightforward. Plasma irradiation is also capable of leading to the formation of some functional groups, like -COOH, at the membrane surface, which can further enhance the hydrophilicity of the membranes. Finally, air exposure for the active membrane during the plasma grafting polymerization process is carried out: a peroxide group was formed on its surface following gas plasma treatment at different energies and exposure times. Monomers with the functional groups carboxyl, amine, and hydroxyl were generated on the membrane surface.

UV

The treatment of polymers, sun exposure-triggered UV emission, has always been considered a key human skin stressor. The three forms of ultraviolet light are distinguished by their spectral range: A (320-400 nm), B (290-320 nm), and C (220-280nm). Of all kinds of UV radiation, UVC has the shortest wavelength and is highly intense; for this reason, it can be fatal. UVB, in turn, is suitable for photodegradation and polymer degradation. Easy operation, a wider industrial application window, and milder reaction conditions are some advantages of UV irradiation grafting techniques. In addition, UV treatment is faster, easier, less expensive, and more effective than other methods [9 - 11].

The advantages and disadvantages of different types of radiation are illustrated in Table **1**.

METHODS:

Here are two main ways to achieve radiation-induced grafting:

- Simultaneous/Mutual Procedure
- Pre-Irradiation

Grafting is often performed in an emulsion or solution with surfactant in low quantity, *e.g.*, Tween 20 and water acting as the reaction medium [14]. The radiation source, the monomer's reactivity, and the polymer that needs to be changed determine the approach to be used. When gamma sources or accelerators are scarce, the pre-irradiation approach comes in very handy. Even after some time has passed for storage, graft polymerization by pre-irradiating a polymer surface (either in air or in vacuum) is initiated.

Table 1. Different sources of radiation and their advantages and disadvantages [12, 13]

S.N.	Method	Source	Species	Advantage	Disadvantage
1	Electron Beam	EB	Electrons	. It can be started using EBs of a variety of energies. . It permits bulk and surface grafting based on the energy of acceleration. . It does not leave any harmful traces behind. . Easy and Quick.	High doses and dosage rates increase the likelihood of mechanical damage to grafted materials. High expense of radiation infrastructure
2.	Gamma	C0-60	Photons	. Permits bulk grafting based on the rate and absorbed dose. Extensively used and ideal for bulk solution simultaneous grafting. . Basic yet not as quick as EB,	Compared to EB, grafting takes a longer time. The origin of Co-60 keeps deteriorating, and thus, the dosage rate slowly decreases and requires modification of the criteria for reactions.
3.	UV	UV	Low-energy photons	Easy to use, affordable, and capable of easily modifying polymer surfaces.	Produces a low grafting level that is limited to the surface and necessitates the use of a photo-initiator. It takes a long time to treat. Too small-scale for big applications.
4.	Plasma	DC glow	Radicals, photons, atoms, molecules, ions, electrons	Easy, pollution-free method to change the surfaces made of polymers without changing their bulk characteristics Permitting functionality combined with portions for entertaining biocide.	The efficiency of surface modification in inhibiting bacterial adherence is challenged by the narrow variety of chemical groups that are available. Not appropriate for extensive use.

Simultaneous/Mutual Procedure

By employing this technique, the polymeric material is subjected to ionizing radiation after being submerged in a monomer solution or pure monomer. Gamma sources are typically used when performing radiation in the air or in an inert atmosphere such as nitrogen. This is the most common and widely used technique for altering the surface of polymeric materials, and it operates efficiently even on substrates that are radiation-sensitive. The following series of equations describes the mechanism of the process:

$$PH \xrightarrow{\gamma} P\bullet + H\bullet \quad (1)$$

$$P\bullet + M \rightarrow PM\bullet \quad (2)$$

$$PM\bullet + nM \rightarrow PM_{n+1}\bullet \quad (3)$$

At the polymeric substrate active sites, PH Eq. (1), ionizing radiation is generated. Graft polymerization is started by the combination of a Monomer molecule (M) and the primary radicals of the polymer backbone (P•) in Eq. (2).

Following the initiation stage, the graft chain grows onto the polymer backbone during the propagation step, following which the monomer is attached to the macroradical centers using Equation (3).

Active sites may form in the solvent, monomer, and polymer backbone of the grafting mixture as a result of exposing the grafting mixture to ionizing radiation. Additional reaction can be seen as radicals other than those found in the backbone of the polymer have the potential to consume monomer, which lowers the degree of grafting. A homopolymer is produced in the mutual approach when the grafting process and monomer homopolymerization occur simultaneously, equations 4-6 represent this [15, 16].

$$M \xrightarrow{\gamma} M\bullet \quad (4)$$

$$M\bullet + nM \rightarrow M_{n+1}\bullet \quad (5)$$

$$M_{n+1}\bullet + M_m\bullet \rightarrow M_{n+m+1}\bullet \quad (6)$$

Homopolymerization raises the viscosity and suppresses the degree of grafting. If produced, it solubilizes in the solvent or monomer in which grafting is done. Monomer finds it more difficult to diffuse to the polymer backbone's reactive areas. Additionally, because of monomer consumption during the homopolymer synthesis process, the availability of monomer for the grafting reaction is less. Inorganic salts can be added to water or another medium used as the solvent for the grafting system to reduce the amount of homopolymer that develops.

Method of Pre-Irradiation

The following steps are included in the pre-irradiation method:

To create active radical sites, the substrate polymer is exposed to radiation in an inert atmosphere or in air. The irradiation of the polymeric substrate starts the monomer reaction. When peroxides are present, heating is used to aid in the propagation of the reaction.

$$PH \xrightarrow{\gamma} P\bullet + H\bullet \tag{7}$$

$$P\bullet + O_2 \longrightarrow POO\bullet \tag{8}$$

$$POO\bullet + PH \longrightarrow POOH + P\bullet \tag{9}$$

$$POOH \xrightarrow{\Delta} PO\bullet + \bullet OH \tag{10}$$

$$PO\bullet + M \longrightarrow POM\bullet \tag{11}$$

$$POM\bullet + nM \longrightarrow POM_{n+1}\bullet \tag{12}$$

Alkyl radicals oxidize to produce peroxide radicals when polymeric substrates are subjected to ionizing radiation in the air (Eqs. **7-8**). These products become hydroperoxides when they come into contact with the polymeric substrate (Eq. **9**). When heated, the hydroperoxides break down into alkoxy radicals (Eq. **10**). In the presence of monomers, radicals produced in this manner can start the graft polymerization (Eqs. **11-12**). Because of the hydroperoxides' durability in this pre-irradiation approach with oxygen present, substrate modification of the graft can be completed after irradiation for a certain period, particularly if the irradiated substrate is held at 0°C or lower. Using radiation-generated non-oxidized reactive radical species is an additional pre-irradiation procedure alternative. In this instance, high dose rates of radiation (such as high-energy electrons) under inert atmospheres or vacuums, followed by grafting shortly after irradiation exposure, are necessary to achieve higher amounts of radicals in the polymer. Fewer homopolymers are produced through the usage of the pre-irradiation approach, which is a significant benefit. Nonetheless, thermal breakdown of hydroperoxide species might result in the production of • OH radicals, which can be used in homopolymerization [16]. Prior to the design process of grafting utilizing a simultaneous/mutual approach, it is vital to compare the G-value (chemical yield by radiation) of the monomer and of the polymeric substrate. The number of product molecules produced or initial molecules altered for every 100 eV of energy received is known as the G-value; the SI unit of radiation-chemical yield is 1 molecule/100 eV = 1.036×10^{-7} mol/J = 0.1036 μmol/J. When G(RP •) of the irradiation polymer is significantly higher than that of the monomer, the reaction continues in favor of graft polymerization. As opposed to homopolymerization, which occurs when the monomer's G(RM •) is greater than that of the polymeric substrate [17]. When the pre-irradiation procedure is used, monomers are not exposed to radiation. Only during the polymeric substrate do radicals arise. As a result, there is not much homopolymer production with this approach.

TYPE OF POLYMER

Commercially available synthetic and natural polymers can undergo surface modification through RIG. Since the grafting degree and final qualities of grafted materials are greatly influenced by the chemical structure and morphology, materials intended for grafting must adhere to specific parameters. It is not advised to employ the high radiation dose that is often used in the pre-irradiation procedure on radiation-degradable polymers. On the other hand, both grafting techniques can modify materials that are resistant to radiation. The degree of grafting is significantly influenced by the number of radicals present in the irradiated polymeric substrate.

Types of Polymers Suitable for Grafting

Fluoropolymers

PTFE and PVDF are examples of this class of polymers, which are characterized by their resistance to chemical attack and high temperature stability. RIGP can modify the surface with functional groups that can enhance its biocompatibility and wettability. These changes have proved useful for applications such as fuel cell membranes or battery separators, where the rate at which ions move through and the permeability need to be increased.

Polyolefins

These thermoplastics, including Polyethylene (PE) and Polypropylene (PP), have wide applications in the packaging industry and in motor vehicle parts. Increasing their surface functionality through RIGP renders them compatible with adhesives and printable. When polar monomers are grafted onto a polyolefin, it enhances liquid spreading on the surface and the adhesion properties thereof. Hence, they contribute to the development of novel coatings or blended materials.

Acrylic Polymers

Acrylic polymers like PAN are often grafted for use in filtration/separation systems. Functional monomers make grafting possible, whereby one can create membranes that selectively adsorb specific ions or molecules from solution. This is important because it helps improve water purification systems and gas separation units.

Cellulosic Polymers

Cellulosic materials, including cellulose and its derivatives, can be modified through RIGP to improve mechanical properties and hydrophilicity. This modification proves particularly useful in biomedical applications. Scaffolds for tissue engineering require enhanced biocompatibility. They also need cell adhesion.

Polyamide and Polyester

Polyamides (such as nylon) and polyesters (like PET) can also be grafted using RIGP. Incorporation of functional groups can enhance thermal stability. It can also boost mechanical strength. Additionally, it improves moisture absorption properties. This is particularly advantageous in textile applications and packaging materials. Performance under varying environmental conditions is critical [18, 19]

Characterization

Radiation-generated paramagnetic species can be identified. Their conversion to subsequent products can be tracked using EPR spectroscopy (electron paramagnetic resonance). The technique is used in predicting the behavior of radicals in radiation-induced grafting. It enables comparison of radical levels in polymers. For example, when ionizing radiation is used to bombard Polystyrene (PS), Polypropylene (PP), and Polyethylene (PE) at the same dose, the concentration of radicals remains steady at room temperature. The radiation yields of radicals (G(RP •)) are in the following order: PE > PP > PS. The degree of radiation-induced grafting of acrylic acid on different polymers was found to have the same relationship under comparable conditions using a mutual method [20].

Advantage Over Conventional Methods

Processes based on radiation have numerous benefits over other traditional techniques. Radiation initiation is not the same as chemical initiation for initiation processes. In radiation processing, the reaction can be started without the need for a catalyst or additives. When using the radiation approach, a free radical reaction is usually started by the backbone polymer absorbing energy. Free radicals are created when an initiator breaks down into smaller pieces and attacks the base polymer, which is how chemical initiation works. When examining the effectiveness of the two methods, it was calculated that at a dose of radiation of 1 rad/s or the use of a chemical initiator, such as 0.01 M benzoyl peroxide, the number of initiating radicals obtained in unit duration is similar. The purity and concentration of the initiators, however, set a limit on chemical initiation. On the other hand, radiation processing allows for a great deal of control over the reaction due to its highly variable radiation dose rate. Additionally, the radiation-induced process is uncontaminated, in contrast to the chemical initiation technique. Chemical initiation frequently results in issues because the initiator overheats locally. However, Temperature is not the determining element in the formation of free radical sites on the polymer during the RI process; rather, it is the polymer matrix's capacity to absorb the high-energy radiation. This indicates the radiation process is not dependent on temperature. To put it another way, its commencement requires zero activation energy. Purity of the processed goods can be preserved as neither a catalyst nor additives are needed. The molecular weight of products can be more precisely controlled using radiation processing. For solid substrates, radiation techniques can also be used for initiation. Moreover, the radiation approach can be used to change the final goods. But despite being incredibly effective at causing chemical reactions, nuclear radiation energy is costly. Installed radiation energy, compared to electrical energy or conventional heat, has a substantially higher cost per unit. Even so, the use of energy by nuclear radiation has proved to be superior in terms of economic viability over other energy sources like heat or electrical energy in a variety of chemical processes [21, 22]

APPLICATIONS IN BIOMEDICINE

Medical technologies are considered one of the greatest intellectual achievements. They are closely linked to advances in polymer research. Polymers, once viewed with skepticism, are now indispensable in medical science. Their use in biomedical applications began with celluloid and bakelite during World War II. These early materials were not bio-stable. They caused strong tissue reactions, leading to their rejection. Over time, more suitable polymers like Polymethylmethacrylate (PMMA) and Poly(Vinyl Alcohol) (PVA) emerged, addressing these limitations [23].

The 1960s marked a significant shift. This shift occurred with the introduction of radiation synthesis and fabrication techniques. They have become crucial in biomedical research. This is particularly true for ensuring bio-compatibility of materials used in contact with blood and tissues. Bio-compatibility is essential to prevent biological disorders. Disorders include changes in blood stability and viscosity. They can affect platelet distribution, clotting, and circulation [24].

Radiation processing is a reliable and straightforward method for modifying polymer properties for bio-compatibility. This process avoids the need for additives and maintains product purity. It also sterilizes materials simultaneously. Research in this field focuses on synthesizing polymers with specific biological functions. It prepares polymeric composites with special biological components *via* radiation immobilization.

Polymers in biomedical applications fall into two categories:
  i. *In vitro,* used in drug forms like solutions, syrups, and blood products
  ii. *In vivo,* temporary materials like sutures and surgical adhesives, and permanent implants, such as vascular grafts and heart valves.

Hydrogels in Medical Applications

Hydrogels are a class of polymers that swell in water without dissolving. They are particularly important due to their non-abrasive feature. The significance of biologically active substances contributes to their wettability and permeability. Hydrogels can be prepared using ionizing radiation, where fresh covalent bonds are created between polymeric chains. It creates insoluble materials ideal for medical use. Hydrogels are three-dimensional networks with a molecular weight of infinite size. They absorb water or solvents in the swollen state. The degree of crosslinking within polymer chains is an indicator of the amount of water absorbed. The more the crosslinking, the less water the hydrogel absorbs. The structure of the hydrogel is often described as a network. The space between crosslinked points is accessible for solute diffusion. Radiation polymerization does not require catalysts or heat. It is a convenient method for hydrogel formation. This process eliminates toxicity and prevents drug decomposition. This makes it ideal for drug delivery systems. Radiation techniques also allow control over the hydrophilicity and porosity of the hydrogel. This is done at low temperatures, enabling the controlled release of drugs [25].

Immobilization Techniques for Hydrogels

The immobilization process by which the mobility of biological species is reduced is crucial for using hydrogels in biomedical applications. This technique fixes or entraps bioactive compounds within the polymer matrix. It provides advantages, such as repeated use and convenient shaping. It also allows controlled release of biological components. Immobilization degree is dependent on the bio-component size. Larger molecules are more challenging to fixate. Radiation immobilization in a supercooled state is a unique technique. It has been found that supercooled acrylates and methacrylates sharply increase in viscosity at low temperatures. This prevents radiation damage to bio-functional components. It also increases porosity in the polymeric matrix. This method differs from typical fixation methods. It enhances the activity of immobilized composites. It does so by concentrating bio-functional components at the interface between frozen water and supercooled monomer phase [26].

Hydrogels in Wound Dressing

Hydrogels, particularly in wound dressings for burns, ulcers, and skin grafts, offer benefits such as prolonged drug release and bacterial barriers. They also provide gas permeability. They enhance adhesion without sticking. PVP poly(acrylamide) agar, polyethylene glycol, and PVA-PVP-based hydrogels have common usage. These hydrogels are superior to traditional bandages. They are especially effective for large wounds, as they effectively relieve pain, control bleeding, and prevent fluid loss.

Therapeutic Application of Hydrogel

PVP-based hydrogels induce labor by locally releasing prostaglandin. They reduce the negative side effects associated with other methods. Devices made from these hydrogels, such as prostaglandin-impregnated rods, are sterilized through radiation. These devices find application in medical procedures like testicular prostheses. They maintain serum testosterone levels effectively. This offers an advantage over injections [27].

Applications of Crosslinked Hydrogels

Crosslinked hydrogels have diverse applications, including in ophthalmic, transdermal, and dental implantation methods. For example, hydrogel materials used to make contact lenses, like poly(HEMA), are preferred. This is due to a controllable network and sterilization through radiation techniques. Hydrogels are also used in glaucoma treatment and cataract surgeries. In dentistry, radiation polymerization of hydrophilic monomers like HEMA fills cavities more effectively. This reduces defects like cracking. Additionally, collagen-based hydrogels are useful in plastic surgery. They assist in repairing cosmetic defects.

Thermoresponsive Gels in Medical Devices

Thermoresponsive gels, which swell and shrink with temperature changes, have gained attention for use in drug delivery systems, artificial muscles, and soft actuators. These gels allow temperature-controlled drug permeation. They also enable enzyme activity. For instance, Maolin and colleagues immobilized enzymes like BSA in thermoresponsive hydrogels. This methodology presents an innovative approach. It enhances efficiency in various applications.

Polymeric Drug Delivery Systems

Polymeric drug delivery systems are designed to control the release of drugs at a desired rate and duration, minimizing secondary reactions. They improve the targeting of affected cells. These systems encapsulate the drug in a polymer capsule. This allows for continuous automatic release. Modern drug delivery devices are composite structures with a sensor and a delivery part. The sensor is often a biochip. It monitors and controls drug release. The delivery part is typically an electrically responsive hydrogel. It regulates drug release through reversible volume changes triggered by an on-off electrical field. This makes it effective as both an actuator and a reservoir. Kaetsu developed an immobilized glucose-oxidase sensor using a thin PMMA membrane on an ISFET gate. This makes the biochip sensitive to glucose. This innovation led to advancements like on-off switching and the release of insulin from electro-responsive hydrogels, which are controlled by an electrical field.

Cancer

Glass-forming monomers and anticancer reagents mixed and mutually dispersed, then g-irradiated at a low temperature (-78°C) to create anticancer drug–polymer composites. Treatment for malignant brain tumors can greatly benefit from the drug's controlled, gradual release. It has several advantages over local chemotherapy, which has some serious drawbacks, including rapid drug removal from the tumor site, harm to healthy tissues, and inadequate anticancer drug intratumoural infiltration to the blood-brain barrier. Immobilization of Anti-a-fetoprotein on a polymeric film is designed to be used for the diagnosis of liver cancer. In addition, proteins, peptides, and anticancer medications created through radiation therapy are incorporated into the poly(2-HEMA) and poly(HEMA) for drug delivery systems. Due to its toxicity, the promising anticancer medication Narcisclasien—which has impressive antimitotic activity—has not yet been put to use. This is how a medication delivery system made from the radiation-induced polymerization of methacrylic monomers at low temperatures has been developed for controlled release. A blood plasma exchange device made of grafted fiber can be utilized to treat acute lymphocytic leukemia [28]. This instance involves immobilizing the asparaginase enzyme on methyl acrylic acid, grafting it onto porous hollow polypropylene fiber, and employing the carbodiimide method to activate the -COOH groups with NHS (N-hydroxy succinimide).

Immobilization of Enzymes

To entrap proteins, peptides, or enzymes, the radiation method is used. The method offers greater benefits than the other methods since it allows the drug molecule to adhere to the surface of the polymer backbone. In order to immobilize active biological compounds, like ferments and medications, on polymeric surfaces intended for use in medicine and biotechnology, radiation-induced graft polymerization has been employed. Horse Radish Peroxidase (HRP) was chemically bound to a Segmented Polyether Urethane Film (SPEU) created *via* grafting acrylamide and acrylic acid. This technique can be used to analyze the peroxide concentration of irradiated wine. Trypsin immobilization is another application of radiation-induced grafting. Trypsin is linked to the nitro groups of p-nitrostyrene, which are converted into isothiocyanate groups on a variety of polymers. The process of treating milk involves the use of immobilized trypsin. The benefit of grafted polymers is that they can also be utilized to make films. Another method for creating temporary artificial skin is to graft hydroxymethylmethacrylate and acrylamide onto a polyurethane film using radiation [29]. Using the grafted polymer offers the advantages of preventing bacterial invasion, improving water permeability, and preventing fluid accumulation. It is painless to remove, translucent, and adheres effectively to the wound.

Sterilization

In addition to the previously mentioned uses, which include the sterilization of medical devices using radiation for the manufacturing of devices for drug delivery, breast implantations, contact lenses, artificial muscles, catheters, diapers, and so forth, it is becoming increasingly popular. Compared to other standard procedures, such as dry heat, autoclaving, and the use of ethylene oxide gas, this method is chosen due to its advantages. While radiation sterilization allows for the simultaneous processing of biomedical materials, conventional sterilization is a laborious operation. Three factors determine how effective the radiation sterilization procedure is:

- The degree of sterility guarantee,
- The resistance of the microorganism, and
- The initial microbiological contamination [30].

Radiation sterilization is successful because of several factors, including:

a. ease of use and dependability;
b. continuous sterilization process;
c. room temperature suitability;
d. single-use medical products sterilization;
e. material choice flexibility; and
f. high level of assurance.

Radiation sterilization process control is an easy task. All that is required is the conveyor and an isotope radiation source. The procedure is solely controlled by the speed of the conveyor. Reliability that surpasses that of the majority of mechanical process equipment is ensured by the simplicity of controlling the output and application of tested mechanical conveyors. The reliabilities in this instance are even higher than 90%. Because ionizing radiation penetrates materials, its sterilizing power is unaffected by the chemical makeup of the materials. Because radiation sterilization is easy to repeat and regulate, it is a very dependable technique that is mostly immune to human mistakes. The primary benefit of radiation over other methods is that, following packing, biomedical items can be sterilized at ambient temperature, thereby preventing recontamination and making the process fully safe.

The technique known as radiation-induced grafting has been used to change polymers in order to improve their characteristics or add new functions. The ecologically friendly elements of this procedure are referred to as "green grafting". One of these features is lowering the usage of dangerous chemicals. They reduce wastage and start the grafting process with radiation, which is a cleaner option. Despite this, radiation-induced green grafting offers a number of exciting prospects. It also has a lot of difficulties, which need to be addressed. We talk about the difficulties and consider the technology's potential for the future.

CHALLENGES OF RADIATION-INDUCED GREEN GRAFTING

- **Regulation Over Grafting Specifications**

Proper control over grafting settings is a major difficulty in radiation-induced green grafting. These consist of the distribution of grafted chains. Grafting density plays an important role. The degree of grafting is also crucial. The process is impacted by radiation dosage. These parameters are affected by monomer and polymer matrix types. Dosage rate contributes to this complexity. Achieving consistent control across several systems can be difficult. This complexity results from the intricacy of interactions. Materials and radiation interact in intricate ways. Changes in these factors may result in grafted polymers with inconsistent characteristics. Certain apps performance gets impacted by discrepancies.

- **Polymers' Sensitivity to Radiation**

The susceptibility of various polymers to radiation varies. Certain polymers may suffer unintended side reactions or deteriorate when exposed to radiation, resulting in unwanted byproducts. Structural alterations can occur, for example, high radiation levels can cause chain scission. It may also cause cross-linking in the polymer backbone. This could weaken the mechanical qualities of the material. Additionally, it may change the intended functionality. Radiation dosages must be carefully chosen. They should be optimized. Due to variability, this procedure can be difficult and may also require time.

- **Environmental and Economic Aspects**

Although radiation-induced green grafting is thought to be more environmentally benign than traditional chemical procedures, there are still issues with process sustainability from an economic and environmental standpoint. The establishment and upkeep of a radiation facility can be quite expensive. This holds true for sources of gamma or electron beam radiation. They tend to be costly. This is especially true in large-scale industrial settings. Furthermore, the production of secondary wastes, such as solvents or unused monomers, presents serious environmental problems. It is essential to develop sustainable and affordable waste management techniques. Recovery of resources is critical for further use of technology.

- **Safety and Regulatory Issues**

The use of radiation in material processing gives rise to serious safety and regulatory issues. Safe handling and storage of radioactive materials or radiation sources necessitate observance of safety regulations. This prevents radiation exposure for employees. It also aims to protect the general public and the environment. Strict guidelines may prevent radiation-induced green grafting from being widely used. This is especially valid in some areas. It holds true for sectors of the economy that lack the required infrastructure, where regulatory frameworks are not well established.

- **Process Scalability**

Another major problem is transferring radiation-induced grafting from laboratory-scale research to industrial-scale production. It can be challenging to ensure radiation exposure uniformity across large-scale materials. This may cause grafting to be uneven, resulting in uneven product quality. Additionally, considerable adjustments to machinery may be necessary for the integration of radiation-induced grafting with current manufacturing processes. These adjustments may be expensive and time-consuming.

- **Mechanistic Pathways: Limited Understanding**

It is unclear exactly how radiation-induced grafting works. This is especially valid for intricate polymer systems. This incomplete knowledge makes it difficult to forecast outcomes. It complicates the management of the results of the grafting procedure. It is imperative to have a clearer understanding of molecular interactions. This involves radiation, monomers, and polymers. It is crucial to optimize the procedure. This will increase applicability to a wider range of materials and functionalities.

FUTURE PERSPECTIVES OF RADIATION-INDUCED GREEN GRAFTING

Despite challenges, radiation-induced green grafting holds significant promise for the future. There are numerous opportunities for advancement in scientific research. It also has implications for industrial applications. Radiation-induced techniques can enhance plant growth. Researchers actively explore these methods. They seek to understand and maximize benefits.

- **Advances in Radiation Technology**

Continued advancements in radiation technology, such as the development of more precise and tunable radiation sources, are expected to enhance control over grafting processes. Innovations in Low-Energy Electron Beam (LEEB) technology, for example, offer potential for more targeted and efficient grafting. This occurs with reduced energy consumption and minimal damage to the polymer matrix. These advancements could make radiation-induced grafting more accessible and cost-effective for a wider range of applications.

- **Development of New Polymer-Monomer Systems**

The exploration of new polymer-monomer systems that are more amenable to radiation-induced grafting is a promising area of research. By designing polymers and monomers with specific radiation-responsive characteristics, researchers can tailor the grafting process. This approach could lead to the development of novel materials with enhanced performance. Areas such as biomedical devices, water treatment, and energy storage are also potential fields impacted by these innovations.

- **Integration with Sustainable Materials**

Radiation-induced green grafting can play a vital function in the development of sustainable products, such as biopolymers and recycled polymers. By modifying these materials, the grafting properties can be enhanced. This enhancement helps meet performance standards required for various applications. Grafting improves mechanical strength as well as enhances thermal stability. Furthermore, grafting can improve barrier properties. This makes biodegradable polymers more competitive with conventional petroleum-based plastics.

- **Application in Emerging Technologies**

Emerging technologies, such as nanotechnology, additive manufacturing, and advanced coatings, present exciting opportunities for the application of radiation-induced green grafting. In nanotechnology, for instance, grafting can be used to functionalize nanoparticles or nanocomposites. This enables integration into advanced materials with tailored properties. In additive manufacturing, radiation-induced grafting can be employed. It modifies the surface properties of 3D-printed objects, enhancing their functionality and durability.

- **Improved Understanding of Mechanistic Pathways**

Ongoing research into the mechanistic pathways of radiation-induced grafting will lead to a more comprehensive understanding of the process at a molecular level. This knowledge will enable the development of more predictive models and simulation tools. It allows for better control. It optimizes grafting processes. Improved mechanistic understanding will also facilitate the design of new materials. It will encourage exploration of novel applications.

- **Regulatory and Safety Innovations**

To address regulatory and safety challenges, there is a need for innovations in radiation protection and safety protocols. The development of more efficient shielding materials is crucial. Real-time monitoring systems must be improved. Automated safety procedures can significantly reduce risks associated with radiation use. Additionally, harmonizing international regulatory standards for radiation-induced processes could streamline approval and adoption of this technology in various regions and industries [31].

CONCLUSION

Radiation-induced green grafting represents a significant advancement in the field of polymer science. It offers numerous benefits over conventional grafting methods. It enhances material properties. It aligns with environmental sustainability. It also provides precise control. This makes it an invaluable tool in various applications. This technique shines particularly in pharmaceuticals. As research and technology continue to evolve, radiation-induced grafting is poised to play a crucial role in the development of innovative materials and solutions, meeting the demands of modern science and industry. In radiation processing, most changes are introduced at the manufacturing stage of the polymer matrix. Further improvements can be achieved through grafting, cross-linking, or additional polymerization using reactive elements of films that have been commercially treated. Specific changes will additionally be introduced after the processing stage. This occurs without affecting the main matrix. Radiation polymer research is advancing towards biodegradability. Biomass conversion addresses environmental pollution by enhancing the solubility of polymers, which is hindered by non-degradable components.

Despite its potential for large-scale applications, radiation processing has been limited due to the duration of the radiation and the associated setup costs. To speed up reactions, the field is expanding. Electron beam processing techniques are being developed. They involve large-volume processes. These processes operate with fast line acceleration. Still, more investigation is required into the smaller-scale applications of radiation grafting. This is particularly true when using Co-60 sources. Dose reduction to a minimum is crucial. This is particularly crucial for relatively inert polymers like poly(tetrafluoroethylene) and macromolecules like cellulose that are sensitive to heat and radiation. The goal of future research is to find new ingredients. These will cut grafting doses even lower than they are now [32].

In a nutshell, considerable progress has been achieved in radiation chemistry research, both in basic and practical fields, over the past three decades. However, continued research is essential. It is necessary to overcome existing challenges. This will fully realize the potential of radiation technology in the polymer field. This method exemplifies integration of scientific progress with environmental consciousness. It opens new avenues. For material design, it also creates applications, paving the way for a more sustainable future and a technologically advanced society.

REFERENCES

[1] Odian G, Chandler HW. Radiation-induced graft polymerization. In: Henley EJ, Kouts H, eds. Advances in nuclear science and technology. Vol. 1. New York: Academic Press; 1962. p. 85–109.
[http://dx.doi.org/10.1016/B978-1-4831-9955-9.50007-6]

[2] Ballantine DS, Glines A, Metz DJ, Beher J, Mesrobian RB, Restaino AJ. G values of gamma-ray initiation of vinyl polymerization and their relation to graft copolymer formation. J Polym Sci 1956; 19(91): 219-24.
[http://dx.doi.org/10.1002/pol.1956.120199128]

[3] Ballantine D, Glines A, Adler G, Metz DJ. Graft copolymerization by pre-irradiation technique. J Polym Sci 1959; 34(127): 419-38.
[http://dx.doi.org/10.1002/pol.1959.1203412732]

[4] Nasef MM, Güven O. Radiation-grafted copolymers for separation and purification purposes: Status, challenges and future directions. Prog Polym Sci 2012; 37(12): 1597-656.
[http://dx.doi.org/10.1016/j.progpolymsci.2012.07.004]

[5] Moad G, Chong YK, Postma A, Rizzardo E, Thang SH. Advances in RAFT polymerization: the synthesis of polymers with defined end-groups. Polymer (Guildf) 2005; 46(19): 8458-68.
[http://dx.doi.org/10.1016/j.polymer.2004.12.061]

[6] Nasef MM, Gupta B, Shameli K, Verma C, Ali RR, Ting TM. Engineered Bioactive Polymeric Surfaces by Radiation Induced Graft Copolymerization: Strategies and Applications. Polymers (Basel) 2021; 13(18): 3102-2.
[http://dx.doi.org/10.3390/polym13183102] [PMID: 34578003]

[7] Nisar S, Pandit AH, Nadeem M, Pandit AH, Rizvi MMA, Rattan S. γ-Radiation induced L-glutamic acid grafted highly porous, pH-responsive chitosan hydrogel beads: A smart and biocompatible vehicle for controlled anti-cancer drug delivery. Int J Biol Macromol 2021; 182: 37-50.
[http://dx.doi.org/10.1016/j.ijbiomac.2021.03.134] [PMID: 33775765]

[8] Sherazi TA. Radiation-Induced Grafting. Springer eBooks. 2014 Jan 1;1–2.
[http://dx.doi.org/10.1007/978-3-642-40872-4_515-1]

[9] Drobny JG. Ionizing Radiation and Polymers: Principles, Technology, and Applications. Oxford: William Andrew 2012; p. 320. https://books.google.co.in/books?id=0b_lfeoDIyoC

[10] Lim SJ, Shin IH. Graft copolymerization of GMA and EDMA on PVDF to hydrophilic surface modification by electron beam irradiation. Nucl Eng Technol 2020; 52(2): 373-80.
[http://dx.doi.org/10.1016/j.net.2019.07.018]

[11] Chen X, Li L, Xu S, et al. Ultraviolet B radiation down-regulates ULK1 and ATG7 expression and impairs the autophagy response in human keratinocytes. J Photochem Photobiol B 2018; 178: 152-64.
[http://dx.doi.org/10.1016/j.jphotobiol.2017.08.043] [PMID: 29154199]

[12] Chen Y, Huang T, Jiang C, et al. Preparation of antifouling poly (ether ether ketone) hollow fiber membrane by ultraviolet grafting of polyethylene glycol. Mater Today Commun 2021; 27: 102326-6.
[http://dx.doi.org/10.1016/j.mtcomm.2021.102326]

[13] Bhattacharya A. Grafting: a versatile means to modify polymersTechniques, factors and applications. Prog Polym Sci 2004; 29(8): 767-814.
[http://dx.doi.org/10.1016/j.progpolymsci.2004.05.002]

[14] Nasef M. Preparation and applications of ion exchange membranes by radiation-induced graft copolymerization of polar monomers onto non-polar films. Prog Polym Sci 2004; 29(6): 499-561.
[http://dx.doi.org/10.1016/j.progpolymsci.2004.01.003]

[15] Makuuchi K, Cheng S. Radiation Processing of Polymer Materials and its Industrial Applications. New York: Wiley 2011; p. 444.

[16] Kumar V, Bhardwaj YK, Rawat KP, Sabharwal S. Radiation-induced grafting of vinylbenzyltrimethylammonium chloride (VBT) onto cotton fabric and study of its anti-bacterial activities. Radiat Phys Chem 2005; 73(3): 175-82.
[http://dx.doi.org/10.1016/j.radphyschem.2004.08.011]

[17] Choi SH, Nho YC. Radiation-induced graft copolymerization of binary monomer mixture containing acrylonitrile onto polyethylene films. Radiat Phys Chem 2000; 58(2): 157-68.
[http://dx.doi.org/10.1016/S0969-806X(99)00367-9]

[18] Estrada-Villegas GM, Bucio E. Comparative study of grafting a polyampholyte in a fluoropolymer membrane by gamma radiation in one or two-steps. Radiat Phys Chem 2013; 92: 61-5.
[http://dx.doi.org/10.1016/j.radphyschem.2013.07.015]

[19] Ferreira HP, Parra DF, Lugao AB. Radiation-induced grafting of styrene into poly(vinylidene fluoride) film by simultaneous method with two different solvents. Radiat Phys Chem 2012; 81(9): 1341-4.

[http://dx.doi.org/10.1016/j.radphyschem.2012.01.047]

[20] Przybytniak G, Kornacka EM, Mirkowski K, Walo M, Zimek Z. Functionalization of polymer surfaces by radiation-induced grafting. NUKLEONIKA 2008; 53(3): 89-95. Available from: http://www.nukleonika.pl/www/back/full/vol53_2008/v53n3p089f.pdf

[21] Chalykh AE, Tverskoy VA, Aliev AD, *et al*. Mechanism of Post-Radiation-Chemical Graft Polymerization of Styrene in Polyethylene. Polymers (Basel) 2021; 13(15): 2512-.
[http://dx.doi.org/10.3390/polym13152512] [PMID: 34372115]

[22] Dubey KA, Bhardwaj YK. Radiation processing of polymers and their industrial applications [Internet]. Mumbai (India): Bhabha Atomic Research Centre; 2021.

[23] Gaylord NG, Ballantine DS. Atomic radiation and polymers. A. C HARLESBY. Pergamon Press, New York, 1960. xiii + 556 pp. $17.50. J Polym Sci 1960; 45(146): 553-3.
[http://dx.doi.org/10.1002/pol.1960.1204514637]

[24] Vo QV, Tram TLB, Phuoc Hoang L, Hoa NT, Mechler A. The alkoxy radical polymerization of *N*-vinylpyrrolidone in organic solvents: theoretical insight into the mechanism and kinetics. RSC Advances 2023; 13(34): 23402-8.
[http://dx.doi.org/10.1039/D3RA03820C] [PMID: 37546223]

[25] Rosiak JM, Olejniczak J. Medical applications of radiation formed hydrogels. Radiat Phys Chem 1993; 42(4-6): 903-6.
[http://dx.doi.org/10.1016/0969-806X(93)90398-E]

[26] Dafader NC, Manir MS, Alam MF, Susmita Paul Swapna, Akter T, Huq D. Effect of kappa-carrageenan on the properties of poly(vinyl alcohol) hydrogel prepared by application of gamma radiation. SOP transactions on applied chemistry. 2015 Jan 31;2(1):1–12.

[27] Rosiak JM, Ulanski P, Leonardo Adamo Pajewski, Yoshii F, Keizo Makuuchi. Radiation formation of hydrogels for biomedical purposes. Some remarks and comments. 1995 Aug 1;46(2):161–8.

[28] Kaetsu I. Radiation synthesis and fabrication for biomedical applications. Radiat Phys Chem 1995; 46(4-6): 1025-30.
[http://dx.doi.org/10.1016/0969-806X(95)00314-N]

[29] Bhattacharya A. Radiation and industrial polymers. Prog Polym Sci 2000; 25(3): 371-401.Available from:
http://oktatas.ch.bme.hu/oktatas/konyvek/fizkem/Fizikai%20kemia_MSc_MuaSzal/2019%20osz_archiv/ea/Sugkh%20Ajanlott%20irodalom/Radiation%20and%20industrial%20polymers.pdf
[http://dx.doi.org/10.1016/S0079-6700(00)00009-5]

[30] I. Kaetsu. Radiation synthesis of polymeric materials for biomedical and biochemical applications. Springer eBooks. 2005 Nov 13;81–97.

[31] Abdel-Ghaffar AM. Radiation synthesis and modification of biopolymers and polymeric composites for biomedical applications. Polym Polymer Compos 2023; 31: 09673911231166636.Available from: https://www-pub.iaea.org/mtcd/publications/pdf/te_1324_web.pdf
[http://dx.doi.org/10.1177/09673911231166636]

[32] Khan MA, Shehrzade S, Sarwar M, Chowdhury U, Rahman MM. Effect of Pretreatment with UV Radiation on Physical and Mechanical Properties of Photocured Jute Yarn with 1,6-Hexanediol Diacrylate (HDDA). J Polym Environ 2001; 9(3): 115-24.
[http://dx.doi.org/10.1023/A:1020450827424]

CHAPTER 7

Plasma-Assisted Green Grafting: Principles, Environmental Benefits, and Uses

Pratiksha Bramhe[1], Prafulla Sabale[1,*], Suchita Waghmare[1], Pramod Khedekar[1], Vidya Sabale[2], Nilesh Rarokar[3] and Mahavir Chougule[4]

[1] *Department of Pharmaceutical Sciences, Rashtrasant Tukadoji Maharaj Nagpur University, Nagpur, Maharashtra 440033, India*

[2] *Department of Pharmaceutics, Dadasaheb Balpande College of Pharmacy, Besa, Nagpur, Maharashtra, 440037 India*

[3] *Department of Pharmaceutics, G. H. Raisoni Institute of Life Sciences, Nagpur 440016, India*

[4] *Department of Pharmaceutical Sciences, Mercer University College of Pharmacy, Atlanta, Georgia 30341, USA*

Abstract: In the field of medicine and pharmacology, PAGG (Plasma-Assisted Green Grafting) is a revolutionary method that uses green chemistry and non-thermal plasma to improve material surfaces in an ecologically friendly way. This novel method utilizes non-thermal plasma for the activation and modification of surface characteristics. This eliminates the need for hazardous chemicals or excessive heat during the grafting of functional groups or polymers. Significant environmental benefits result from this, including reduced energy consumption, lower environmental pollution, and a decrease in the use of hazardous chemicals. PAGG has demonstrated tremendous promise in the pharmaceutical and healthcare industries. By enhancing the biocompatibility of medical implants, PAGG ensures that the body will absorb these devices more easily, thereby reducing the likelihood of rejection and complications. Furthermore, PAGG enhances medication delivery mechanisms, enabling more effective and targeted therapeutic administration that can reduce adverse effects and improve treatment outcomes. Another important achievement that contributes to infection prevention and enhanced patient safety in clinical settings is the creation of antimicrobial surfaces made possible by PAGG. The technique provides an environmentally friendly alternative to conventional grafting procedures by enhancing the mechanical stability, antibacterial properties, and biocompatibility of implants, medical devices, and drug delivery systems. As a result, PAGG aligns perfectly with the growing demand for sustainable practices in the medical and pharmaceutical industries. By reducing the environmental footprint and improving the safety and efficacy of medical treatments, PAGG drives innovation and sustainability, offering

---

** **Corresponding author Prafulla M. Sabale:** Department of Pharmaceutical Sciences, Rashtrasant Tukadoji Maharaj Nagpur University, Nagpur, Maharashtra 440033, India; E-mail: prafullasable@yahoo.com*

Kuldeep Vinchurkar, Satish Polshettiwar, Nilesh Mahajan &Yogeshwar Bachhav (Eds.)
All rights reserved-© 2026 Bentham Science Publishers

substantial benefits for both patient care and environmental health. This technique exemplifies how cutting-edge technology can contribute to sustainable development in critical fields.

Keywords: Biocompatibility, Drug delivery system, Environmentally friendly, Non-thermal plasma, Plasma-assisted green grafting, Sustainability.

INTRODUCTION

The goal of the scientific discipline of "green chemistry" is to create chemical products and processes that minimize or completely do away with the need for dangerous materials. Green chemistry aims to develop more environmentally friendly and sustainable chemical processes by using waste reduction, energy conservation, and renewable resource utilization strategies. This strategy places a strong emphasis on enhancing safety and efficiency while minimizing the adverse environmental impacts of chemical manufacturing and use. Green chemistry aims to preserve the environmental and public health benefits of chemical advancements while promoting safer chemical development, utilizing benign solvents, and developing energy-efficient processes [1, 2]. Green grafting is a polymer grafting method that is environmentally friendly and aims to mitigate the pollution that polymer synthesis has historically caused. Through the attachment of side chains to a polymer backbone, this approach modifies the mechanical strength, solubility, and thermal stability of the polymer. The environmentally friendly feature of this process is highlighted by the substitution of sustainable monomers made from plant-based materials, natural oils, or sugars for conventional petroleum-based compounds [3]. In green grafting, green solvents are essential. These eco-friendly substitutes for traditional solvents are derived from plant- or bio-based resources that are replenishable [4]. Green solvents help reduce pollutants and greenhouse gas emissions during the grafting process, as they are less harmful to the environment, often biodegradable, and have a smaller environmental footprint. Green solvents ensure that the grafting procedure complies with green chemistry principles, in contrast to conventional solvents, which are typically petroleum-based and can cause significant environmental contamination [5]. Green grafting stresses energy efficiency by utilizing procedures that operate at lower pressures, temperatures, or energy inputs, in addition to employing green solvents. To reduce energy usage, technologies such as microwave or ultrasonic energy are utilized in place of conventional heating techniques [6]. Enzymes that are biocompatible, reusable, or non-toxic are also preferable catalysts, which further aligns the grafting process with green chemistry concepts. There are major health and environmental benefits when green solvents and green grafts are combined. For example, due to their decreased volatility and toxicity, green solvents are typically safer, as they expose users to fewer hazardous compounds and reduce the risk of respiratory difficulties, skin irritation, and other health concerns associated with solvent usage. Contrarily, a lot of common solvents are combustible, volatile, and poisonous, which pose serious health hazards, including the possibility of long-term consequences like cancer or neurological damage. In terms of economics, green grafting techniques and solvents may occasionally incur higher upfront costs, but total expenditures may be lower due to lower waste disposal costs, safer handling, and less stringent regulations [7]. Moreover, implementing green practices can enhance a business's sustainability profile and potentially provide it with competitive advantages. On the other hand, traditional solvents may be quite expensive to handle, dispose of, and comply with environmental regulations, even if they are initially less expensive. In terms of performance, green solvents used in green grafting yield outcomes that are on par with or even better than those of conventional solvents in various applications, including coatings, cleaning products, and pharmaceuticals [8].

Green chemistry innovations continue to become increasingly effective and widely adopted. Green grafting techniques include, for example, grafting cellulose with biodegradable polymers such as Poly(Lactic Acid) (PLA) using green solvents and enzymatic catalysts, or grafting natural rubber with polymers derived from plant oils without the need for solvents. Grafting involves attaching polymer chains or side groups to an existing polymer and is achieved through methods such as free radical or living polymerization [9]. Polymer grafting has emerged as a promising alternative in response to the growing global demand for natural products and the depletion of available resources. By using this technique, it is possible to create segments of polymers that are covalently attached to the main chains of polymers, resulting in the development of copolymers and the integration of special properties from the grafted materials. Grafted polymers are extremely advantageous in a variety of applications because this method adds additional functions while maintaining the extended conjugated structure of the original chain, as shown in Fig. (**1**) [10].

Green grafting is a significant advancement in polymer chemistry, facilitated by the use of environmentally friendly solvents. This method offers a safer, more environmentally friendly, and often more cost-effective option for polymer grafting compared to older techniques, aligning with the growing emphasis on green chemistry [11]. Green grafting reduces the use and manufacture of harmful compounds by employing sustainable processes and green solvents, thereby promoting an industry that is more ecologically and health-conscious. Using different plasma systems, plasma-induced graft copolymerization is an appealing method for creating antibacterial and antifouling coatings. With the use of monomer precursors (silicon, hydrocarbons, and fluorocarbons) or polymerizable gases (NH_3, N_2, O_2, CO_2 or H_2O), plasma treatment offers a special technique for grafting on the surface of polymer substrates [12]. This process results in energy gain, activation, and initiation of grafting. Either a direct grafting of activated monomers onto polymer surfaces or the generation of a radical on polymer surfaces, followed by contact with monomers, is one of the two ways the reaction occurs. The plasma treatment parameters, including monomer reactivity, flow rate, system pressure, discharge power, excitation signal frequency, and substrate temperature, all influence the degree of grafting [13]. Some commonly used green solvents used in green synthesis are illustrated in Table **1**.

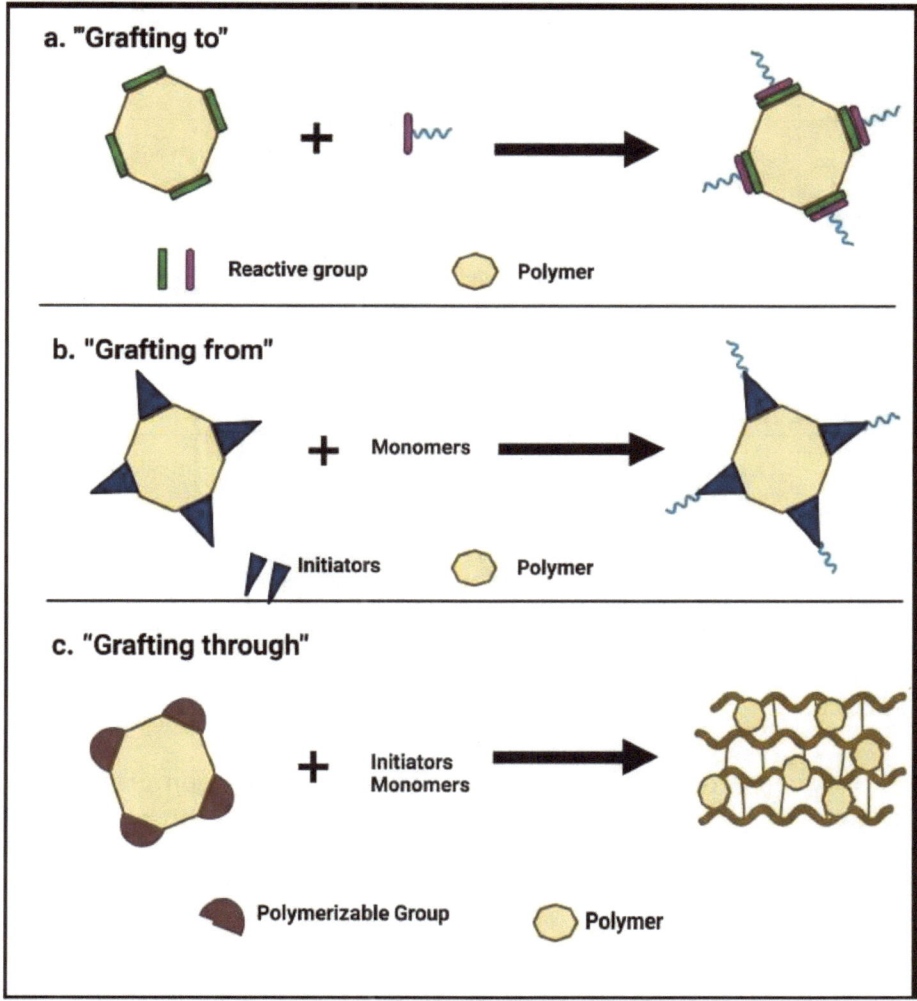

Fig. (1). Diagrammatic representation of grafting.

Table 1. Common green solvents used in green chemistry.

Green Solvents	Source	Properties	Applications	Advantages
Water	Ubiquitous, renewable	Non-toxic, abundant, high heat capacity	Cleaning, chemical reactions, and extraction.	Safe, environment-friendly, cost-effective
Ionic liquids	Synthetic or biobased	Non-volatile, tunable properties	Catalysis, extraction, and electrochemistry.	Recyclable, customizable for specific use
Supercritical CO_2	Carbon dioxide	Non-toxic, non-flammable, easily separable	Extraction, decaffeination, and cleaning.	Low toxicity, easily recoverable
Bioethanol	Fermentation of biomass	Biodegradable, lower toxicity, renewable	Solvent in coatings, pharmaceuticals, and fuels.	Renewable, low toxicity
Ethyl lactate	Fermentation of sugars	Biodegradable, non-toxic, high solvency	Cleaning, paint formulations, and degreasing.	Biodegradable, derived from renewable resources
Glycerol	By-product of biodiesel production	High boiling point, low volatility	Pharmaceuticals, cosmetics, and the food industry.	Non-toxic, biodegradable, renewable
Limonene	Extracted from citrus peels	Biodegradable, pleasant odor, high solvency	Cleaning agents, degreasing, and the formation of fragrances	Renewable, low toxicity, pleasant aroma
Propylene Carbonate	Derived from propylene oxide	Low toxicity, high boiling point	Paints, coatings, adhesives, and battery electrolytes.	Biodegradable, low VOC content
2-Methyl Tetrahydrofuran	Derived from hemicellulose	Low toxicity, hydrophobic, good solubility	Reactions, extractions, and coatings.	Renewable, good substitute for THF
Dimethyl Carbonate (DMC)	Methanol and carbon dioxide	Low toxicity, biodegradable, high solvency	Paints, coatings, and batteries.	Low toxicity, low environmental impact

A recently developed technique called "plasma-assisted green grafting" utilizes plasma technology to modify the surfaces of materials in an eco-friendly and sustainable manner. The ionized gases that make up plasma, the fourth state of matter, include electrons, ions, and radicals, among other reactive species. Without the need for strong chemicals or extremely hot temperatures, these reactive species may start and drive chemical reactions on material surfaces, making the procedure "green" and environmentally beneficial. The material surface is exposed to plasma during plasma-assisted grafting, which breaks chemical bonds to produce active sites. Grafting molecules or functional groups onto these active sites enhances the material's qualities, such as adhesion, wettability, or biocompatibility [14]. This approach is very useful in materials science and biomedical engineering, where it may be used to improve the characteristics of textiles, polymers, and other substrates, as well as to improve the performance of implants and other medical devices. Because of its efficiency, adaptability, and environmental friendliness, plasma-assisted green grafting is a viable strategy for developing sustainable material innovations. This technique's environmental friendliness stems from its ability to operate at lower temperatures and decrease or eliminate the need for harmful chemicals. Its flexibility also stems from its ability to be applied to a wide range of materials and introduce a variety of functions [15]. Green grafting with plasma assistance enables regulated, surface-specific alterations, resulting in consistent coatings even on intricate geometries. Applications are found in many different sectors, including textiles, packaging, environmental protection, and biomedical engineering. Preparing the substrate, creating plasma, activating the surface, grafting, and post-treatment are usually steps in the procedure. Plasma-assisted green grafting is a potent and environmentally friendly surface modification approach that offers several benefits over conventional techniques and has significantly advanced various sectors [16, 17]. The different types of plasma gases are illustrated in Table **2**.

Table 2. Illustrative examples of plasma gases used in plasma-assisted green grafting.

Plasma Gas	Characteristics	Role in Plasma-Assisted Green Grafting	Advantages	Refs.
Argon (Ar)	Inert, non-reactive, widely available	Provides an inert atmosphere and helps to activate surfaces without chemical reactions.	Prevents unwanted reactions and stabilizes the process.	[18, 19]
Oxygen (O_2)	Reactive strong oxidizer	Used to oxidize and activate surfaces, and introduces oxygen-containing functional groups	Enhances surface wettability, promotes adhesion.	[20]
Nitrogen (N_2)	Inert, non-reactive, moderate activation energy	Used to introduce nitrogen-containing functional groups like amines or nitriles.	Improves surface functionality, and enhances adhesion.	[21]
Carbon Dioxide (CO_2)	Non-toxic, moderate reactivity	Used to introduce carboxyl or ester groups, and contributes to surface activation.	Environmentally friendly, introduces polar groups	[22]
Hydrogen (H_2)	Highly reactive, reducing agent	Used to reduce surface oxides, clean and prepare surfaces before grafting	Effective surface cleaning, and removes contaminants	[23]
Helium (He)	Inert, low reactivity, high thermal conductivity	Used as a carrier gas to stabilize plasma, aids in uniform surface treatment.	Provides a stable plasma environment, enhances control	[24]

(Table 2) cont.....

Plasma Gas	Characteristics	Role in Plasma-Assisted Green Grafting	Advantages	Refs.
Ammonia (NH_3)	Reactive, introduces nitrogen	Used to introduce amine groups onto surfaces, and increases surface reactivity	Enhances surface reactivity and promotes functionalization.	[25]
Methane (CH_4)	Reactive, carbon source	Used to introduce hydrocarbon chains onto surfaces, and increases hydrophobicity.	Enhances hydrophobicity and modifies surface energy.	[26]
Air	Readily available, mixture of O_2 and N_2	Provides a combination of oxidation and nitrogenation effects, and activates surfaces.	Cost-effective and enhances surface wettability.	[27]
Fluorine-based Gases (*e.g.*, CF_4)	Highly reactive, introduces fluorine atoms	Used to introduce fluorinated groups, enhances hydrophobicity, and chemical resistance.	Improves chemical resistance, and enhances hydrophobicity.	[28]

PRINCIPLES OF PLASMA-ASSISTED GREEN GRAFTING:

The basic principles of green chemistry are integrated into plasma-assisted green grafting, providing a safe, effective, and environmentally friendly method for surface modification. This method not only enhances material performance but also advances the broader objectives of sustainable chemical manufacturing and environmental protection. Green chemistry and plasma technology are utilized in plasma-assisted green grafting to modify surfaces in an environmentally friendly manner. This technique offers accurate and effective surface alterations while minimizing its negative environmental effects [29].

Minimization of Environmental Impact

Plasma-assisted grafting offers significant advantages over conventional techniques by minimizing chemical use and waste generation. This method utilizes plasma to activate surfaces, thereby reducing or eliminating the need for hazardous chemicals and solvents. Additionally, since plasma serves as the primary energy source, rather than chemical reagents that might produce harmful byproducts, the process generates minimal waste [30, 31].

Energy Efficiency

Plasma-assisted grafting processes offer the advantage of being conducted at low temperatures and pressures, in contrast to traditional methods that often require high-temperature conditions. This not only reduces energy consumption but also enhances the overall efficiency of the process. Additionally, plasma activation of surfaces typically requires less energy compared to methods that rely on heat or chemicals, making it a more energy-efficient alternative [32].

Use of Environmentally Benign Plasma Gases

The process reduces the possibility of harmful emissions or poisonous byproducts by using inert, non-toxic gases, such as argon or nitrogen. Green plasma gases, such as carbon dioxide or oxygen, are also used where feasible to add functional groups, which reduces the environmental impact and prevents the production of hazardous waste [33].

Selective Surface Modification

Plasma-assisted grafting allows for precise control over surface functionalization, ensuring that only the desired modifications are made while preserving the bulk properties of the material. The process is typically confined to the material's surface, preventing unnecessary chemical reactions in the bulk and thereby minimizing waste [34].

Reduction of Toxic Catalysts and Solvents

Traditional catalysts, which can be hazardous to human health or the environment, are often not required in plasma-assisted grafting procedures, as the plasma itself provides the activation energy necessary for grafting. Furthermore, many of these methods reduce the environmental impact associated with the use of hazardous and Volatile Organic Compounds (VOCs) by using green solvents or by being completely solvent-free [35].

Sustainable Feedstocks and Monomers

Aligned with sustainability principles, plasma-assisted grafting can utilize monomers derived from renewable resources, such as plant- or bio-based materials. Additionally, by grafting biodegradable polymers onto surfaces, the process enhances the environmental friendliness of the resulting products [36].

Scalability and Industrial Relevance

Plasma-assisted grafting is a scalable technique suitable for industrial applications, making it a viable strategy for green manufacturing while maintaining its environmental benefits. The process adheres to the 12 principles of green chemistry, ensuring compliance with legal and environmental regulations while producing high-performance materials [37].

PLASMA TECHNOLOGY AND SURFACE ACTIVATION/ MODIFICATION

Graft polymerization generated by plasma is a process that forms specific groups on the surface of modified polymers using non-polymeric gases. All chemical bonds may be broken by the nearly infinite energy of plasma, which will cause a chemical reaction on any surface that is exposed. Distinct gases give distinct chemical characteristics to plasma [38]. In green chemistry, plasma-assisted methods are gaining recognition for their ability to modify surfaces and produce functionalized materials without the use of harsh chemicals or extreme environments. Specifically, plasma-assisted grafting utilizes plasma, a partially ionized gas comprising ions, electrons, and neutral species, to modify a material's surface, thereby facilitating the grafting of polymers or other molecules. Green chemistry principles align with this ecologically beneficial technique, which often requires minimal energy, little to no solvents, and minimal waste generation [39]. One of the primary characteristics of plasma-assisted grafting is surface activation, where a material's surface is treated with plasma to introduce reactive functional groups, such as hydroxyl, carboxyl, or amine groups, thereby increasing the material's reactivity and facilitating grafting. Because the process runs at very low temperatures and doesn't require hazardous chemicals or solvents, it is environmentally benign. A variety of materials, including polymers, metals, ceramics, and composites, can be altered using plasma, demonstrating the material's adaptability in many grafting applications [40]. Furthermore, because plasma parameters, such as power, gas type, and exposure duration, can be precisely regulated to provide homogeneous and stable grafting with desired qualities, plasma-assisted grafting also allows for controlled grafting. In comparison to conventional chemical grafting processes, this approach often yields fewer by-products, making it a safer and cleaner choice in various industrial applications [41]. Through the use of plasma, reactive sites are created on a surface to facilitate the grafting of different functional groups or molecules. This process is known as plasma-assisted grafting. Because it can be performed at low temperatures, it frequently uses less energy, and doesn't include hazardous chemicals, this process is regarded as ecologically benign [42]. A diagrammatic representation of plasma technology is shown in Fig. (**2**).

Modification of Polymer Surfaces for Water Purification

The introduction of hydrophilic groups onto polymer surfaces using plasma-assisted grafting has been demonstrated to be an efficient method for modifying polymer surfaces and enhancing their water-purification properties. For example, functional carboxyl and hydroxyl groups were added to the surface of polypropylene fibers using a low-pressure oxygen plasma treatment. After poly(acrylic acid) was grafted onto these altered fibers, a material with better heavy metal adsorption properties from wastewater was produced. This method is especially beneficial since it reduces the need for harmful solvents and high temperatures, which are normally needed for surface changes of this kind. As a result, the process is more sustainable and environmentally friendly [43]. Slow discharge plasma radiations offer about the same grafting probabilities as with ionizing radiation. The main events in plasma radiation-induced grafting are electron-induced excitation, ionization, and dissociation. Thus, the high-energy accelerated electrons from the plasma induce the cleavage of chemical bonds in the polymeric structure, which subsequently form macromolecule radicals and initiate graft copolymerization [44]. Acrylic acid-grafted polyethylene terephthalate has been successfully used to reduce the antithrombogenic property of polyethylene terephthalate. Heparin immobilized over the grafted polymer was found to be blood-compatible [45].

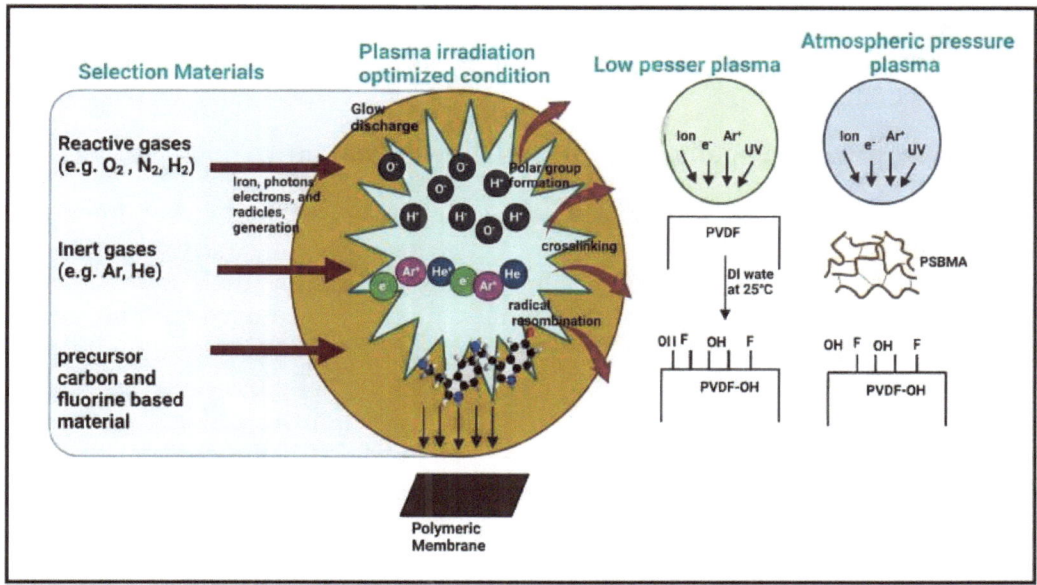

Fig. (2). Diagrammatic representation of plasma technology.

Grafting of Biocompatible Coatings on Medical Devices

Since plasma alters surfaces without altering the material's intrinsic qualities, it is a quick and flexible process with several benefits (bulk properties). Additionally, this method of altering polymer surfaces is highly industrially significant and environmentally beneficial. Due to the surface's adjustable qualities, it can function in various ways without requiring the replacement of the entire plasma system, such as by simply swapping gas or opening up new application possibilities [46]. Juliana M. Vaz *et al* in the study, plasma grafting was employed to functionalize polytetrafluoroethylene (PTFE) surfaces with a chitosan. Chitosan was grafted onto PTFE surfaces using two different spacer molecules: poly(ethylene glycol) bis(carboxymethyl) ether (PEG) and poly(ethylene-al--maleic anhydride) (PEMA). The results showed that PTFE-Plasma-PA-Chitosan exhibited a higher amount of chitosan attachment and superior antibacterial performance compared to PTFE-Plasma-PEG-Chitosan, as demonstrated in the study [47]. The inventive method of plasma-assisted grafting has gained traction due to the growing need for medical equipment to have surfaces that are both biocompatible and antibacterial. An illustration of this is the application of plasma treatment to activate the polydimethylsiloxane (PDMS) surface, which was subsequently grafted with the well-known biocompatible polymer polyethylene glycol (PEG). With this alteration, PDMS's blood compatibility is significantly improved, making it more suitable for use in medical devices such as implants and catheters. Compared to conventional surface modification techniques, which usually require the use of harsh chemicals and higher temperatures, the plasma-assisted procedure is especially beneficial because it is solvent-free and operates at low temperatures [48].

Functionalization of Cellulose for Sustainable Packaging

The growing demand for sustainable packaging solutions has sparked interest in natural polymers, such as cellulose, which provide an eco-friendly alternative to synthetic materials. One innovative approach to enhancing the performance of cellulose in packaging is plasma treatment, which has been used to graft acrylic acid onto cellulose fibers. This modification significantly improves the material's barrier properties against moisture and oxygen, making cellulose more suitable for use in biodegradable packaging applications. The plasma-assisted process is particularly beneficial because it reduces the need for chemical reagents and generates fewer byproducts, aligning with the principles of green chemistry and sustainability [49].

Grafting of Antifouling Layers on Marine Equipment

Biofilm generation and corrosion are common problems for marine equipment, which can shorten its lifespan and raise maintenance expenses. It has been demonstrated that grafting antifouling polymers, such as poly(sulfobetaine methacrylate), onto stainless steel surfaces with plasma assistance is a highly successful method for overcoming these obstacles. By drastically reducing the buildup of biofilms and other fouling organisms, this technique prolongs the usable life of maritime equipment and lowers maintenance frequency and expenses. Because the plasma-assisted method eliminates the need for harmful antifouling paints and coatings, it is regarded as an ecologically beneficial approach that offers a more sustainable and environmentally friendly option for maritime applications [50]. A study by *Soolmaz et al.* demonstrates that plasma-assisted green grafting was applied to develop graphene/polydimethylsiloxane (PDMS)-based coatings aimed at preventing the adhesion of fouling organisms. A nanocomposite, GOH@Ag, was synthesized by reducing graphene oxide (GO) using Avicennia marina and silver, thereby leveraging the natural and metallic components for enhanced functionality [51].

Enhancement of Catalytic Surfaces for Green Chemistry

Catalysis is vital in chemical industries, essential for producing many key chemicals. As petroleum resources dwindle and focus on renewable energy and environmental concerns grows, the role of catalysis is set to increase. However, traditional methods of preparing heterogeneous catalysts are environmentally taxing, resulting in pollution and consuming significant amounts of materials and energy. Although catalysis aligns with green chemistry principles, its conventional preparation methods are often not environmentally friendly [52]. This is where plasma-assisted green grafting comes into play, offering a sustainable alternative for catalyst development. By utilizing plasma technology, green grafting can reduce the environmental footprint associated with catalyst preparation, minimizing pollution and energy consumption. This approach not only aligns with green chemistry principles but also addresses the challenges of traditional catalyst preparation, thereby paving the way for more sustainable and eco-friendly catalytic processes in the chemical industry [53]. For marine equipment, biofilm formation and corrosion are frequent issues that can reduce equipment lifespan and increase maintenance costs. It has been demonstrated that a highly effective method for addressing these issues is to graft antifouling polymers, such as poly(sulfobetaine methacrylate), onto stainless steel surfaces with the aid of plasma. This method minimizes maintenance costs and frequency while also extending the usable life of marine equipment by significantly minimizing the accumulation of biofilms and other fouling organisms. The plasma-assisted technology is considered an environmentally advantageous method, offering a more sustainable and environmentally friendly solution for marine applications, as it eliminates the need for toxic antifouling paints and coatings [54].

MECHANISM OF PLASMA ASSISTED GRAFTING

A flexible surface modification method known as "plasma-assisted grafting" uses plasma to graft desired molecules or polymers onto a material's surface by introducing functional groups. The procedure is a useful tool in green synthesis, as it is highly effective, often carried out in temperate environments, and is considered ecologically benign [55]. A schematic representation of polymer grafting is shown in Fig. (3).

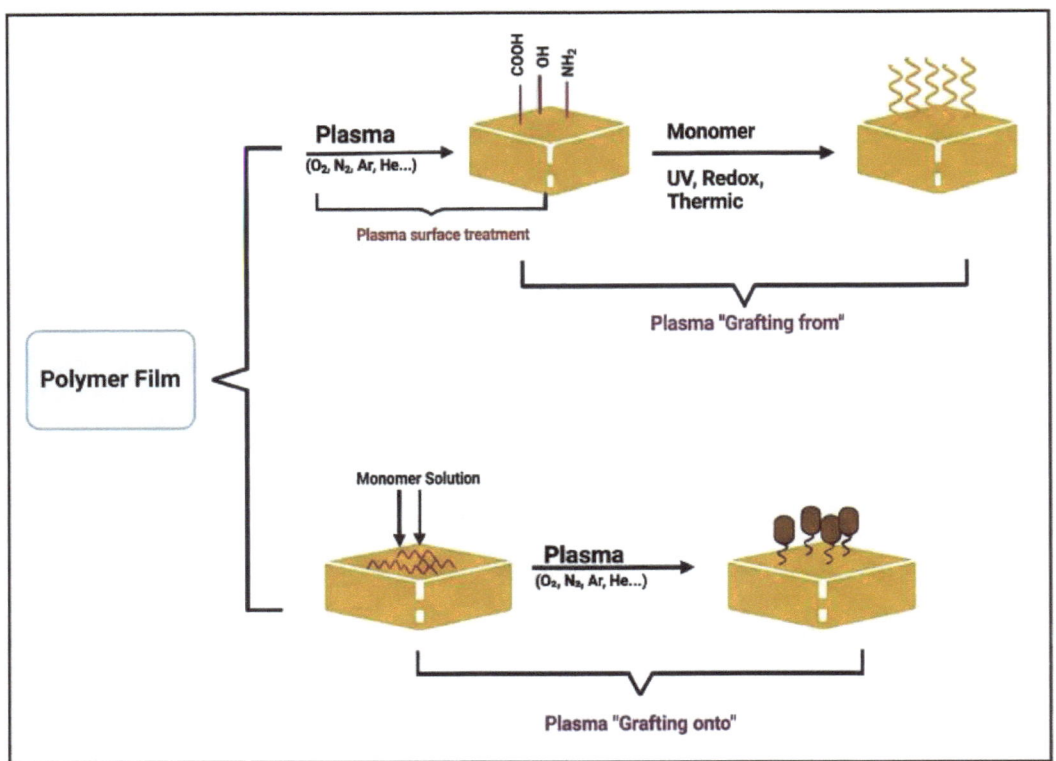

Fig. (3). Schematic representation of polymer grafting.

Plasma Generation

The process of applying an electric field to a gaseous medium causes the gas molecules to become ionized, producing a quasi-neutral ensemble of charged particles, including ions, electrons, radicals, and excited atomic or molecular species. This is how plasma is formed. The type of gas used—typically oxygen, nitrogen, argon, or air—as well as the variables controlling plasma creation, such as pressure, applied power, and frequency, have a significant impact on the specific composition and characteristics of the plasma [56]. These parameters are carefully calibrated to elicit specific surface topographies and chemistries, providing accurate control over the surface alterations that follow. The intricate relationships present in the plasma environment enable the production of very reactive species, which are necessary to start surface activation and subsequent grafting processes [57].

Surface Activation

Bombardment by Plasma Species: When a substrate is exposed to the plasma environment, its surface undergoes intensive bombardment by energetic plasma species, including ions, electrons, and radicals. This bombardment leads to a series of critical surface modifications [58, 59]

Surface Cleaning: The high-energy impact of plasma species efficiently eliminates loosely adherent layers, weakly bonded molecules, and organic impurities, leaving behind a surface that is both highly reactive and immaculate. For the grafting procedure to be consistent and effective thereafter, this cleaning step is particularly important [60].

Formation of Reactive Sites: Reactive sites, such as free radicals, ionized groups, and other chemically active functionalities (*e.g.*, carboxyl, hydroxyl, or amine groups), are formed when energetic species interact with the substrate surface. By selecting the appropriate plasma gas and operating parameters, these sites function as nucleation centers for the grafting process, enabling precise control over the type and density of reactive sites. Introducing specific functional groups is essential for customizing surface characteristics to meet application needs [61, 62].

Grafting of Functional Groups or Polymers

Initiation of Grafting: The grafting process begins when reactive sites are present on the substrate surface. Covalent bonds between the substrate and the grafted species are formed as a result of interactions between monomers or functional molecules and these reactive sites in the plasma environment. The grafting species' and substrate's chemical makeup, as well as the plasma circumstances that affect these sites' activation and stability, all play a significant role in this process [63].

Polymerization: When monomers are used, they can become polymer chains that are covalently grafted onto the substrate by directly undergoing plasma-induced polymerization on the surface. In addition to initiating the polymerization process, the plasma environment facilitates simultaneous grafting, resulting in a homogeneous, conformal coating with specific chemical functions. It is possible to engineer the resulting polymer layer to possess certain characteristics, such as hydrophilicity, biocompatibility, or improved adhesion [64].

Cross-Linking: Cross-linking processes between the grafted polymer chains and the substrate surface, or between the grafted chains themselves, can also be induced under plasma conditions. By strengthening the grafted layer's mechanical stability, chemical resistance, and durability, this cross-linking makes it suitable for demanding applications such as environmental protection or biomedical devices [65].

Control of Grafting Density and Thickness

Parameter Optimization: By adjusting the plasma working conditions, two crucial factors that may be carefully controlled are the grafting density and the thickness of the grafted layer. When assessing the degree of surface alteration, variables including exposure duration, applied power, gas composition, and pressure are important considerations. Longer exposure durations or greater power levels often increase the density of grafted chains and the subsequent layer thickness. To prevent unfavourable outcomes, such as over-etching or excessive cross-linking, which could jeopardize the intended surface qualities, these parameters must be properly adjusted [66].

Selective Grafting: Spatial selectivity in plasma-assisted grafting enables the deliberate alteration of particular areas of the substrate surface. This selectivity can be achieved by carefully regulating plasma exposure to specific regions or by employing masking strategies. For applications such as microelectronics or biomedical implants that require patterned surfaces or targeted functionalization, this level of accuracy is crucial [67].

Post-Grafting Treatment

Stabilization and Purification: To stabilize the grafted layer and ensure its long-term performance, the substrate may undergo post-treatment procedures after the grafting process. These treatments may involve washing to remove unreacted monomers or by-products, or thermal annealing to strengthen cross-linking or eliminate any weakly attached species. To ensure that the modified surface meets the stringent requirements of cutting-edge technological applications, post-grafting procedures are necessary to achieve the requisite mechanical strength, chemical stability, and performance characteristics of the grafted layer [68, 69].

COMPARATIVE ANALYSIS OF PLASMA-ASSISTED SYNTHESIS AND TRADITIONAL SYNTHESIS METHODS IN THE FRAMEWORK OF GREEN CHEMISTRY

Within the framework of green chemistry, plasma-assisted synthesis offers several advantages over conventional synthesis techniques, including increased energy efficiency, reduced use of hazardous chemicals, enhanced reaction selectivity, and a lower environmental impact. Even if scaling remains a barrier, continuous developments in plasma technology are making it an increasingly attractive substitute for sustainable chemical synthesis. Even if they are scalable and well-established, traditional methods often fall short in terms of energy efficiency and environmental impact, highlighting the potential of plasma-assisted methods to promote greener industrial practices in the future [70].

Energy Efficiency

Since plasma-assisted synthesis processes may function at low temperatures and air pressure, they are often considered to be energy-efficient. Highly reactive species are generated in a plasma environment, which facilitates chemical processes without requiring substantial thermal energy inputs. As a consequence, less energy is used than in conventional high-temperature synthesis procedures. Furthermore, by reducing the need for external reagents or catalysts, the in situ production of reactive species reduces energy requirements and improves overall process efficiency [71]. To overcome activation barriers, however, conventional synthesis methods typically require high temperatures, prolonged reaction periods, and substantial energy inputs. Particularly, energy-intensive processes such as thermal decomposition, calcination, and sintering have a significant negative impact on the environment. Moreover, the production of solvents, catalysts, and reagents required for these processes is often energy-intensive, exacerbating the overall energy footprint of conventional synthesis methods [72].

Use of Hazardous Chemicals

One major benefit of plasma-assisted synthesis is that it can reduce or even eliminate the need for hazardous ingredients. Without the need for hazardous reagents or solvents, the reactive species produced in the plasma phase can efficiently initiate and maintain chemical reactions, resulting in cleaner operations with reduced formation of hazardous waste [73]. This improves the environmental safety of the synthesis process while also streamlining waste management. Furthermore, the ability to carefully adapt plasma processes to specific reactions enables more selective and effective synthesis routes. Traditional synthesis techniques, on the other hand, sometimes rely on hazardous substances, such as strong acids, bases, and organic solvents, which pose serious threats to the environment and human health [74].

Reaction Selectivity and Yield

Plasma-assisted synthesis is advantageous due to its non-equilibrium nature, which enables highly selective reactions by controlling the energy distribution among plasma species to favour specific chemical pathways. This selectivity results in higher yields of desired products with minimal by-products, aligning it well with green chemistry principles [75]. Additionally, plasma-assisted processes offer fine-tuning capabilities through adjustments to parameters like gas composition, power input, and pressure, providing a level of control that is often difficult to achieve with traditional methods. In contrast, traditional synthesis methods often suffer from lower selectivity, particularly in complex reactions where multiple pathways are possible. This often results in lower yields and the formation of unwanted by-products, necessitating additional separation and purification steps that reduce overall efficiency and contribute to increased waste and resource consumption [76].

Scalability and Practical Application

Scalability is a challenge in plasma-assisted synthesis, as maintaining consistent plasma conditions over large volumes can be difficult, limiting its application in industrial production despite its high efficiency in laboratory settings. However, advancements in continuous-flow plasma systems and plasma reactor design are addressing these issues, thereby enhancing the potential for plasma-assisted synthesis in large-scale applications [77]. In contrast, conventional synthesis techniques are widely accepted and extensively used in large-scale manufacturing due to decades of optimization for scalability. However, this scalability often comes with increased waste output, higher energy consumption, and more significant environmental impacts, making these traditional processes less sustainable despite their industrial viability [78].

Environmental Impact

Plasma-assisted synthesis typically has a smaller environmental impact compared to conventional techniques due to its high selectivity of reactions, the elimination of hazardous chemicals, and reduced energy consumption. The ability to perform plasma operations under ambient conditions further reduces the need for stringent safety measures, thereby lowering the overall environmental impact. On the other hand, conventional synthesis methods have a significant environmental impact due to their reliance on hazardous chemicals, high energy demands, and substantial waste generation. These processes often require complex waste management and purification systems, which increase the overall carbon footprint and resource consumption, exacerbating their environmental impact [79, 80].

ENVIRONMENTAL BENEFITS OF PLASMA-ASSISTED GREEN GRAFTING

Plasma-Assisted Green Grafting (PAGG) represents an innovative approach that aligns with the principles of green chemistry and sustainable development. It utilizes plasma, an ionized gas containing reactive species such as ions, radicals, and electrons, to activate or functionalize material surfaces, particularly polymers, without relying on traditional chemical grafting agents [81, 82].

Elimination of Hazardous Chemicals

Traditional grafting techniques frequently employ hazardous chemicals—including poisonous, volatile, and environmentally persistent initiators, solvents, and monomers. By using radicals produced by plasma to trigger grafting, PAGG eliminates the requirement for these substances. This complies with strict environmental laws and reduces the ecological footprint of industrial operations by significantly lowering the formation of hazardous byproducts and Volatile Organic Compounds (VOCs) [83, 84].

Energy Efficiency and Process Optimization

Unlike thermal grafting methods, which need significant energy inputs to reach the required activation energies, PAGG operates at ambient or near-ambient temperatures. In addition to saving energy, PAGG's low-temperature operation shields delicate substrates from thermal deterioration, enabling the processing of a wider variety of materials. Due to the energy savings, PAGG is a more environmentally friendly choice for large-scale industrial applications, as it leaves a smaller total carbon imprint [85, 86].

Minimal Environmental Impact and Waste Reduction

The majority of PAGG's byproducts are safe, readily controlled, non-toxic gases, such as nitrogen or oxygen. In contrast to chemical grafting, which produces a significant amount of liquid and solid waste that needs to be disposed of and treated, PAGG produces very little waste, allowing for cleaner manufacturing conditions. This waste reduction aligns with the tenets of the circular economy, which prioritize resource conservation and waste minimization [85, 87].

Enhanced Material Properties and Recyclability

Without altering the material's bulk characteristics, plasma treatment enables the precise functionalization of material surfaces, thereby enhancing properties such as hydrophilicity, adhesion, and biocompatibility. Due to this accuracy, high-performance materials with improved recyclability can be produced. To enhance compatibility in recycling streams, for example, the selective modification of polymer surfaces can lessen the deterioration of recovered materials and encourage closed-loop recycling systems [88, 89].

Water Conservation and Process Sustainability

Water is used extensively in conventional grafting processes to remove unreacted monomers, initiators, and solvents through lengthy washing stages. Since PAGG is a dry process, it uses no water, thereby saving this vital resource and reducing the environmental impact of wastewater treatment. This benefit aligns with global sustainability goals centered on water conservation and is particularly relevant in areas where water shortages are an issue [90, 91].

Reduced Environmental Footprint in Surface Engineering

Reduced material use and environmental impact are further benefits of PAGG's ability to functionalize surfaces at the nanoscale without the need for bulk chemicals or high temperatures. Through the facilitation of surface alterations that enhance material performance (*e.g.*, increased resilience, improved barrier qualities), PAGG facilitates the creation of materials with a longer lifespan and a lower environmental footprint during both use and disposal stages [92, 93]. Plasma-Assisted Green Grafting (PAGG) provides a multifaceted approach to surface modification, significantly reducing the environmental impact of manufacturing processes. Its alignment with green chemistry principles, energy efficiency, waste minimization, and enhanced material recyclability makes it a compelling technology for advancing sustainable manufacturing and material engineering practices [94].

APPLICATION OF PAGG IN THE PHARMACEUTICAL INDUSTRY

Plasma-Assisted Green Grafting (PAGG) holds significant potential in the healthcare industry due to its ability to modify and enhance the surface properties of medical devices and materials in an environmentally friendly manner [95].

Enhance Biocompatibility of Medical Implants

Specifically for implants such as dental implants, stents, or joint replacements, PAGG offers a long-term way to enhance the integration of medical implants with surrounding biological tissue. By grafting biocompatible molecules onto the surfaces of these implants, PAGG improves their biocompatibility and lowers the likelihood of rejection, inflammation, and other complications—all of which have a positive impact on patient outcomes [96]. Different surface treatments have been developed over the last 20 years to address issues with the biocompatibility of medical implants. PAGG distinguishes itself by providing a more environmentally friendly method, especially for implants that are subjected to severe mechanical stress, such as knee and hip prostheses, or those with specialized needs for biocompatibility, like coronary stents. By enhancing the surface characteristics of materials such as PEEK and titanium alloys, PAGG addresses key concerns of durability, infection prevention, and inflammation, representing a significant advancement in biomaterial surface engineering [97].

Development of Antimicrobial Surfaces

By administering antimicrobial chemicals to the surfaces of catheters, surgical instruments, medical equipment, and wound dressings, a technique known as Plasma-Assisted Green Grafting (PAGG) plays a vital role in the fight against infections associated with medical devices. Hospital-Acquired Infections (HAIs) are a significant problem in healthcare settings, and antimicrobial surfaces play a crucial role in reducing the incidence of HAIs [98]. PAGG is a cutting-edge strategy for utilizing environmentally friendly techniques to enhance the antibacterial efficacy of metals, polymers, and other biocompatible materials. Through the use of non-equilibrium reactive plasma chemistries, PAGG enables the creation of high-end, complex coatings with antifungal or antibacterial properties. This groundbreaking method not only minimizes stages in the production process but also drastically reduces the need for costly and hazardous chemicals, aligning with the healthcare industry's growing emphasis on environmentally friendly and sustainable practices [99].

Improved Drug Delivery Systems

The objective is to enhance the effectiveness and precision of drug delivery. Plasma-assisted grafting and functionalization can be utilized to modify drug delivery vehicles, such as nanoparticles, microneedles, or patches, thereby enhancing their biocompatibility and targeting capabilities. This leads to more efficient and localized drug release, thereby reducing side effects and improving therapeutic outcomes [100]. By modifying the surface of the IOLs with 2-hydroxyethyl methacrylate (HEMA) using argon plasma-assisted grafting, the system integrated MFX by trapping inside the grafted polyHEMA layer as well as through soaking [101]. Produced at temperatures below 40°C, cold atmospheric plasma (CAP) has shown great promise in biological applications, particularly in the field of medication administration. Building on the ideas of CAP, Plasma-Assisted Green Grafting (PAGG) provides an efficient and environmentally friendly approach to enhancing medication delivery systems. Using UV light, small temperature changes, and reactive species such as free radicals generated by CAP, PAGG modifies the surface characteristics of biological tissues and drug carriers [102].

CONCLUSION

Plasma-Assisted Green Grafting (PAGG) represents a significant advancement in surface modification techniques, offering a more environmentally friendly and efficient alternative to traditional grafting methods. The principle of PAGG relies on the use of plasma technology to activate and modify surfaces, thereby facilitating the precise grafting of functional groups and polymers. This method not only enhances the surface properties of materials but also aligns with green chemistry principles by minimizing the use of hazardous chemicals and reducing waste. Compared to conventional grafting techniques, PAGG stands out due to its ability to achieve high grafting efficiency under mild conditions, thereby preserving the integrity of sensitive substrates and contributing to more sustainable manufacturing practices. The environmental benefits of PAGG, including reduced chemical consumption and lower environmental impact, underscore its potential as a key technology in the shift towards greener industrial processes. In the pharmaceutical industry, PAGG offers promising applications, from improving drug delivery systems to enhancing the functionality of medical devices. By leveraging the precision and environmental advantages of PAGG, the pharmaceutical sector can advance towards more sustainable and effective solutions, ultimately benefiting both technological progress and environmental stewardship.

ABBREVIATIONS

PAGG	Plasma-Assisted Green Grafting
PLA	Poly(Lactic Acid)
DMC	Dimethyl Carbonate'
VOC	Volatile Organic Compounds
PTFE	Polytetrafluoroethylene
PA	Poly(ethylene-alt-maleic anhydride
PEG	polyethylene glycol
PDMS	Polydimethylsiloxane
GO	Graphene Oxide
HAIs	Hospital-Acquired Infections
HEMA	2-Hydroxyethyl methacrylate
CAP	Cold Atmospheric Plasma
PEEK	Polyether ether ketone
MFX	Moxifloxacin
IOLs	Intraocular lens

ACKNOWLEDGEMENTS

All of the Authors are thankful to the Department of Pharmaceutical Sciences, Rashtrasant Tukadoji Maharaj Nagpur University, Nagpur, for providing all necessary facilities and guidance for this work.

REFERENCES

[1] Pleissner D, Kümmerer K. Green Chemistry and Its Contribution to Industrial Biotechnology. 2018; pp. 281-98.
[http://dx.doi.org/10.1007/10_2018_73]

[2] Zimmerman JB, Anastas PT, Erythropel HC, Leitner W. Designing for a green chemistry future. Science 2020; 367(6476): 397-400.
[http://dx.doi.org/10.1126/science.aay3060] [PMID: 31974246]

[3] Matlin SA, Mehta G, Hopf H, Krief A. The role of chemistry in inventing a sustainable future. Nat Chem 2015; 7(12): 941-3.
[http://dx.doi.org/10.1038/nchem.2389] [PMID: 26587703]

[4] Du X, Gao F, Hua Y, Zhang X, Li H, Di L. Green, mild, and rapid plasma synthesis of nitrogen/amino-cofunctionalized MWCNTs-supported Pd catalyst in solution for boosting formic acid dehydrogenation. Fuel 2024; 377: 132751.
[http://dx.doi.org/10.1016/j.fuel.2024.132751]

[5] Ardila-Fierro KJ, Hernández JG. Sustainability Assessment of Mechanochemistry by Using the Twelve Principles of Green Chemistry. ChemSusChem 2021; 14(10): 2145-62.
[http://dx.doi.org/10.1002/cssc.202100478] [PMID: 33835716]

[6] Kümmerer K, Clark J. Green and Sustainable Chemistry Sustain Sci. Dordrecht: Springer Netherlands 2016; pp. 43-59.

[7] Mohd Makhtar SNN, Ahmad Sohaimi KSB, Salleh NKM, Fajrina N. Grafting process on photocatalytic membrane Advanced Ceramics for Photocatalytic Membranes. Elsevier 2024; pp. 157-78.
[http://dx.doi.org/10.1016/B978-0-323-95418-1.00003-3]

[8] Constable DJC, Jimenez-Gonzalez C, Henderson RK. Perspective on Solvent Use in the Pharmaceutical Industry. Org Process Res Dev 2007; 11(1): 133-7.
[http://dx.doi.org/10.1021/op060170h]

[9] Shen Y, Qi R, Liu Q, Wang Y, Mao Y, Yu J. Grafting of maleic anhydride onto polyethylene through a green chemistry approach. J Appl Polym Sci 2008; 110(4): 2261-6.
[http://dx.doi.org/10.1002/app.28789]

[10] Bramhe P, Waghmare S, Rarokar N, et al. Polymer blends innovation: Advancement in novel drug delivery. Int J Polym Mater Polym Biomater 2024; 1-18.

[11] Wang C, Chen L, Li J, et al. Enhancing the interfacial strength of carbon fiber reinforced epoxy composites by green grafting of poly(oxypropylene) diamines. Compos, Part A Appl Sci Manuf 2017; 99: 58-64.
[http://dx.doi.org/10.1016/j.compositesa.2017.04.003]

[12] Dridi D, Bouaziz A, Gargoubi S, et al. Enhanced Antibacterial Efficiency of Cellulosic Fibers: Microencapsulation and Green Grafting Strategies. Coatings 2021; 11(8): 980.
[http://dx.doi.org/10.3390/coatings11080980]

[13] Zainol Abidin MN, Nasef MM, Matsuura T. Fouling Prevention in Polymeric Membranes by Radiation Induced Graft Copolymerization. Polymers (Basel) 2022; 14(1): 197.
[http://dx.doi.org/10.3390/polym14010197] [PMID: 35012218]

[14] Karimah A, Ridho MR, Munawar SS, et al. A review on natural fibers for development of eco-friendly bio-composite: characteristics, and utilizations. J Mater Res Technol 2021; 13: 2442-58.
[http://dx.doi.org/10.1016/j.jmrt.2021.06.014]

[15] Walden R, Goswami A, Scally L, McGranaghan G, Cullen PJ, Pillai SC. Nonthermal plasma technologies for advanced functional material processing and current applications: Opportunities and challenges. J Environ Chem Eng 2024; 12(5): 113541.
[http://dx.doi.org/10.1016/j.jece.2024.113541]

[16] Wang Z, Chen J, Sun S, et al. Plasma-enabled synthesis and modification of advanced materials for electrochemical energy storage. Energy Storage Mater 2022; 50: 161-85.
[http://dx.doi.org/10.1016/j.ensm.2022.05.018]

[17] Zhu X, Dai L, Guan X, et al. A manual operated and eco-friendly plasma-assisted polymerization method for enhancing surface weather resistance of insulation. Appl Surf Sci 2024; 675: 160970.
[http://dx.doi.org/10.1016/j.apsusc.2024.160970]

[18] Chandwani N, Jain V, Dave P, Dave H, Jhala PB, Nema SK. Experimental Studies on Applications of Atmospheric Pressure Air Plasma for Eco-friendly Processing of Textiles and Allied Material. J Inst Eng India Ser E 2021; 102: 203-13.
[http://dx.doi.org/10.1007/s40034-021-00219-z]

[19] Vesel A. Deposition of Chitosan on Plasma-Treated Polymers—A Review. Polymers (Basel) 2023; 15(5): 1109.
[http://dx.doi.org/10.3390/polym15051109] [PMID: 36904353]

[20] Hossain MF, Park JY, Kim J. Surface modification of polymer materials using atmospheric pressure plasma for enhanced functional performance. Polymers (Basel) 2021; 13(17): 2899.

[21] Pillai RR, Thomas V. Plasma Surface Engineering of Natural and Sustainable Polymeric Derivatives and Their Potential Applications. Polymers (Basel) 2023; 15(2): 400.
[http://dx.doi.org/10.3390/polym15020400] [PMID: 36679280]

[22] Pustahija L, Kern W. Surface Functionalization of (Pyrolytic) Carbon—An Overview. C (Basel). 2023; 9: 38.

[23] Ma C, Nikiforov A, Hegemann D, De Geyter N, Morent R, Ostrikov KK. Plasma-controlled surface wettability: recent advances and future applications. Int Mater Rev 2023; 68(1): 82-119.
[http://dx.doi.org/10.1080/09506608.2022.2047420]

[24] Luo Y, Wu Y, Huang C, Menon C, Feng SP, Chu PK. Plasma modified and tailored defective electrocatalysts for water electrolysis and hydrogen fuel cells. EcoMat 2022; 4(4): e12197.
[http://dx.doi.org/10.1002/eom2.12197]

[25] Ahmed HS, Yahya Z, Ali khan W, Faraz A. Sustainable pathways to ammonia: a comprehensive review of green production approaches. Clean Energy 2024; 8(2): 60-72.
[http://dx.doi.org/10.1093/ce/zkae002]

[26] Sun X, Bao J, Li K, et al. Advance in Using Plasma Technology for Modification or Fabrication of Carbon-based Materials and Their Applications in Environmental, Material, and Energy Fields. Adv Funct Mater 2021; 31(7): 2006287.
[http://dx.doi.org/10.1002/adfm.202006287]

[27] Singh A, Bhowmick AK, Rane AV. Atmospheric pressure plasma surface modification of polymers for improved adhesion and functional properties. Surf Coat Technol 2022; 433: 128153.

[28] Primc G. Recent Advances in Surface Activation of Polytetrafluoroethylene (PTFE) by Gaseous Plasma Treatments. Polymers (Basel) 2020; 12(10): 2295.
[http://dx.doi.org/10.3390/polym12102295] [PMID: 33036423]

[29] Hati S, Patel M, Yadav D. Food bioprocessing by non-thermal plasma technology. Curr Opin Food Sci 2018; 19: 85-91.

[http://dx.doi.org/10.1016/j.cofs.2018.03.011]

[30] Di L, Zhang J, Zhang X, *et al.* Cold plasma treatment of catalytic materials: a review. J Phys D Appl Phys 2021; 54(33): 333001.
[http://dx.doi.org/10.1088/1361-6463/ac0269]

[31] Zaborniak I, Chmielarz P. Polymer-modified regenerated cellulose membranes: following the atom transfer radical polymerization concepts consistent with the principles of green chemistry. Cellulose 2023; 30(1): 1-38.
[http://dx.doi.org/10.1007/s10570-022-04880-4]

[32] Soltys L, Olkhovyy O, Tatarchuk T, Naushad M. Green Synthesis of Metal and Metal Oxide Nanoparticles: Principles of Green Chemistry and Raw Materials. Magnetochemistry 2021; 7(11): 145.
[http://dx.doi.org/10.3390/magnetochemistry7110145]

[33] Kolen'ko B, Mafla C, Richards K, et al. Plasma-Induced Graft Polymerization for the In Situ Synthesis of Cross-Linked Nanocoatings. ACS Appl Eng Mater 2024;2(3):563-73.
[http://dx.doi.org/10.1021/acsaenm.3c00536]

[34] Palma V, Cortese M, Renda S, Ruocco C, Martino M, Meloni E. A Review about the Recent Advances in Selected NonThermal Plasma Assisted Solid–Gas Phase Chemical Processes. Nanomaterials (Basel) 2020; 10(8): 1596.
[http://dx.doi.org/10.3390/nano10081596] [PMID: 32823944]

[35] Yusuf A, Amusa HK, Eniola JO, *et al.* Hazardous and emerging contaminants removal from water by plasma-based treatment: A review of recent advances. Chem Eng J Adv 2023; 14: 100443.
[http://dx.doi.org/10.1016/j.ceja.2023.100443]

[36] Maraveas C, Bayer IS, Bartzanas T. Recent Advances in Antioxidant Polymers: From Sustainable and Natural Monomers to Synthesis and Applications. Polymers (Basel) 2021; 13(15): 2465.
[http://dx.doi.org/10.3390/polym13152465] [PMID: 34372069]

[37] Laureano-Anzaldo CM, González-López ME, Pérez-Fonseca AA, Cruz-Barba LE, Robledo-Ortíz JR. Plasma-enhanced modification of polysaccharides for wastewater treatment: A review. Carbohydr Polym 2021; 252: 117195.
[http://dx.doi.org/10.1016/j.carbpol.2020.117195] [PMID: 33183635]

[38] Wang Y, Zhu L, Guo P, Zhang Y, Lan X, Xu W. Research progress of All-in-One PCR tube biosensors based on functional modification and intelligent fabrication. Biosens Bioelectron 2024; 246: 115824.
[http://dx.doi.org/10.1016/j.bios.2023.115824] [PMID: 38029707]

[39] Levchenko I, Xu S, Baranov O, Bazaka O, Ivanova E, Bazaka K. Plasma and Polymers: Recent Progress and Trends. Molecules 2021; 26(13): 4091.
[http://dx.doi.org/10.3390/molecules26134091] [PMID: 34279431]

[40] Pustahija L, Kern W. Surface Functionalization of (Pyrolytic) Carbon—An Overview. C (Basel). 2023; 9: 38.

[41] Andresen J, Lim Xiao Y. Pyrolysis processes and technology for the conversion of hydrocarbons and biomass Advances in Clean Hydrocarbon Fuel Processing. Elsevier 2011; pp. 186-98.
[http://dx.doi.org/10.1533/9780857093783.2.186]

[42] Chu P. Plasma-surface modification of biomaterials. Mater Sci Eng Rep 2002; 36(5-6): 143-206.
[http://dx.doi.org/10.1016/S0927-796X(02)00004-9]

[43] Wang J, Yu W, Graham NJD, Jiang L. Evaluation of a novel polyamide-polyethylenimine nanofiltration membrane for wastewater treatment: Removal of Cu^{2+} ions. Chem Eng J 2020; 392: 123769.
[http://dx.doi.org/10.1016/j.cej.2019.123769]

[44] Sakhare MS, Rajput HH. Polymer grafting and applications in pharmaceutical drug delivery systems -

a brief review. Asian J Pharm Clin Res. 2017; 10(6): 59.
[http://dx.doi.org/10.22159/ajpcr.2017.v10i6.18072]

[45] Kim YJ, Kang IK, Huh MW, Yoon SC. Surface characterization and in vitro blood compatibility of poly(ethylene terephthalate) immobilized with insulin and/or heparin using plasma glow discharge. Biomaterials 2000; 21(2): 121-30.
[http://dx.doi.org/10.1016/S0142-9612(99)00137-4] [PMID: 10632394]

[46] Ogino A, Kral M, Yamashita M, Nagatsu M. Effects of low-temperature surface-wave plasma treatment with various gases on surface modification of chitosan. Appl Surf Sci 2008; 255(5): 2347-52.
[http://dx.doi.org/10.1016/j.apsusc.2008.07.119]

[47] Vaz JM, Taketa TB, Hernandez-Montelongo J, et al. Antibacterial noncytotoxic chitosan coatings on polytetrafluoroethylene films by plasma grafting for medical device applications. J Coat Technol Res 2022; 19(3): 829-38.
[http://dx.doi.org/10.1007/s11998-021-00560-3]

[48] Roh S, Jang Y, Yoo J, Seong H. Surface Modification Strategies for Biomedical Applications: Enhancing Cell–Biomaterial Interfaces and Biochip Performances. Biochip J 2023; 17(2): 174-91.
[http://dx.doi.org/10.1007/s13206-023-00104-4]

[49] Ranjha MMAN, Shafique B, Aadil RM, Manzoor MF, Cheng JH. Modification in cellulose films through ascent cold plasma treatment and polymerization for food products packaging. Trends Food Sci Technol 2023; 134: 162-76.
[http://dx.doi.org/10.1016/j.tifs.2023.03.011]

[50] Nikiforov A, Ma C, Choukourov A, Palumbo F. Plasma technology in antimicrobial surface engineering. J Appl Phys. 2022; 131(1): 1-35.

[51] Soleimani S, Jannesari A, Yousefzadi M, Ghaderi A, Shahdadi A. Eco-friendly foul release coatings based on a novel reduced graphene oxide/Ag nanocomposite prepared by a green synthesis approach. Prog Org Coat 2021; 151: 106107.
[http://dx.doi.org/10.1016/j.porgcoat.2020.106107]

[52] Liu C, Li M, Wang J, et al. Plasma methods for preparing green catalysts: Current status and perspective. Chin J Catal 2016; 37(3): 340-8.
[http://dx.doi.org/10.1016/S1872-2067(15)61020-8]

[53] Xu J, Yang B, Chen K, et al. Cold plasma and ethanol coupling-activated Ni/MCM-41 catalysts with highly stability and activity for dry reforming of methane. Microporous Mesoporous Mater 2024; 364: 112847.
[http://dx.doi.org/10.1016/j.micromeso.2023.112847]

[54] Dufour T. From Basics to Frontiers: A Comprehensive Review of Plasma-Modified and Plasma-Synthesized Polymer Films. Polymers (Basel) 2023; 15(17): 3607.
[http://dx.doi.org/10.3390/polym15173607] [PMID: 37688233]

[55] Konchekov EM, Kolik LV, Danilejko YK, et al. Enhancement of the Plant Grafting Technique with Dielectric Barrier Discharge Cold Atmospheric Plasma and Plasma-Treated Solution. Plants 2022; 11(10): 1373.
[http://dx.doi.org/10.3390/plants11101373] [PMID: 35631800]

[56] Cheng CY, Chung FY, Chou PY, Huang C. Surface Modification of Polytetrafluoroethylene by Atmospheric Pressure Plasma-Grafted Polymerization. Plasma Chem Plasma Process 2020; 40(6): 1507-23.
[http://dx.doi.org/10.1007/s11090-020-10112-z]

[57] Hegemann D, Gaiser S. Plasma surface engineering for manmade soft materials: a review. J Phys D Appl Phys 2022; 55(17): 173002.
[http://dx.doi.org/10.1088/1361-6463/ac4539]

[58] Kyzioł A, Kyzioł K. Surface Functionalization With Biopolymers via Plasma-Assisted Surface Grafting and Plasma-Induced Graft Polymerization—Materials for Biomedical Applications Biopolymer Grafting: Applications. Elsevier 2018; pp. 115-51.
[http://dx.doi.org/10.1016/B978-0-12-810462-0.00004-1]

[59] Qin L, Ishizaki T, Takeuchi N, Takahashi K, Kim KH, Li OL. Green Sulfonation of Carbon Catalysts *via* Gas–Liquid Interfacial Plasma for Cellulose Hydrolysis. ACS Sustain Chem& Eng 2020; 8(15): 5837-46.
[http://dx.doi.org/10.1021/acssuschemeng.9b07156]

[60] Gleissner C, Landsiedel J, Bechtold T, Pham T. Surface Activation of High Performance Polymer Fibers: A Review. Polym Rev (Phila Pa) 2022; 62(4): 757-88.
[http://dx.doi.org/10.1080/15583724.2022.2025601]

[61] Sun W, Liu W, Wu Z, Chen H. Chemical Surface Modification of Polymeric Biomaterials for Biomedical Applications. Macromol Rapid Commun 2020; 41(8): 1900430.
[http://dx.doi.org/10.1002/marc.201900430] [PMID: 32134540]

[62] Purohit P, Bhatt A, Mittal RK, *et al.* Polymer Grafting and its Chemical Reactions. Front Bioeng Biotechnol 2023;11:1044927.
[http://dx.doi.org/10.3389/fbioe.2022.1044927]

[63] Kumar M, Gehlot PS, Parihar D, Surolia PK, Prasad G. Promising grafting strategies on cellulosic backbone through radical polymerization processes – A review. Eur Polym J 2021; 152: 110448.
[http://dx.doi.org/10.1016/j.eurpolymj.2021.110448]

[64] Suresh D, Goh PS, Ismail AF, Hilal N. Surface Design of Liquid Separation Membrane through Graft Polymerization: A State of the Art Review. Membranes (Basel) 2021; 11(11): 832.
[http://dx.doi.org/10.3390/membranes11110832] [PMID: 34832061]

[65] Asadian M, Chan KV, Norouzi M, *et al.* Fabrication and Plasma Modification of Nanofibrous Tissue Engineering Scaffolds. Nanomaterials (Basel) 2020; 10(1): 119.
[http://dx.doi.org/10.3390/nano10010119] [PMID: 31936372]

[66] Mishra H, Bolouki N, Hsieh ST, Li C, Wu W, Hsieh JH. Application of Spectroscopic Analysis for Plasma Polymerization Deposition onto the Inner Surfaces of Silicone Tubes. Coatings 2022; 12(6): 865.
[http://dx.doi.org/10.3390/coatings12060865]

[67] Sundriyal P, Pandey M, Bhattacharya S. Plasma-assisted surface alteration of industrial polymers for improved adhesive bonding. Int J Adhes Adhes 2020; 101: 102626.
[http://dx.doi.org/10.1016/j.ijadhadh.2020.102626]

[68] Bhattarai DP, Pokharel P, Xiao D. Correction to: Surface Functionalization of Polymers Reactive and Functional Polymers Volume Four. Cham: Springer International Publishing 2020; pp. C1-1.

[69] Suresh D, Goh PS, Ismail AF, Hilal N. Surface Design of Liquid Separation Membrane through Graft Polymerization: A State of the Art Review. Membranes 2021;11(11):832.
[http://dx.doi.org/10.3390/membranes11110832]

[70] Palumbo F, Lo Porto C, Fracassi F, Favia P. Recent Advancements in the Use of Aerosol-Assisted Atmospheric Pressure Plasma Deposition. Coatings 2020; 10(5): 440.
[http://dx.doi.org/10.3390/coatings10050440]

[71] Lee J, Kim J, Hyeon T. Recent Progress in the Synthesis of Porous Carbon Materials. Adv Mater 2006; 18(16): 2073-94.
[http://dx.doi.org/10.1002/adma.200501576]

[72] Vandenabeele CR, Lucas S. Technological challenges and progress in nanomaterials plasma surface modification – A review. Mater Sci Eng Rep 2020; 139: 100521.
[http://dx.doi.org/10.1016/j.mser.2019.100521]

[73] Taaca KLM, Prieto EI, Vasquez MR Jr. Current Trends in Biomedical Hydrogels: From Traditional Crosslinking to Plasma-Assisted Synthesis. Polymers 2022;14(13):2560.
[http://dx.doi.org/10.3390/polym14132560]

[74] Rouwenhorst KHR, Engelmann Y, van 't Veer K, Postma RS, Bogaerts A, Lefferts L. Plasma-driven catalysis: green ammonia synthesis with intermittent electricity. Green Chem 2020; 22(19): 6258-87.
[http://dx.doi.org/10.1039/D0GC02058C]

[75] Oliveira Palm M, Luchetti Alves de Freitas Barbosa S, Wilgen Gonçalves M, Duarte DA, de Camargo Catapan R, Silva de Carvalho Pinto CR. Plasma-assisted catalytic route for transesterification reactions at room temperature. Fuel 2022; 307: 121740.
[http://dx.doi.org/10.1016/j.fuel.2021.121740]

[76] Evans EH, Pisonero J, Smith CMM, Taylor RN. Atomic spectrometry update: review of advances in atomic spectrometry and related techniques. J Anal At Spectrom 2020; 35(5): 830-51.
[http://dx.doi.org/10.1039/D0JA90015J]

[77] Salimi P, Aroujalian A, Iranshahi D. Development of PES-based hydrophilic membranes *via* corona air plasma for highly effective water purification. J Environ Chem Eng 2022; 10(3): 107775.
[http://dx.doi.org/10.1016/j.jece.2022.107775]

[78] Özmen F, Korpayev S, Kavaklı PA, Kavaklı C. Activation of inert polyethylene/polypropylene nonwoven fiber (NWF) by plasma-initiated grafting and amine functionalization of the grafts for Cu(II), Co(II), Cr(III), Cd(II) and Pb(II) removal. React Funct Polym 2022;174:105234.
[http://dx.doi.org/10.1016/j.reactfunctpolym.2022.105234]

[79] Dehbashi Nia N, Lee S-W, Bae S, Kim T-H, Hwang Y. Surface modification of polypropylene non-woven filter by O2 plasma/acrylic acid enhancing Prussian blue immobilization for aqueous cesium adsorption. Appl Surf Sci 2022;590:153101.
[http://dx.doi.org/10.1016/j.apsusc.2022.153101]

[80] Dastgheib SA, Karanfil T, Cheng W. Tailoring activated carbons for enhanced removal of natural organic matter from natural waters. Carbon 2004; 42(3): 547-57.
[http://dx.doi.org/10.1016/j.carbon.2003.12.062]

[81] Asadollahi M, Bastani D, Musavi SA. Enhancement of surface properties and performance of reverse osmosis membranes after surface modification: A review. Desalination 2017; 420: 330-83.
[http://dx.doi.org/10.1016/j.desal.2017.05.027]

[82] Kalia S, Thakur K, Celli A, Kiechel MA, Schauer CL. Surface modification of plant fibers using environment friendly methods for their application in polymer composites, textile industry and antimicrobial activities: A review. J Environ Chem Eng 2013; 1(3): 97-112.
[http://dx.doi.org/10.1016/j.jece.2013.04.009]

[83] Jha MK, Joshi S, Sharma RK, *et al.* Surface Modified Activated Carbons: Sustainable Bio-Based Materials for Environmental Remediation. Nanomaterials (Basel) 2021; 11(11): 3140.
[http://dx.doi.org/10.3390/nano11113140] [PMID: 34835907]

[84] Wei P, Lou H, Xu X, *et al.* Preparation of PP non-woven fabric with good heavy metal adsorption performance via plasma modification and graft polymerization. Appl Surf Sci 2021;539:148195.
[http://dx.doi.org/10.1016/j.apsusc.2020.148195]

[85] Basak S, Annapure US. Recent trends in the application of cold plasma for the modification of plant proteins - A review. Future Foods 2022; 5: 100119.
[http://dx.doi.org/10.1016/j.fufo.2022.100119]

[86] Nguyen HT, Huynh LG, Wang YF, Bui XT, You SJ. Environment-friendly fluoride-free membranes from plasma-activated hydrophilic PES and alkylsilanes applied in MD: grafting optimization, surface properties, and performance. Environ Sci Water Res Technol 2023; 9(10): 2706-24.
[http://dx.doi.org/10.1039/D3EW00249G]

[87] Babiker DMD, Li L. Plasma-assisted grafting of functional layers on battery electrodes for enhanced

performance. J Power Sources 2023;564:232828.
[http://dx.doi.org/10.1016/j.jpowsour.2023.232828]

[88] Turkoglu Sasmazel H, Alazzawi M, Kadim Abid Alsahib N. Atmospheric Pressure Plasma Surface Treatment of Polymers and Influence on Cell Cultivation. Molecules 2021; 26(6): 1665.
[http://dx.doi.org/10.3390/molecules26061665] [PMID: 33802663]

[89] Coudane J, Van Den Berghe H, Mouton J, Garric X, Nottelet B. Poly(Lactic Acid)-Based Graft Copolymers: Syntheses Strategies and Improvement of Properties for Biomedical and Environmentally Friendly Applications: A Review. Molecules 2022; 27(13): 4135.
[http://dx.doi.org/10.3390/molecules27134135] [PMID: 35807380]

[90] Hu J, Cavalcante J, Abdellah M, Szekely G. Green surface modification methods and coating techniques for polymer membranes Green Membrane Technologies towards Environmental Sustainability. Elsevier 2023; pp. 209-39.

[91] Brunengo E, Conzatti L, Utzeri R, *et al.* Chemical modification of hemp fibres by plasma treatment for eco-composites based on biodegradable polyester. J Mater Sci 2019; 54(23): 14367-77.
[http://dx.doi.org/10.1007/s10853-019-03932-8]

[92] Mitchell S, Qin R, Zheng N, Pérez-Ramírez J. Nanoscale engineering of catalytic materials for sustainable technologies. Nat Nanotechnol 2021; 16(2): 129-39.
[http://dx.doi.org/10.1038/s41565-020-00799-8] [PMID: 33230317]

[93] Naebe M, Haji A. Grafting and polymerization of antimicrobial compounds onto textiles using plasma techniques. Engineering 2022;8(1):139-58.
[http://dx.doi.org/10.1016/j.eng.2021.03.022]

[94] He B, Gu Z. Sustainable design synthesis for product environmental footprints. Des Stud 2016; 45: 159-86.
[http://dx.doi.org/10.1016/j.destud.2016.04.001]

[95] Pawar HA. An Overview of Natural Polysaccharides as Biological Macromolecules: Their Chemical Modifications and Pharmaceutical Applications. Biology and Medicine. 2014; 07.

[96] Chu PK. Enhancement of surface properties of biomaterials using plasma-based technologies. Surf Coat Tech 2007; 201(19-20): 8076-82.
[http://dx.doi.org/10.1016/j.surfcoat.2005.12.053]

[97] García JA, Rivero PJ, Ortiz R, Quintana I, Rodríguez RJ. Advanced Surface Treatments for Improving the Biocompatibility of Prosthesis and Medical Implants. Advanced Surface Engineering Research. InTech 2018.
[http://dx.doi.org/10.5772/intechopen.79532]

[98] Bazaka K, Jacob MV, Chrzanowski W, Ostrikov K. Anti-bacterial surfaces: natural agents, mechanisms of action, and plasma surface modification. RSC Advances 2015; 5(60): 48739-59.
[http://dx.doi.org/10.1039/C4RA17244B]

[99] Yang H, Wang H. Plasma technique for green surface modification of composites in sustainable applications. Compos Part B Eng 2019;160:448-56.
[http://dx.doi.org/10.1016/j.compositesb.2018.12.048]

[100] Abdulkareem A, Kasak P, Nassr MG, *et al.* Surface Modification of Poly(lactic acid) Film *via* Cold Plasma Assisted Grafting of Fumaric and Ascorbic Acid. Polymers (Basel) 2021; 13(21): 3717.
[http://dx.doi.org/10.3390/polym13213717] [PMID: 34771274]

[101] Filipe HP, Bozukova D, Pimenta A, *et al.* Moxifloxacin-loaded acrylic intraocular lenses: In vitro and in vivo performance. J Cataract Refract Surg 2019; 45(12): 1808-17.

[http://dx.doi.org/10.1016/j.jcrs.2019.07.016] [PMID: 31856994]

[102] Adhikari M, Adhikari B, Adhikari A, Yan D, Soni V, Sherman JH, *et al.* Combination drug delivery using cold atmospheric plasma technology Nanocarriers for the Delivery of Combination Drugs. Elsevier 2021; pp. 393-423.
[http://dx.doi.org/10.1016/B978-0-12-820779-6.00008-6]

CHAPTER 8

Enzymatic Green Grafting: Role of Enzymes and their Sustainable Advantages

Kishor Danao[1], **Deweshri Nandurkar**[1], **Vijayshri Rokde**[1], **Atul Bendale**[2,*], **Nilesh Karande**[3], **Akhil Nagar**[4] and **Viren Soni**[5]

[1] *Department of Pharmaceutical Chemistry, Dadasaheb Balpande College of Pharmacy, Nagpur, Maharashtra 440037, India*

[2] *Department of Pharmaceutical Chemistry, Mahavir Institute of Pharmacy, Nashik 422202, Maharashtra, India*

[3] *Department of Pharmaceutical Chemistry, Institute of Pharmaceutical Science and Education, Wardha 442001, Maharashtra, India*

[4] *Department of Pharmaceutical Chemistry, R.C. Patel Institute of Pharmaceutical Science and Education, Shirpur 425405 Maharashtra, India*

[5] *Department of Pharmacology and experimental Therapeutics, Thomas Jefferson University, Philadelphia, PA 19107, USA*

Abstract: Enzymatic grafting of biopolymers has lately become a focus of green chemistry technologies due to increased environmental concerns and resulting legislative constraints. Over the past decade, polymer science has witnessed a surge in research on enzymes such as laccases and lipases. The goal of this research is to use these enzymes to graft multifunctional polymers for various applications. In this context, a number of bio-composites, such as bacterial cellulose (BC), poly-3-hydroxybutyrate grafted ethyl cellulose, and keratin-g-ethyl cellulose, were effectively synthesized using enzyme-based grafting, with laccase and lipase as model bio-catalysts. Wood preservative made by creating covalent connections between bioactive compounds and wood using the laccase enzyme. The free-radical polymerization of aromatic substances, including gallate esters and lignins, is catalyzed by horseradish peroxidase. To increase the hydrophobicity of the fibers, Dodecyl Gallate (DG) was grafted onto the surfaces of lignin-rich jute textiles using HRP-mediated oxidative polymerization. By using the laccase enzyme, cellulose grafting with ferulic acid is expected to enhance the mechanical properties of the resultant biocomposites. Moreover, the low molecular weight phenol and Pycnoporus cinnabarinus laccase enzyme's ability to biomodify high-content cellulose fibre for use in making paper. These low-molecular-weight phenols, which are covalently bonded to flax fibers by laccase treatment, can function as antibacterial agents, resulting in antimicrobial handsheets. This economical and resilient method offers enormous potential in the production of cellulose-based functional polymers.

---

[*] **Corresponding author Atul Bendale:** Department of Pharmaceutical Chemistry, Mahavir Institute of Pharmacy, Nashik 422202, Maharashtra, India; E-mail: atulbendale123@gmail.com

Kuldeep Vinchurkar, Satish Polshettiwar, Nilesh Mahajan &Yogeshwar Bachhav (Eds.)
All rights reserved-© 2026 Bentham Science Publishers

Keywords: Covalent, Enzymatic grafting, Green chemistry, Laccase, Lipase, Polymerization.

INTRODUCTION

An inventive and eco-friendly method of surface or material modification is enzymatic green grafting, which involves attaching bioactive chemicals or polymers *via* enzymatic processes [1, 2]. Because this process can produce functionalized materials with improved qualities and a reduced environmental impact compared to typical chemical procedures, it offers significant potential in several sectors, including environmental science, biotechnology, materials science, and medicine [3]. Enzymatic green grafting is based on the concept of enzymatic catalysis, which utilizes enzymes as biocatalysts to accelerate specific chemical reactions [4, 5]. Due to their exceptional specificity in recognizing substrates and executing reactions, enzymes provide precise control over the modification process. Enzymatic green grafting operates under mild conditions, such as neutral pH and moderate temperatures, utilizing water as the solvent, which reduces energy consumption and waste generation [6, 7]. This is in contrast to conventional chemical grafting methods, which often require harsh conditions and organic solvents and produce hazardous waste. Grafting is the process of affixing molecules or polymers to a substrate surface in order to impart particular characteristics or functions [8]. The selection of both the enzyme and the substrate determines the type of alteration and final material qualities in enzymatic green grafting. This method can be used with a variety of substrates, including metals, ceramics, biomaterials, natural polymers (such as cellulose and chitosan), synthetic polymers (like polyethylene and polystyrene), and cellulose [9 - 11]. The selection of enzymes with high specificity for the intended substrate and reaction conditions is crucial to the effectiveness of enzymatic green grafting. Catalysts generated from biology, such as enzymes, accelerate chemical reactions without being consumed in the process [12]. Generally, these enzymes identify specific functional groups or chemical structures that are present on both the grafting molecule (also known as the "grafting agent") and the substrate. For example, lipases and esterases are frequently utilized in grafting operations involving ester bonds, and proteases and peptidases are used to change proteins or peptides on surfaces [13, 14]. Similarly, carbohydrates are grafted onto substrates by glycosyltransferases. It is frequently necessary to prepare the substrate surface to expose reactive functional groups or create an environment conducive to enzyme activity before enzymatic grafting can occur [15, 16]. Pre-treatment procedures, such as cleaning, functionalization, or activation, may be necessary for this. Techniques for activation include chemical etching, plasma treatment, and surface functionalization using linker molecules that can communicate with both the grafting agent and the substrate. The molecule or polymer chain that will be affixed to the substrate surface is known as the grafting agent. It is essential to consider factors such as the grafting agent's stability under grafting conditions, compatibility with the enzyme, and the desired features it imparts to the substrate when designing an efficient grafting agent [17, 18]. Small compounds, oligomers, or polymers containing specific functional groups that can participate in enzymatic activity and form covalent bonds with the substrate are known as grafting agents.

It is possible to apply enzymatic green grafting to modify the surfaces of various materials, thereby acquiring specific characteristics or functions. Typical methods encompass immobilization on polymers, bioconjugation, and functional coatings. Immobilization on polymer surfaces is a technology used to improve catalytic stability and efficiency [19, 20]. It includes immobilizing enzymes on polymer surfaces. This method is helpful in biomedical settings where tissue engineering, medication administration, and biosensing can all be accomplished with immobilized enzymes. Enzymes can be coupled with biomolecules (such as peptides or antibodies) in a process known as bioconjugation to produce functionalized surfaces for use in targeted drug delivery systems, biosensors, or diagnostic tests. The process of applying functional coatings to surfaces to enhance adhesion, antibacterial characteristics, or biocompatibility is known as enzymatic grafting [21 - 23]. To reduce the risk of infections, antimicrobial peptides, for example, can be grafted onto medical devices. Biopolymers modified by natural polymers, such as cellulose, chitosan, and starch, can be enzymatically modified to enhance their properties for various applications, including chitosan grafting and cellulose modification [24, 25]. Cellulases and hemicellulases can be utilized in cellulose modification to alter cellulose fibers for use in papermaking, textile applications, or biocomposite materials. Surface functionalization or degradation to change the mechanical characteristics is one type of modification. In contrast, chitosan grafting is a biopolymer derived from chitin that can be enzymatically grafted with molecules or polymers to enhance its antibacterial activity, biocompatibility, or drug delivery properties. Enzymatic modification of synthetic polymers, such as polyethylene, polypropylene, or polystyrene, can introduce particular functional groups or enhance their compatibility with other materials, a process known as polymer functionalization [26 - 28]. By inserting reactive groups, such as hydroxyl or carboxyl groups, enzymes like oxidases or peroxidases can be employed to activate polymer surfaces. Subsequent grafting reactions with functional compounds or polymers can be carried out on these activated surfaces. Moreover, enzymatic grafting can help mix polymers with various characteristics together by altering their interfacial interactions or surface chemistry. This method works well for combining polymers to create composite, coating, or packaging materials. The capacity of enzymatic green grafting to modify biomaterials with enhanced biocompatibility, bioactivity, or improved drug transport capabilities makes it a promising technique for biomedical applications. Implant surfaces can be altered by enzymatic grafting to enhance osseointegration, reduce inflammation, or prevent bacterial adhesion. To improve tissue integration, peptides with cell-adhesive motifs, for instance, can be grafted onto titanium implants [29 - 31]. Therapeutic substances can be released under regulated conditions by grafting polymers onto nanoparticles or micelles with the application of enzymes. This strategy lessens systemic toxicity, increases targeted effectiveness, and stabilizes drugs. For the sensitive and targeted identification of analytes in biological samples, biomolecules (such as enzymes or antibodies) can be functionalized onto the surfaces of biosensors or diagnostic equipment by the process of enzymatic grafting. Sustainable approaches to waste management, resource recovery, and environmental remediation are provided *via* enzymatic green grafting. Enzymatic modification of biopolymers, such as cellulose or starch, can improve their biodegradability and suitability for use in agriculture or as sustainable packaging materials [32 - 34]. Adsorbent materials, such as activated carbon and zeolites, can enhance their ability to remove contaminants from wastewater streams by using enzymes to graft functional groups onto them. This method enhances the selectivity and efficiency of wastewater treatment operations [35]. Enzymatic green grafting can enhance the qualities of fibers, textiles, or paper goods in the paper and textile industries. Textile fibers can be modified through enzymatic grafting to add hydrophobic coatings, flame retardants, or antimicrobial compounds, thereby imparting the desired qualities of stain resistance or durability. Paper surfaces can be modified with enzymes to enhance water resistance, ink adhesion, and printability [36 - 38]. This method enhances the performance and quality of paper products for various applications. There are several benefits that enzymatic green grafting offers over chemical grafting techniques. When compared to chemical processes, enzymatic reactions produce less hazardous waste and use less energy when they take place in mild environments (such as room temperature and neutral pH). They also use water as a solvent. Elevated efficiency and reduced by-products are achieved through the selective reactions that enzymes catalyse [39, 40]. Enhancing the repeatability of modified materials, this specificity enables precise control over the grafting process. Biomolecules, such as proteins, peptides, and carbohydrates, can be functionalized for biomedical applications using enzymatic green grafting without compromising their biological activity. Enzymatic grafting is an adaptable technique for modifying a variety of materials, as it can be applied to a broad range of substrates, including natural and synthetic polymers, metals, ceramics, and biomaterials, as shown in Fig. (**8**). The industrial applications of enzymatic green grafting techniques can be expanded, potentially yielding financial benefits in industries such as environmental remediation, healthcare, textiles, and packaging [41, 42]. Notwithstanding its encouraging benefits, enzymatic green grafting still has several issues that must be resolved before it can be widely used and made commercially viable. Certain conditions (such as high temperatures or severe pH) can cause denaturation or loss of activity in enzymes, which limits their reusability in continuous industrial processes. Research is underway to develop methods for enhancing the stability and recyclability of enzymes [43]. The compatibility of enzymes, substrates, and grafting agents is essential for the successful use of enzymatic grafting. To obtain effective grafting results, reaction conditions and substrate preparation techniques must be optimized. When compared to conventional chemical catalysts, enzymes used in grafting reactions may be more expensive. The creation of affordable enzyme supplies, strategies for enzyme immobilization, and process optimization are crucial for enhancing the economic feasibility of enzymatic green grafting. Enzyme use in industrial settings may require regulatory permission due to safety concerns, environmental impacts, and compliance with standards [44, 45]. To support the commercialization of enzymatic green grafting technologies, specific standards and regulations are necessary. Enzymatic green grafting, all things considered, is a promising method of surface modification.

METHODS OF GRAFTING [46]

The primary goal of surface modification is to enhance a polymer's wettability, biological compatibility, and physical properties. While various grafting techniques are illustrated in Fig. (**1**), this particular approach is based on the concept of using an enzyme to initiate the chemical or electrochemical grafting process. By removing the need for reactive chemicals, enzyme application can provide a safer, more cost-effective, and environmentally friendly method of grafting procedures. Moreover, the specificity of enzymes may provide a possibility for precisely customizing desired macromolecular characteristics.

MECHANISMS OF ENZYME-INITIATED GREEN GRAFTING

The mechanisms include sustainable wood preserving methods, Radical polymerization methods, and oxidative polymerization methods.

Sustainable Wood Preserving Method

Utilizing an enzyme-mediated reaction to adhere to the wood surface, natural extractives are suggested as viable, eco-friendly components that are bioactive for wood preservative compositions [47]. This green method requires the laccase enzyme's activity, which causes covalent bonds to form between these substances and the wood in its natural form. Enzyme-based grafting is a process that covalently attaches target molecules to a chosen substrate. A key enzyme involved in this process is laccase (EC 1.10.3.2), which catalyzes the oxidation of phenols and amines into phenoxy and amino free radicals, respectively. This approach offers a sustainable method for creating durable wood protection solutions that resist leaching. For example, extracts from *Cryptomeria japonica* and *Pinus* spp. have been successfully used in this method to treat wood. Antioxidants may enhance the effectiveness of these natural biocides. The grafting process, catalyzed by laccases, involves activating the phenolic structures in both the target compound and the wood's lignin. This treatment is environmentally friendly, effective under mild conditions, and has the key advantage of enabling the covalent attachment of the substrates to the wood surface. An eco-friendly and reasonably priced preservation product for wood is prepared from the residue left over after making soymilk and tofu, known as okara (OK). It also examines the efficacy of preservatives using enzymatically hydrolyzed okara against decomposing fungi [48]. The important thing is that hot-water leaching did not affect OK-based wood preservatives. Furthermore, because of its chemical composition, it shows promise for use as a component in a wood preservative system. Specifically, by oven-dry weight, OK includes approximately 27 percent protein, 53 percent fiber and sugar, and 12% petroleum products. Enzymatically hydrolyzed OK may also be utilized in the formulation of OK-wood preservatives [49, 50].

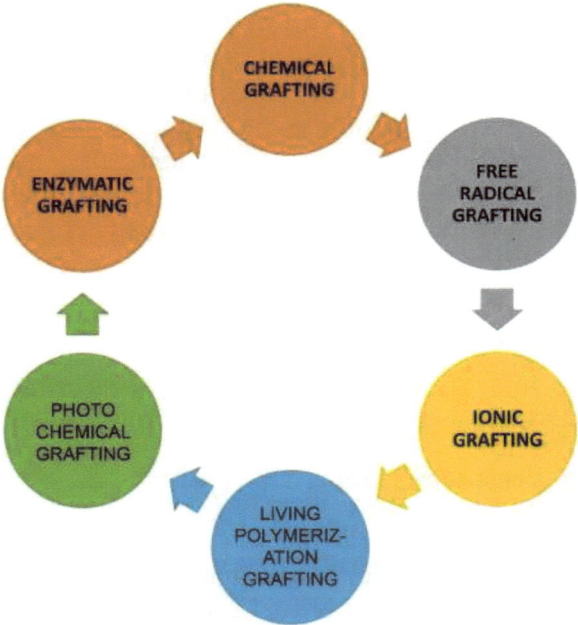

Fig. (1). The different techniques of grafting.

Radical Polymerization Method [50 - 54]

The fields of biocatalytic polymerization of radicals are dominated by two types of enzymes: laccases and peroxidases, having Enzyme Commission numbers 1.10.3.2 and 1.11.1, respectively. Both enzyme groups primarily catalyze proton-depletion processes, which generate radical forms that initiate the polymerization reaction, while exhibiting quite distinct catalytic mechanisms and active site structures. Only a small part is played by other enzyme types, such as lipoxygenases and oxidases. Heme peroxidases are widely used as catalysts to initiate enzymatic radical polymerization, utilizing aromatic and vinyl monomers. The peroxidase enzyme contains protoporphyrin IX and Fe(III) as a prosthetic group in its resting state, as shown in Fig. (2).

Fig. (2). Structure of protoporphyrin IX (A) and Fe^{III} as prosthetic group in the resting state (B).

As shown in Fig. (3), Hydrogen peroxide replaces the water ligand throughout the catalytic process, creating a peroxo complex. When the O-O A bond molecule undergoes heterolytic cleavage; it is then returned to its resting state by two separate hydrogen removals from the reduction (In-H). This produces two radical types, which initiate the polymerization process.

Fig. (3). Mechanism of peroxidases containing protoporphyrin IX with Fe^{III} and L as histidine ligand.

The typical method of preventing excess hydrogen peroxide involves adding it in small increments, which can be laborious and increase the risk of some irreproducibility. The authors described a bi-enzymatic system for creating polymers of phenols that includes glucose oxidase enzyme, as evident in Fig. **(4)**.

Fig. (4). Polymerization reaction using bienzymes.

OXIDATIVE POLYMERISATION [55 - 60]

By Horseradish peroxidase (HRP) enzymes grafted dodecyl gallate (DG) containing groups that were hydrophobic to the outermost layer of lignin-rich jute cloth by oxidative polymerisation mechanism, as shown in Fig. (**5**).

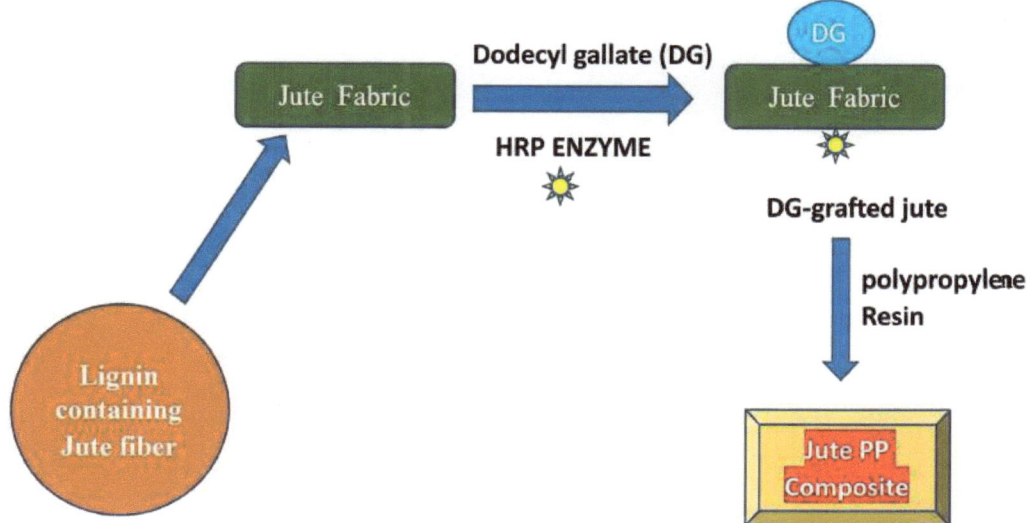

Fig. (5). Oxidative polymerisation mechanism by using HRP enzyme.

ROLE OF ENZYMATIC GREEN GRAFTING

Green Approaches for Starch Modification

Enzymatic green grafting offers approaches to designing a starch with a modified structure, such as gelatinized starch, as depicted in Fig. (**6**). The enzymes treated with gelatinized starch modified the molecular mass, branch chain length, and altered the ratio of amylopectin [61]. Enzymatic modifications can significantly alter the properties of starch, improving the freeze-thaw stability of gels and hindering retrogradation during storage. These modified starches have been successfully used as ingredients in the food industry to enhance product quality and improve food processing efficiency [62].

Polymer-grafting from a Metallo-Centered Enzyme

The surface of laccase was modified by grafting a polymer, in which a metalloenzyme with a copper residue (Fig. **7**) active site was adapted to polymerization species ligated to an amino acid residue, lysine [63]. For grafting, regeneration of activators and atom transfer radical polymerization were used to create a tiny library of polymers from proteins. Enzyme activity is altered by changes in pH function and temperature of the solvent, impacting polymer grafting. Moreover, polymer grafting from laccase increased enzymatic activity, as evidenced by the high molecular weight of dimethyl amino ethyl methacrylate grafted, which depicted a significant enhancement of catalytic activity [64]. Hence, it was proved that polymer grafting enhances enzyme laccase activity in both native and non-native conditions.

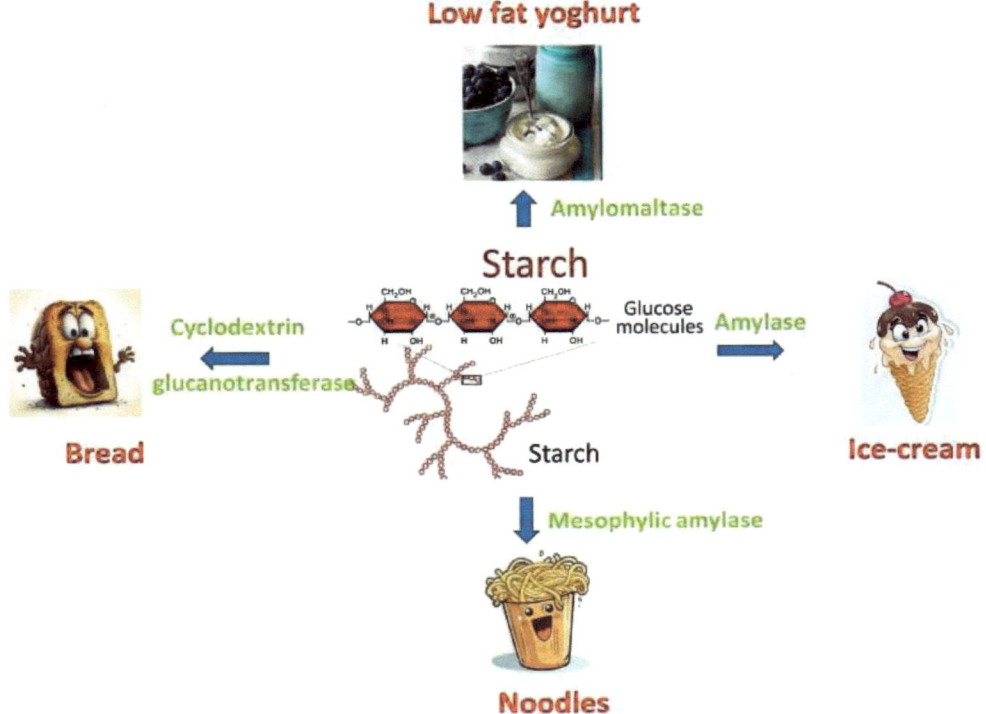

Fig. (6). Enzymatic modification of starch in food processing,

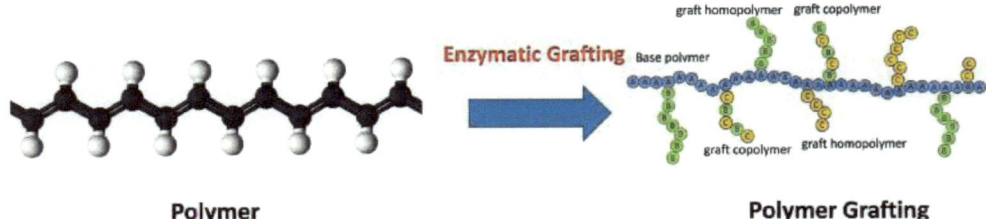

Fig. (7). Polymer-grafting of copper residue active site.

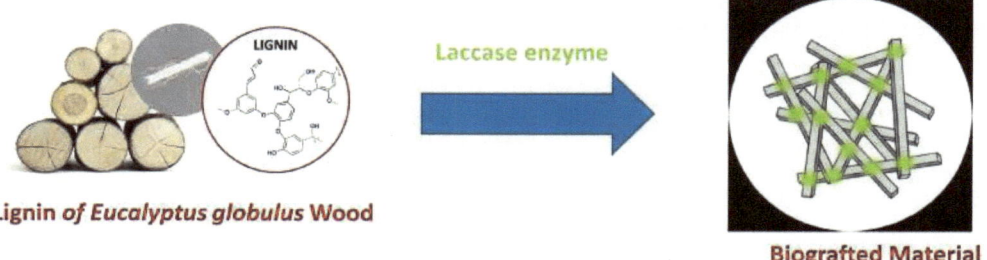

Fig. (8). Enzymatic grafting on *Eucalyptus globulus* wood by laccase.

ENZYMATIC GRAFTING ON *EUCALYPTUS GLOBULUS* WOOD BY LACCASE

Eucalyptus wood with laccase has enzymatic grafting from the basidiomycetes Marasmiellus palmivorus, as shown in Fig. (**8**). The treated reaction yielded three phenols and an amine substrate, along with two mediator compounds, through oxidative grafting, which demonstrated enzymatic potency [65]. There were better results with laccase-aided grafting of octyl and lauryl gallate. Subsequently, enzymatic grafting of these compounds increased the hydrophobicity of eucalyptus veneers, which was beneficial for developing the durability and dimensional stability of wood in the building and furniture industries [66].

RAFT POLYMERIZATION FOR COMPLICATED POLYMER ARCHITECTURE

The broad applicability to the majority of vinyl monomers, good tolerance to functional groups, and compatibility with a variety of system conditions have made Reversible Addition-fragmentation chain Transfer (RAFT) polymerization an almost universal synthetic approach [67]. Without the aid of external degassing, oxygen-tolerant Enz-RAFT polymerization —a GOx or P2Ox oxidase deoxygenation approach with outstanding controllability and high efficiency —has also been created.

Enz-RAFT polymerization in open vessels has been demonstrated to be remarkably efficient in producing both UHMW polymers and precise multi-block copolymers, thereby eliminating the need for additional deoxygenation operations during the process [68]. RAFT-mediated polymerization-induced self-assembly (RAFT-PISA), photo-enzymatic RAFT, cascade-catalyzed RAFT (Fig. (**9**), and "grafting from" RAFT provide an overview of the growing role that enzymes are playing in the high-throughput, environmentally friendly synthesis of precision RAFT polymers with high efficiency, specificity, and selectivity under mild conditions. It also provides detailed information about the opportunities and insights that arise from using enzymatic catalysis to produce novel polymers for biomedical applications through green polymer chemistry [69 - 71].

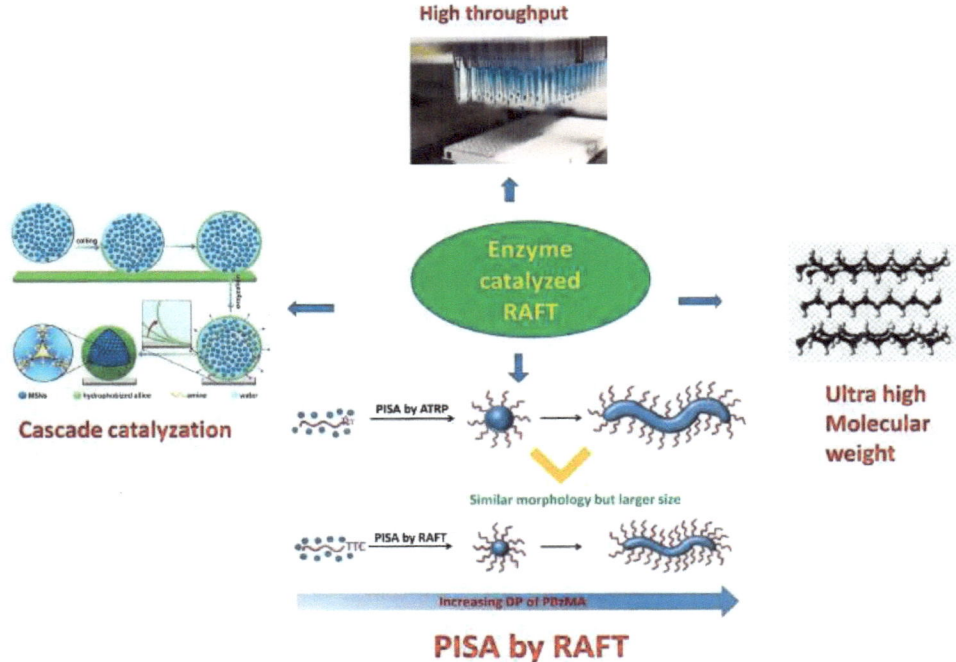

Fig. (9). RAFT polymerization.

ENZYMATIC GRAFTING OF NATURAL PHENOLS TO FLAX FIBRES

The laccase phenol treatment could result in the grafting of these phenols onto the fibres. These enzyme treatments might enable flax fibers to acquire new or improved characteristics. The treated papers exhibit strong antibacterial activity, resulting in the creation of bioactive papers shown in Fig. (**10**) [72]. The grafted antibacterial compounds in the flax fibers become immobile. To test the possibility of biomodifying high cellulose content fibers, unbleached flax fibers used in paper manufacture were treated with low-molecular-weight phenols (p-coumaric acid, PCA), acetosyringone, and syringaldehyde from *Pycnoporus cinnabarinus*. Following the phenols' enzymatic treatment, a rise in kappa number was observed, most likely as a result of the phenoxy radicals' covalent attachment to the fibers. Pulps treated with PCA showed increased grafting (an increase of 4 kappa number points relative to the laccase control was achieved).

Fig. (10). Enzymatic grafting of natural phenols to flax fibres.

Paper handsheets made from treated pulps exhibited antimicrobial activity against the bacteria tested: Staphylococcus aureus, Pseudomonas aeruginosa, and *Klebsiella pneumoniae*. Following the incubation of liquid cultures of the bacteria with grafted handsheets, a significant decrease in the microbial count was observed. Grafted fibers with AS and PCA exhibited strong antibacterial action against *K. pneumoniae*, resulting in almost complete growth suppression. Additionally, AS fibers significantly reduced the *P. aeruginosa* bacterial population (97% reduction) [73 - 75].

NATURAL GAS CONVERSION

The utilization of methane monooxygenase (MMO) as a biocatalyst to convert methane to methanol has garnered attention in the context of growing natural gas production. Due to its numerous benefits over the chemical conversion pathway, MMO has been studied for its scalability in methane conversion. It has been demonstrated to convert methane to methanol at ambient temperatures with selectivity approaching 100% [76 - 78]. The MMO-catalyzed methanol synthesis exhibits great selectivity, which facilitates product separation and may lead to a significant reduction in the number of stages needed for the conversion process depicted in Fig. (**11**). Moreover, the moderate reaction conditions of the enzymatic process may reduce the expenses related to heating, pressurization, and other feedstock conditioning steps [79, 80].

ADVANTAGES OF ENZYMATIC GREEN GRAFTING

Enzymatic green grafting is an innovative methodology that provides numerous advantages due to its environmentally friendly approach to altering materials. The potential of enzymatic green grafting is particularly promising in various sectors, including pharmaceuticals, food production, agriculture, and energy, due to its capacity to enhance sustainability, productivity, quality, and ecological compatibility [81].

Fig. (11). Natural gas conversion *via* biocatalyst.

This new technology has numerous advantages due to its various applications, including those in the pharmaceutical industry. This technology can aid in the enzymatic synthesis of biological materials for therapeutics, making it a sustainable approach that can meet the rapidly growing global demand for therapeutics to treat various diseases. In addition, the pharmaceutical industry utilizes toxic catalysts to carry out numerous reactions. The introduction of this technology will provide an avenue for using enzymes and water, which will reduce the production of hazardous by-products that have a negative environmental impact. Scalability is another advantage that enzymatic green grafting can provide, as this technology can be scaled up efficiently, which can be used to meet the demand for products that are needed as potential therapeutics for patients [82, 83]. This process follows the principles of green chemistry, where enzymatic green grafting aims to reduce waste and conserve energy. To this end, enzymatic green grafting features a continuous flow of synthesis, which enhances energy efficiency and minimizes the manufacturing footprint. Lastly, compared to traditional chemical processes, the use of enzymes in this technology provides a safer alternative, thereby increasing safety for scientists and enabling the construction of complex structures with increased specificity, as this technology has a high degree of selectivity [84].

Enzymatic green grafting has applications in the food industry, and its use offers numerous advantages. It provides high efficiency and specificity, as enzymes, by nature, are efficient and specific in their functions, which allows them to be precise in modifying food products to enhance texture, nutrition, and flavor [85]. This technological innovation enables the improvement of food standards to meet the growing market demands, yielding substantial benefits for the food sector. This enzymatic methodology provides a path for innovation in food processing, such as supercritical technology (scCO2), in which certain enzymes are inactivated while others are activated [86, 87]. This process lowers energy consumption compared to ultra-high temperature processing and achieves the goal of preserving food quality for the companies that market these products [88].

The agricultural industry has recently adopted the use of this technology, as it provides numerous advantages for crops and produce. Enzymatic green grafting technology can be utilized to enhance plant vigor, leading to robust growth, better crops for farmers, and improved food quality for consumers [89]. The harvesting period for many agricultural products could be extended using this technology, resulting in longer seasons of crop production. This would yield a greater production of various agricultural products and improve their overall quality [90]. Grafted plants produced through this methodology would exhibit characteristics conducive to withstanding challenging environmental conditions, including extreme temperatures, salinity, heavy metal exposure, and drought periods, as well as facilitating the effective utilization of water in regions with restricted water supply. These aspects are impacted by climate change, and enzymatic green grafting can help address this issue. Finally, this technology enables agricultural products to deal with and fend off soil-borne and foliar pathogens, as well as various pests, and contributes to weed control [91 - 93].

An emerging field of nanotechnology has adopted enzymatic green grafting, as this technology offers a significant advantage for further advances in the field. This innovative technology is ecologically sound and can be employed in the nanoparticle synthesis process to reduce the need for and minimize the use of harmful chemicals in their production, while also mitigating environmental pollution [94]. On a similar note, this method is safer for the environment and human health, as toxic by-products are not produced throughout the production process, thus lowering toxicity compared to conventional synthetic methodologies. Enzymatic green grafting paves the way for reducing the production cost of nanotechnology products, making them economically viable by lowering energy requirements and increasing the use of renewable resources [95]. Additionally, it promotes sustainability as the materials used are derived from biodegradable or natural sources. Enzymes, enabled by this technology, allow for tailoring the synthesis of nanoparticles with properties that enhance their therapeutic potential in biomedicine.

Lastly, enzymatic green grafting technology has been utilized in the energy sector due to its advantageous applications. The technology offers a more energy-efficient approach compared to traditional chemical processes, particularly when applied to large-scale production of various products, by reducing the energy required for the manufacturing processes involved [96]. There are also environmental benefits, as the enzymes used in these processes aid in reducing greenhouse gas emissions. Implementing this enzymatic methodology will result in reduced costs due to lower energy requirements and decreased waste production. Enzymatic green grafting opens the door to new avenues of recycling and upcycling, which can have a positive impact both economically and socioeconomically [97].

For the development of enzyme-polymer hybrids, five primary synthetic approaches are employed: covalent bonding, ionic and non-ionic interaction, encapsulation, physical entrapment, and affinity-based interaction. The method, commonly employed in the manufacture of Enzyme-Polymer Conjugates (EPCs), involves the covalent attachment of pre-prepared polymers to a target enzyme [98]. In a process commonly referred to as the grafting-to approach, synthetic polymerization can be engineered to produce a range of topologies and functional end-groups, ultimately allowing it to react with several residues on the enzyme surface [99].

As shown in Fig. (**12**), the protein surface can have one or more polymer chains attached to it (Fig. **12A**). The number of attached polymers depends on the type and quantity of amino acids targeted, as well as any potential intrinsic steric problems in specific polymer chains (*e.g.*, bulky polymers or dendrimers).

The stability of single enzymes can be increased by covalently modifying them from many unique points through the use of branching polymers or networks. Fig. (**12B**) shows a multipoint approach that is crucial for stabilizing and immobilizing enzymes with multiple subunits. For the multipoint covalent immobilization of enzymes, polymeric supports functionalized by glutaraldehyde (GA) or glyoxal groups, such as agarose, epoxy resins, or polymethacrylate, are often utilized. This grafting technique is utilized in various reactions, including carboxylic acid–amine group reactions *via* carbodiimide chemistry, nucleophilic sulfhydryl side chain reactions, and alkylation, among others.

A high degree of control over the hybrid is necessary for the effective use of biorthogonal chemistry, accurate chain localization on the enzyme surface, and management of the grafting density. Because they are absent from biological systems, successful examples of this kind of grafting have been accomplished using the thiol-ene, alkyne-azide, and diels-alder reactions in the presence of nucleophiles, electrophiles, reductants, oxidants, and water, without altering or affecting the reaction's evolution. The covalent bonds that form are stable, and the byproducts are safe. Using this method, numerous hybrids have been successfully synthesized, including polymer brushes, polymer monoliths, enzyme-polymer nanoconjugates, and enzyme-MOF conjugates. Moreover, obtaining a high grafting density using this method is typically challenging (Fig. **12C**).

A second strategy, called grafting from, arose to target bulky polymers on the protein surface and increase the grafting density. With this method, the polymer grows in situ, beginning with chain transfer agents or reaction initiators that have already been conjugated to the enzyme. The grafting-from approach requires mild polymerization conditions while preserving the catalytic properties of the enzymes to maintain chain-end functionality and accurately control the molecular weight of the polymer. In this regard, Atom Transfer Radical Polymerization (ATRP) and Reversible Addition-Fragmentation chain Transfer Polymerization (RAFT) are the most widely used techniques that meet the previously described requirements (Fig. **12D**). Because the monomers in the grafting-from approach are significantly smaller than the conjugate itself, downstream purification is easier than in the grafting-to approach [100 - 105].

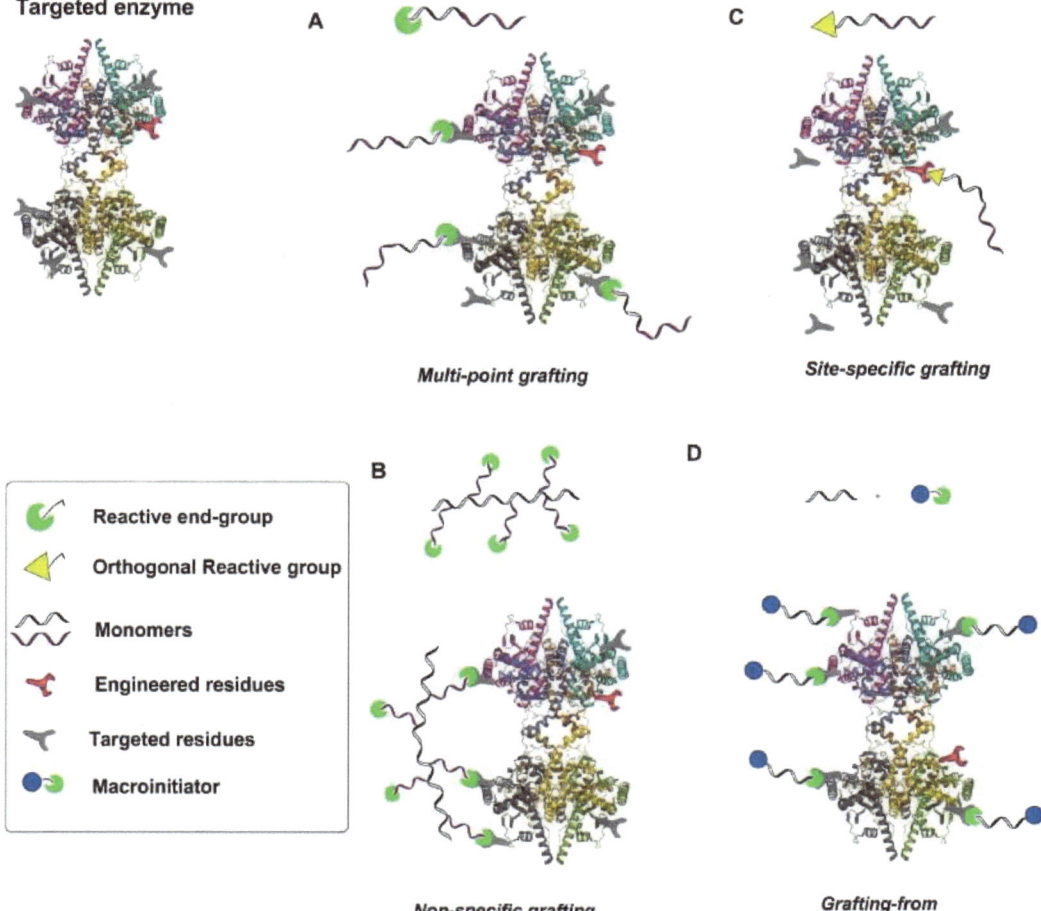

Fig. (12). Different forms of grafted enzymes.

Overall, enzymatic green grafting is a novel technology that offers various advantages (Table 1) to multiple industries, thanks to its environmentally friendly approach to modifying and enhancing various products. It has advantages for the pharmaceutical, food, agricultural, and energy industries, with the key objective of improving each sector's viability, efficiency, and quality.

Table 1. Table describing the advantages of green grafting in various fields [106 - 110].

Industry/Sector	Advantages	Description
Pharmaceutical	Sustainable Therapeutics	This technology can aid in the enzymatic synthesis of biological materials for therapeutics, making it a sustainable approach that can meet the rapidly growing global demand for therapeutics to treat various diseases patients may have.
	Reduced Environmental Impact	The pharmaceutical industry utilizes toxic catalysts to facilitate numerous chemical reactions. The introduction of this technology will provide an avenue for utilizing enzymes and water, thereby reducing the production of hazardous byproducts.
	Scalability	Enzymatic green grafting can be scaled up efficiently, which can be used to meet the demand for products that are needed as potential therapeutics for patients.
	Green Chemistry	This process adheres to the principles of green chemistry, where enzymatic green grafting seeks to minimize waste and conserve energy.
	Safety & Selectivity	In comparison to traditional chemical processes, the use of enzymes in this technology offers a safer alternative and enables the construction of complex structures with increased specificity.
	Energy Efficiency & Net-Zero Carbon Footprint	Enzymatic green grafting has a continuous synthesis process, which enhances energy efficiency and minimizes the manufacturing footprint. The pharmaceutical industry is moving towards this technology, which will lead to sustainable practices that reduce waste and mitigate its negative environmental impact.
Food	High Efficiency & Specificity	The nature of enzymes to be efficient and specific in their functions allows them to be precise in modifying food products to enhance texture, nutrition, and flavor.
	Quality Improvement	This technology will enhance the quality of food to meet the growing demand, resulting in significant benefits for the food industry.
	Innovation of Food Processing	Enzymatic green grafting can be employed in food processing, such as supercritical technology (scCO2), in which certain enzymes are inactivated while others are activated, resulting in lower energy consumption compared to ultra-high temperature processing. This achieves the goal of preserving food quality for the companies that market these products.

(Table 1) cont.....

Industry/Sector	Advantages	Description
Agricultural	Plant Vigor	Enzymatic green grafting technology can be utilized to enhance plant vigor, leading to robust growth and improved crops for farmers, as well as better food quality for consumers.
	Harvesting Period, Yield & Quality	The harvesting period for many agricultural products could be extended using this technology, resulting in longer seasons of crop production. This would yield a greater production of various agricultural products and an overall improvement in their quality.
	Increased Stress Tolerance & Efficient Water Use	Grafted plants using this technology would possess qualities that enable them to endure harsh environments, such as extreme temperatures, salinity, heavy metal stress, and periods of drought. This also enables the efficient use of water in areas where water resources are scarce. Both of these aspects are impacted by climate change, and enzymatic green grafting can help in this regard.
	Disease, Pest Management & Weed Control	The technology can provide agricultural products with the ability to combat and defend against soil-borne and foliar pathogens, as well as various pests, and contribute to effective weed control.
Nanotechnology	Environmental Friendliness	Enzymatic green grafting technology can be utilized in the process of synthesizing nanoparticles, as it decreases the need for and use of hazardous chemicals and reduces environmental contamination.
	Low Toxicity	The method is safer for the environment and human health, as it does not produce toxic byproducts throughout the production process, unlike conventional synthetic methodologies.
	Cost Effectiveness & Sustainability	This technology paves the way for reducing the production costs of nanotechnology products, making them economically viable by lowering energy requirements and increasing the use of renewable resources. Additionally, it promotes sustainability as the materials used are derived from biodegradable or natural sources.
	Customizable Properties	Enzymes, made possible by this technology, allow for tailoring the synthesis of nanoparticles with properties that enhance their therapeutic potential in biomedicine.
Energy	Energy Efficiency	Enzymatic green grafting technology provides a more energy-efficient method compared to traditional chemical processes. This applies to making different products on a large scale by reducing the energy required to execute the manufacturing processes involved.
	Environmental Benefits	The use of enzymes in this technology helps reduce greenhouse gas emissions. This can help reduce plastic pollution by recycling it enzymatically.
	Economic Socioeconomic Impact	Implementing this enzymatic methodology will lead to reduced costs, resulting from lower energy requirements and decreased waste production. This method also opens the door to new avenues of recycling and upcycling, which can have a positive impact both economically and socioeconomically.

Future Scope of Enzymatic Green Grafting

Enzymatic green grafting is a burgeoning technological advancement that holds considerable promise for diverse applications owing to its eco-friendly methodology for material modification. The pharmaceutical sector foresees a promising trajectory for enzymatic green grafting, given its capacity to enhance drug delivery efficiency, improve drug durability, and promote the development of greener manufacturing practices [111 - 116].

Here are some specific areas where enzymatic green grafting can make a significant impact:

1. Drug Delivery Systems
- **Controlled Release:** Enzymes can graft polymers onto drug molecules, enabling the development of controlled-release formulations. This can improve therapeutic outcomes by maintaining optimal drug concentrations in the body over extended periods of time.
- **Targeted Delivery:** Enzymatic grafting enables the attachment of targeting ligands to drug carriers, facilitating site-specific drug delivery. This process has the potential to mitigate adverse reactions and enhance the effectiveness of therapeutic interventions.

2. Drug Stability and Solubility
- **Improved Stability**: The enhancement of stability in drug molecules can be achieved through the grafting of protective groups, thereby mitigating degradation caused by environmental conditions like light, temperature, and pH.
- **Enhanced Solubility**: The use of enzymatic grafting has the potential to improve the solubility of drugs with low solubility, ultimately enhancing their solubility and bioavailability, which are essential factors for the efficacy of numerous pharmaceutical compounds.

3. Biocompatible Materials
- **Biodegradable Polymers:** Enzymatic grafting can be used to create biocompatible and biodegradable polymers for use in drug delivery systems, implants, and medical devices.
- **Functional Coatings:** Enzymatic grafting can produce functional coatings on medical instruments, such as stents and catheters, to enhance their functionality and reduce the likelihood of infections.

4. Personalized Medicine
- **Customized Drug Formulations:** Enzymatic grafting allows for the customization of drug formulations to meet individual patient needs, such as grafting specific therapeutic agents onto carriers tailored to a patient's unique

physiology.
5. **Vaccine Development**
   - **Adjuvant-free Vaccines:** Enzymatic grafting can be used to attach antigens to delivery systems that enhance the immune response without the need for traditional adjuvants.
   - **Stable Vaccine Formulations:** Enzymatic grafting can stabilize vaccine components, improving their shelf life and efficacy.
6. **Nanomedicine**
   - **Functionalized Nanoparticles:** Enzymatic grafting enables the functionalization of nanoparticles with therapeutic agents, targeting moieties, or imaging agents, thereby creating multifunctional nanomedicines for diagnosis and treatment.
   - **Enhanced Penetration:** Enzymatically linked nanoparticles have the potential to increase drug penetration into cells and tissues, thereby boosting therapeutic efficacy in the treatment of a range of diseases, including cancer.
7. **Sustainable Pharmaceutical Manufacturing**
   - **Green Chemistry:** Enzymatic grafting offers a greener alternative to traditional chemical grafting methods, reducing the use of toxic solvents and harsh chemicals.
   - **Efficient Synthesis:** Enzymatic processes can be more efficient and selective, potentially lowering production costs and reducing waste.
8. **Combination Therapies**
   - **Multi-drug Conjugates:** Enzymatic grafting offers the potential to produce conjugates involving multiple drugs, allowing for the application of combination therapies that can target various pathways simultaneously. This approach proves particularly beneficial in the treatment of complex diseases, such as cancer.

Applications of Enzymatic Green Grafting

1. **Textile Industry**
   - **Eco-friendly Dyeing and Finishing:** Enzymatic grafting can be used to attach functional groups to textile fibers, improving dye uptake and fixation without the use of harmful chemicals.
   - **Antimicrobial Fabrics:** By integrating antimicrobial agents through enzymatic grafting, it is possible to enhance the hygienic characteristics of textiles.
2. **Biomedical Applications**
   - **Drug Delivery Systems:** Enzymes can graft polymers onto drug molecules or carriers, allowing for controlled release and targeted delivery.
   - **Tissue Engineering:** Enzymatic grafting can modify scaffolds to enhance cell attachment and growth, improving tissue regeneration.

3. Food Industry
- **Packaging Materials:** Enzymatic grafting can create biodegradable and antimicrobial packaging materials, extending the shelf life of food products.
- **Functional Foods:** The enzyme grafting of bioactive compounds can be used in food components to enhance the nutritional values.

4. Environmental Applications
- **Water Treatment:** Enzymatic grafting can be used to modify materials for more effective removal of contaminants from water.
- **Biodegradable Plastics:** The biodegradability of plastic can be increased by incorporating natural polymers into synthetic plastics *via* enzyme grafting.

5. Agriculture
- **Enhanced Fertilizers:** Enzymatic grafting can create slow-release fertilizers, improving nutrient use efficiency and reducing environmental impact.
- **Pesticide Delivery:** Enzymes can help in creating controlled-release pesticide formulations, minimizing environmental contamination.

6. Nanotechnology
- **Nanomaterial Functionalization:** Enzymes can graft functional groups onto nanoparticles, enhancing their properties for applications in sensors, catalysis, and electronics.

7. Cosmetics and Personal Care
- **Active Ingredient Delivery:** To increase the efficacy and stability of active ingredients during drug delivery, enzymatic grafting can be used.
- **Sustainable Formulations:** Incorporating enzymes into formulation processes can reduce the need for harsh chemicals, making products more environmentally friendly.

8. Energy Sector
- **Biofuel Production:** Enzymatic grafting can modify substrates to improve their conversion into biofuels.
- **Energy Storage:** Superconductors, capacitors, and battery performance can be enhanced by grafting functional groups onto materials using enzyme technology.

Challenges and Considerations of Enzymatic Green Grafting

Enzymatic green grafting holds promise for developing sustainable and high-performance materials across various industries, addressing the growing demand for environmentally friendly and effective technologies. The application of enzymatic green grafting has a considerable capacity to transform the pharmaceutical sector through the introduction of novel approaches to drug delivery, stability, and production [117]. With the advancement of research and technology, an increase in the use of this method is anticipated in the development of safer, more efficient, and environmentally conscious pharmaceutical products.

Scalability: Developing Cost-Effective Methods to Scale Up Enzymatic Grafting Processes

Scaling up enzymatic grafting processes for large-scale pharmaceutical manufacturing involves several critical steps. The primary objective is to achieve cost-effectiveness while maintaining the quality of the procedure. Initially, it is crucial to select enzymes that are highly specific and efficient, which reduces the amount needed and thus lowers costs. High-throughput screening (HTS) methodologies can be utilized to detect these enzymes, thereby verifying their suitability for use in industrial settings.

Process optimization is another key factor. This entails optimizing reaction parameters, including temperature, pH, and substrate concentrations, to enhance enzyme effectiveness and increase product output. Utilizing computational modeling can help predict optimal conditions and reduce the number of experimental trials needed [118]. Additionally, bioreactor design plays a significant role in scaling up. Continuous flow reactors, for instance, can offer better control over reaction parameters and facilitate consistent product quality. The integration of in-line monitoring systems allows real-time adjustments, enhancing efficiency and reducing waste.

Another cost-effective strategy is enzyme immobilization. Immobilizing enzymes on solid supports can enhance their stability and reusability, thereby reducing the overall cost of enzymes. Various methods, including covalent bonding, adsorption, and encapsulation, can be employed based on the specific needs of the procedure. Reusability not only reduces enzyme costs but also simplifies the purification process, as immobilized enzymes can be easily separated from the reaction mixture.

Enzyme Stability: Ensuring Enzymes Remain Active and Stable During the Grafting Process

Ensuring enzyme stability throughout the grafting process is essential for maintaining efficiency and product consistency. Choosing the right enzymes involves considering their intrinsic stability and compatibility with the reaction conditions. Enzymes that are naturally stable or have been engineered for enhanced stability are preferable. Protein engineering methodologies, including directed evolution and rational design, can enhance enzyme stability by incorporating mutations that strengthen their structural resilience.

Stabilizing agents, such as polyols, sugars, and specific salts, can be added to the reaction mixture to protect enzymes from denaturation. Additionally, immobilization techniques not only enhance reusability but also stabilize enzymes by providing a supportive microenvironment that prevents unfolding and aggregation. The selection of support material and the immobilization technique employed may have a notable impact on both the stability and activity of the enzyme [119].

Storage conditions play a crucial role in maintaining the stability of enzymes. It is imperative to store enzymes under specific conditions that uphold their catalytic activity, including low temperatures and optimal pH levels. A widely used technique to prolong the shelf-life of enzymes is lyophilization, also known as freeze-drying [120]. This method enables enzymes to be preserved and delivered with minimal loss of functionality.

Regulatory Approvals: Meeting Regulatory Standards for New Materials and Applications

Meeting regulatory standards is crucial for the successful implementation of enzymatically grafted pharmaceuticals. Regulatory authorities, exemplified by the Food and Drug Administration and European Medicines Agency, impose rigorous criteria pertaining to safety, effectiveness, and quality. Ensuring compliance involves a thorough understanding of these regulations and integrating them into the development process from the outset.

The initial stage involves conducting thorough preclinical investigations to establish the safety and effectiveness of the enzymatically modified products, which encompass evaluations such as toxicity studies, pharmacokinetic assessments, and pharmacodynamic assessments. Ensuring that the grafting process does not introduce harmful contaminants or alter the pharmacological properties of the drug is vital [121 - 123].

Documentation, compliance, and traceability are crucial in adhering to regulatory requirements. All these parameters are essential to keep thorough documentation of the production process, covering aspects such as the origin and management of raw materials, process variables, and quality assurance steps. Adherence to the Good Manufacturing Practice guidelines is imperative to uphold uniformity in product quality.

Engaging with regulatory bodies early in the development process can facilitate smoother approval pathways [124, 125]. This can include seeking advice on study designs and regulatory expectations through meetings and consultations. Furthermore, the implementation of quality by design principles can play a significant role in identifying and managing crucial process parameters, thereby ensuring the reliability and consistency of manufacturing processes.

In summary, scaling up enzymatic grafting processes in pharmaceutical manufacturing requires careful selection and stabilization of enzymes, optimization of process conditions, and strict adherence to regulatory standards. These efforts ensure cost-effective, stable, and compliant production of high-quality pharmaceuticals.

CONCLUSION

Enzymatic green grafting of biopolymers has recently been a focus of green chemistry technologies, owing to rising environmental trepidations and related legal constraints. Over the last decade, there has been a substantial increase in research, particularly in polymer science, on the numerous applications of enzymes. There are five basic synthetic techniques used to create enzyme-polymer hybrids: covalent bonding, ionic and non-ionic interaction, encapsulation, physical entrapment, and affinity-based interactions. In a procedure known as the grafting-to approach, synthetic polymerization may be created with a variety of topologies and functional end-groups to react with several residues on the enzyme surface. Specific requirements and constraints are necessary to facilitate the commercialization of enzymatic green grafting methods. Enzymatic green grafting is, overall, a promising way of surface alteration.

REFERENCES

[1] Buller R, Lutz S, Kazlauskas RJ, Snajdrova R, Moore JC, Bornscheuer UT. From nature to industry: Harnessing enzymes for biocatalysis. Science 2023; 382(6673): eadh8615.
[http://dx.doi.org/10.1126/science.adh8615] [PMID: 37995253]

[2] Anwar A, Imran M, Iqbal HMN. Smart chemistry and applied perceptions of enzyme-coupled nano-engineered assemblies to meet future biocatalytic challenges. Coord Chem Rev 2023; 493: 215329.
[http://dx.doi.org/10.1016/j.ccr.2023.215329]

[3] Payen PA, Persoz JF. Memoir on diastase, the principal products of its reactions, and their applications to the industrial arts. Ann Chim Phys 1833; 53: 73-92.

[4] Kühne W. Über das Verhalten verschiedener organisirter und sog. Ungeformter Fermente. *Verh. Heidelb. Naturhist.-.* Med Ver Neue Folge 1877; 1: 190-3.

[5] Sumner JB. The Isolation and Crystallization of the Enzyme Urease. J Biol Chem 1926; 69(2): 435-41.
[http://dx.doi.org/10.1016/S0021-9258(18)84560-4]

[6] Liu J, Ren H, Tang T, *et al.* The Biocatalysis in Cancer Therapy. ACS Catal 2023; 13(12): 7730-55.
[http://dx.doi.org/10.1021/acscatal.3c01363]

[7] Gurung N, Ray S, Bose S, Rai V. A broader view: microbial enzymes and their relevance in industries, medicine, and beyond. BioMed Res Int 2013; 2013: 1-18.
[http://dx.doi.org/10.1155/2013/329121] [PMID: 24106701]

[8] Mustafa A, Faisal S, Ahmed IA, *et al.* Has the time finally come for green oleochemicals and biodiesel production using large-scale enzyme technologies? Current status and new developments. Biotechnol

Adv 2023; 69: 108275.
[http://dx.doi.org/10.1016/j.biotechadv.2023.108275] [PMID: 39492461]

[9] Westley J. Enzymic Catalysis. New York, NY, USA: Harper & Row 1969; pp. 5-15.

[10] Koshland DE Jr. Application of a Theory of Enzyme Specificity to Protein Synthesis. Proc Natl Acad Sci USA 1958; 44(2): 98-104.
[http://dx.doi.org/10.1073/pnas.44.2.98] [PMID: 16590179]

[11] Johnson KA, Goody RS, Johnson KA, Goody RS. The original Michaelis constant: translation of the 1913 Michaelis-Menten paper. Biochemistry 2011; 50(39): 8264-9.
[http://dx.doi.org/10.1021/bi201284u] [PMID: 21888353]

[12] Reetz MT, Sun Z, Qu G. Enzyme Engineering: Selective Catalysts for Applications in Biotechnology, Organic Chemistry, and Life Science. Weinheim, Germany: Wiley-VCH GmbH 2023.
[http://dx.doi.org/10.1002/9783527836895]

[13] Reetz MT, Qu G, Sun Z. Engineered enzymes for the synthesis of pharmaceuticals and other high-value products. Nat Synth 2024; 3(1): 19-32.
[http://dx.doi.org/10.1038/s44160-023-00417-0]

[14] Arnodo D, Maffeis E, Marra F, Nejrotti S, Prandi C. Combination of enzymes and deep eutectic solvents as powerful toolbox for organic synthesis. Molecules 2023; 28(2): 516.
[http://dx.doi.org/10.3390/molecules28020516] [PMID: 36677575]

[15] Kries H, Trottmann F, Hertweck C. Novel Biocatalysts from Specialized Metabolism. Angew Chem Int Ed 2024; 63(4): e202309284.
[http://dx.doi.org/10.1002/anie.202309284] [PMID: 37737720]

[16] Sicard C. In Situ Enzyme Immobilization by Covalent Organic Frameworks. Angew Chem Int Ed 2023; 62(1): e202213405.
[http://dx.doi.org/10.1002/anie.202213405] [PMID: 36330829]

[17] Johansson AC, Mosbach K. Acrylic copolymers as matrices for the immobilization of enzymes. Biochimica et Biophysica Acta (BBA) - Enzymology 1974; 370(2): 348-53.
[http://dx.doi.org/10.1016/0005-2744(74)90095-3] [PMID: 4374238]

[18] Miwa N, Ohtomo K. Enzyme immobilization in the presence of substrates and inhibitors. Jpn Kokai Tokkyo Koho 1975; 56: 591.

[19] Chen N, Chang B, Shi N, Yan W, Lu F, Liu F. Cross-linked enzyme aggregates immobilization: preparation, characterization, and applications. Crit Rev Biotechnol 2023; 43(3): 369-83.
[http://dx.doi.org/10.1080/07388551.2022.2038073] [PMID: 35430938]

[20] Costa IO, Morais JRF, de Medeiros Dantas JM, Gonçalves LRB, dos Santos ES, Rios NS. Enzyme immobilization technology as a tool to innovate in the production of biofuels: A special review of the Cross-Linked Enzyme Aggregates (CLEAs) strategy. Enzyme Microb Technol 2023; 170: 110300.
[http://dx.doi.org/10.1016/j.enzmictec.2023.110300] [PMID: 37523882]

[21] Rodrigues RC, Ortiz C, Berenguer-Murcia Á, Torres R, Fernández-Lafuente R. Modifying enzyme activity and selectivity by immobilization. Chem Soc Rev 2013; 42(15): 6290-307.
[http://dx.doi.org/10.1039/C2CS35231A] [PMID: 23059445]

[22] Isanapong J, Lohawet K, Kumnorkaew P. Optimization and characterization of immobilized laccase on titanium dioxide nanostructure and its application in removal of Remazol Brilliant Blue R. Biocatal Agric Biotechnol 2021; 37: 102186.
[http://dx.doi.org/10.1016/j.bcab.2021.102186]

[23] Heilmann A, Teuscher N, Kiesow A, Janasek D, Spohn U. Nanoporous aluminum oxide as a novel support material for enzyme biosensors. J Nanosci Nanotechnol 2003; 3(5): 375-9.
[http://dx.doi.org/10.1166/jnn.2003.224] [PMID: 14733146]

[24] Huckel M, Wirth HJ, Hearn MTW. Porous zirconia: a new support material for enzyme

immobilization. J Biochem Biophys Methods 1996; 31(3-4): 165-79.
[http://dx.doi.org/10.1016/0165-022X(95)00035-P] [PMID: 8675959]

[25] Stranix B, Darling G. Functional polymers from (vinyl)polystyrene. Enzyme immobilization through a cysteinyl-S-ethyl spacer. Biotechnol Tech 1995; 9(2): 75-80.
[http://dx.doi.org/10.1007/BF00224401]

[26] Shinde P, Musameh M, Gao Y, Robinson AJ, Kyratzis IL. Immobilization and stabilization of alcohol dehydrogenase on polyvinyl alcohol fibre. Biotechnol Rep (Amst) 2018; 19: e00260.
[http://dx.doi.org/10.1016/j.btre.2018.e00260] [PMID: 30003052]

[27] Grotzky A, Altamura E, Adamcik J, *et al.* Structure and enzymatic properties of molecular dendronized polymer-enzyme conjugates and their entrapment inside giant vesicles. Langmuir 2013; 29(34): 10831-40.
[http://dx.doi.org/10.1021/la401867c] [PMID: 23895383]

[28] Biró E, Németh ÁS, Sisak C, Feczkó T, Gyenis J. Preparation of chitosan particles suitable for enzyme immobilization. J Biochem Biophys Methods 2008; 70(6): 1240-6.
[http://dx.doi.org/10.1016/j.jprot.2007.11.005] [PMID: 18155771]

[29] Hanachi P, Jafary F, Jafary F, Motamedi S. Immobilization of the Alkaline Phosphatase on Collagen Surface *via* Cross-Linking Method. Iran J Biotechnol 2015; 13(3): 32-8.
[http://dx.doi.org/10.15171/ijb.1203] [PMID: 28959297]

[30] Vega Erramuspe IB, Fazeli E, Näreoja T, *et al.* Advanced Cellulose Fibers for Efficient Immobilization of Enzymes. Biomacromolecules 2016; 17(10): 3188-97.
[http://dx.doi.org/10.1021/acs.biomac.6b00865] [PMID: 27575620]

[31] Yu CC, Kuo YY, Liang CF, *et al.* Site-specific immobilization of enzymes on magnetic nanoparticles and their use in organic synthesis. Bioconjug Chem 2012; 23(4): 714-24.
[http://dx.doi.org/10.1021/bc200396r] [PMID: 22424277]

[32] Zahidah KA, Kakooei S, Ismail MC, Bothi Raja P. Halloysite nanotubes as nanocontainer for smart coating application: A review. Prog Org Coat 2017; 111: 175-85.
[http://dx.doi.org/10.1016/j.porgcoat.2017.05.018]

[33] Liu X, Gao Z, Wang D, Yu F, Du B, Gitsov I. Improving the Protection Performance of Waterborne Coatings with a Corrosion Inhibitor Encapsulated in Polyaniline-Modified Halloysite Nanotubes. Coatings 2023; 13(10): 1677.
[http://dx.doi.org/10.3390/coatings13101677]

[34] Tully J, Yendluri R, Lvov Y. Halloysite Clay Nanotubes for Enzyme Immobilization. Biomacromolecules 2016; 17(2): 615-21.
[http://dx.doi.org/10.1021/acs.biomac.5b01542] [PMID: 26699154]

[35] Kumar-Krishnan S, Hernandez-Rangel A, Pal U, *et al.* Surface functionalized halloysite nanotubes decorated with silver nanoparticles for enzyme immobilization and biosensing. J Mater Chem B Mater Biol Med 2016; 4(15): 2553-60.
[http://dx.doi.org/10.1039/C6TB00051G] [PMID: 32263278]

[36] Wang D, Ding W, Zhou K, Guo S, Zhang Q, Haddleton DM. Coating Titania Nanoparticles with Epoxy-Containing Catechol Polymers *via* Cu(0)-Living Radical Polymerization as Intelligent Enzyme Carriers. Biomacromolecules 2018; 19(7): 2979-90.
[http://dx.doi.org/10.1021/acs.biomac.8b00544] [PMID: 29738234]

[37] Gómez JL, Bódalo A, Gómez E, Bastida J, Hidalgo AM, Gómez M. Immobilization of peroxidases on glass beads: An improved alternative for phenol removal. Enzyme Microb Technol 2006; 39(5): 1016-22.
[http://dx.doi.org/10.1016/j.enzmictec.2006.02.008]

[38] Kim MI, Ham HO, Oh SD, Park HG, Chang HN, Choi SH. Immobilization of *Mucor javanicus* lipase on effectively functionalized silica nanoparticles. J Mol Catal, B Enzym 2006; 39(1-4): 62-8.

[http://dx.doi.org/10.1016/j.molcatb.2006.01.028]

[39] Gao Y, Kyratzis I. Covalent immobilization of proteins on carbon nanotubes using the cross-linker 1-ethyl-3-(3-dimethylaminopropyl)carbodiimide--a critical assessment. Bioconjug Chem 2008; 19(10): 1945-50.
[http://dx.doi.org/10.1021/bc800051c] [PMID: 18759407]

[40] Koneracká M, Kopčanský P, Antalík M, et al. Immobilization of proteins and enzymes to fine magnetic particles. J Magn Magn Mater 1999; 201(1-3): 427-30.
[http://dx.doi.org/10.1016/S0304-8853(99)00005-0]

[41] Velasco-Lozano S, López-Gallego F, Vázquez-Duhalt R, Mateos-Díaz JC, Guisán JM, Favela-Torres E. Carrier-free immobilization of lipase from Candida rugosa with polyethyleneimines by carboxyl-activated cross-linking. Biomacromolecules 2014; 15(5): 1896-903.
[http://dx.doi.org/10.1021/bm500333v] [PMID: 24720524]

[42] Mohamad NR, Marzuki NHC, Buang NA, Huyop F, Wahab RA. An overview of technologies for immobilization of enzymes and surface analysis techniques for immobilized enzymes. Biotechnol Biotechnol Equip 2015; 29(2): 205-20.
[http://dx.doi.org/10.1080/13102818.2015.1008192] [PMID: 26019635]

[43] Zimmermann JL, Nicolaus T, Neuert G, Blank K. Thiol-based, site-specific and covalent immobilization of biomolecules for single-molecule experiments. Nat Protoc 2010; 5(6): 975-85.
[http://dx.doi.org/10.1038/nprot.2010.49] [PMID: 20448543]

[44] Heck T, Faccio G, Richter M, Thöny-Meyer L. Enzyme-catalyzed protein crosslinking. Appl Microbiol Biotechnol 2013; 97(2): 461-75.
[http://dx.doi.org/10.1007/s00253-012-4569-z] [PMID: 23179622]

[45] Minamihata K, Goto M, Kamiya N. Site-specific protein cross-linking by peroxidase-catalyzed activation of a tyrosine-containing peptide tag. Bioconjug Chem 2011; 22(1): 74-81.
[http://dx.doi.org/10.1021/bc1003982] [PMID: 21142129]

[46] Sakhare MS, Rajput HH. Polymer grafting and applications in pharmaceutical drug delivery systems-a brief review. Asian J Pharm Clin Res 2017; 10(6): 59.
[http://dx.doi.org/10.22159/ajpcr.2017.v10i6.18072]

[47] Marney DCO, Russell LJ. Combined fire retardant and wood preservative treatments for outdoor wood applications–a review of the literature. Fire Technol 2008; 44(1): 1-14.
[http://dx.doi.org/10.1007/s10694-007-0016-6]

[48] Fernández-Costas C, Palanti S, Charpentier JP, Sanromán MÁ, Moldes D. A sustainable treatment for wood preservation: enzymatic grafting of wood extractives. ACS Sustain Chem& Eng 2017; 5(9): 7557-67.
[http://dx.doi.org/10.1021/acssuschemeng.7b00714]

[49] Ahn SH, Oh SC, Choi I, et al. Environmentally friendly wood preservatives formulated with enzymatic-hydrolyzed okara, copper and/or boron salts. J Hazard Mater 2010; 178(1-3): 604-11.
[http://dx.doi.org/10.1016/j.jhazmat.2010.01.128] [PMID: 20153107]

[50] A. S., Oh SeiChang, O. S., Choi InGyu, C. I., Han GyuSeong, H. G., Jeong HanSeob, J. H., Kim KiWoo, K. K., & Yang In. Y 2010; I. [Environmentally friendly wood preservatives formulated with enzymatic-hydrolyzed okara, copper and/or boron salts.].

[51] Hollmann F, Arends IWCE. Enzyme initiated radical polymerizations. Polymers (Basel) 2012; 4(1): 759-93.
[http://dx.doi.org/10.3390/polym4010759]

[52] Sigg SJ, Seidi F, Renggli K, Silva TB, Kali G, Bruns N. Horseradish peroxidase as a catalyst for atom transfer radical polymerization. Macromol Rapid Commun 2011; 32(21): 1710-5.
[http://dx.doi.org/10.1002/marc.201100349] [PMID: 21842510]

[53] Trivedi V, Chand P, Maulik PR, Bandyopadhyay U. Mechanism of horseradish peroxidase-catalyzed

heme oxidation and polymerization (β-hematin formation). Biochim Biophys Acta, Gen Subj 2005; 1723(1-3): 221-8.
[http://dx.doi.org/10.1016/j.bbagen.2005.02.005] [PMID: 15780996]

[54] Rodriguez KJ, Pellizzoni MM, Chadwick RJ, Guo C, Bruns N. Enzyme-initiated free radical polymerizations of vinyl monomers using horseradish peroxidase. Methods Enzymol 2019; 627: 249-62.
[http://dx.doi.org/10.1016/bs.mie.2019.08.013] [PMID: 31630743]

[55] Dadashi-Silab S, Matyjaszewski K. Iron catalysts in atom transfer radical polymerization. Molecules 2020; 25(7): 1648.
[http://dx.doi.org/10.3390/molecules25071648] [PMID: 32260141]

[56] Xie, W., Zhao, L., Wei, Y., & Yuan, J. (2021). Advances in enzyme-catalysis-mediated RAFT polymerization. Cell Reports Physical Science, 2(7).

[57] Kyomuhimbo HD, Feleni U, Haneklaus NH, Brink H. Recent advances in applications of oxidases and peroxidases polymer-based enzyme biocatalysts in sensing and wastewater treatment: a review. Polymers (Basel) 2023; 15(16): 3492.
[http://dx.doi.org/10.3390/polym15163492] [PMID: 37631549]

[58] Shen C, Wang Y. Recent Progress on Peroxidase Modification and Application. Appl Biochem Biotechnol 2024; 196(9): 5740-64.
[http://dx.doi.org/10.1007/s12010-023-04835-w] [PMID: 38180646]

[59] Liu R, Dong A, Fan X, *et al.* Enzymatic Hydrophobic Modification of Jute Fibers *via* Grafting to Reinforce Composites. Appl Biochem Biotechnol 2016; 178(8): 1612-29.
[http://dx.doi.org/10.1007/s12010-015-1971-x] [PMID: 26754422]

[60] Zhang W, Hollmann F. Synthesis of vinyl polymers *via* enzymatic oxidative polymerisation. In: Hori K, Kobayashi S, Rinaudo M, eds. Enzymatic polymerization towards green polymer chemistry. Singapore: Springer; 2019. p. 343-56.

[61] Bangar SP, Ashogbon AO, Singh A, Chaudhary V, Whiteside WS. Enzymatic modification of starch: A green approach for starch applications. Carbohydrate polymers. 2022, 1;287:119265.

[62] Xie SX, Liu Q, Cui SW. Starch modification and applications. Food carbohydrates: Chemistry, physical properties, and applications. 2005, 23:357-405.

[63] Kalia S, Thakur K, Kumar A, Celli A. Laccase-assisted surface functionalization of lignocellulosics. J Mol Catal B Enzym 2014; 102: 48-58.
[http://dx.doi.org/10.1016/j.molcatb.2014.01.014]

[64] Schneider WDH, Bolaño Losada C, Moldes D, *et al.* A sustainable approach of enzymatic grafting on Eucalyptus globulus wood by laccase from the newly isolated white-rot basidiomycete *Marasmiellus palmivorus* VE111. ACS Sustain Chem& Eng 2019; 7(15): 13418-24.
[http://dx.doi.org/10.1021/acssuschemeng.9b02770]

[65] Schneider WDH, Fontana RC, Baudel HM, *et al.* Lignin degradation and detoxification of eucalyptus wastes by on-site manufacturing fungal enzymes to enhance second-generation ethanol yield. Appl Energy 2020; 262: 114493.
[http://dx.doi.org/10.1016/j.apenergy.2020.114493]

[66] Xie W, Zhao L, Wei Y, Yuan J. Advances in enzyme-catalysis-mediated RAFT polymerization. Cell Reports Physical Science. 2021 Jul 21; 2(7).
[http://dx.doi.org/10.1016/j.xcrp.2021.100487]

[67] Kumar S, Gaddala R, Thomas S, Schumacher J, Schönherr H. Green synthesis of polymer materials *via* enzyme- initiated RAFT polymerization. Polym Chem 2024; 15(20): 2011-27.
[http://dx.doi.org/10.1039/D4PY00294F]

[68] Monajati M, Tamaddon A, Yousefi G, Abolmaali SS, Dinarvand R. Applications of RAFT polymerization for chemical and enzymatic stabilization of L-asparaginase conjugates with well-

defined poly(HPMA). New J Chem 2019; 43(29): 11564-74.
[http://dx.doi.org/10.1039/C9NJ01211G]

[69] Xu J, Jung K, Atme A, Shanmugam S, Boyer C. A robust and versatile photoinduced living polymerization of conjugated and unconjugated monomers and its oxygen tolerance. J Am Chem Soc 2014; 136(14): 5508-19.
[http://dx.doi.org/10.1021/ja501745g] [PMID: 24689993]

[70] Chapman R, Gormley AJ, Herpoldt KL, Stevens MM. Highly controlled open vessel RAFT polymerizations by enzyme degassing. Macromolecules 2014; 47(24): 8541-7.
[http://dx.doi.org/10.1021/ma5021209]

[71] Filat A, Gallardo O, Vidal T, Pastor FIJ, Díaz P, Roncero MB. Enzymatic grafting of natural phenols to flax fibres: Development of antimicrobial properties. Carbohydr Polym 2012; 87(1): 146-52.
[http://dx.doi.org/10.1016/j.carbpol.2011.07.030] [PMID: 34662943]

[72] Garcia-Ubasart J, Esteban A, Vila C, Roncero MB, Colom JF, Vidal T. Enzymatic treatments of pulp using laccase and hydrophobic compounds. Bioresour Technol 2011; 102(3): 2799-803.
[http://dx.doi.org/10.1016/j.biortech.2010.10.020] [PMID: 21050744]

[73] Elegir G, Kindl A, Sadocco P, Orlandi M. Development of antimicrobial cellulose packaging through laccase-mediated grafting of phenolic compounds. Enzyme Microb Technol 2008; 43(2): 84-92.
[http://dx.doi.org/10.1016/j.enzmictec.2007.10.003]

[74] Strong PJ, Kalyuzhnaya M, Silverman J, Clarke WP. A methanotroph-based biorefinery: Potential scenarios for generating multiple products from a single fermentation. Bioresour Technol 2016; 215: 314-23.
[http://dx.doi.org/10.1016/j.biortech.2016.04.099] [PMID: 27146469]

[75] Stone KA, Hilliard MV, He QP, Wang J. A mini review on bioreactor configurations and gas transfer enhancements for biochemical methane conversion. Biochem Eng J 2017; 128: 83-92.
[http://dx.doi.org/10.1016/j.bej.2017.09.003]

[76] Strong PJ, Xie S, Clarke WP. Methane as a resource: can the methanotrophs add value? Environ Sci Technol 2015; 49(7): 4001-18.
[http://dx.doi.org/10.1021/es504242n] [PMID: 25723373]

[77] Chapman J, Ismail AE, Dinu CZ. Industrial applications of enzymes: recent advances, techniques, and outlooks. Catalysts 2018; 8(6): 238.
[http://dx.doi.org/10.3390/catal8060238]

[78] Eedin-Dahlström J, Wu J, Koivula A, et al. Green Synthesis of Polymer Materials via Enzyme-Initiated RAFT Polymerization. Polym Chem. 2024;15(20):2094-2105.
[http://dx.doi.org/10.1039/D4PY00294F]

[79] Yang D, Huang J, Hu Z, et al. Catalytic conversion of lignin into monoaromatic hydrocarbons over a Ni/Al hydrotalcite-derived catalyst. Fuel 2024; 357: 129982.
[http://dx.doi.org/10.1016/j.fuel.2023.129982]

[80] Fei Q, Guarnieri MT, Tao L, Laurens LML, Dowe N, Pienkos PT. Bioconversion of natural gas to liquid fuel: Opportunities and challenges. Biotechnol Adv 2014; 32(3): 596-614.
[http://dx.doi.org/10.1016/j.biotechadv.2014.03.011] [PMID: 24726715]

[81] Friščić T, Mottillo C, Titi HM. Mechanochemistry for Synthesis. Angew Chem Int Ed 2020; 59(3): 1018-29.
[http://dx.doi.org/10.1002/anie.201906755] [PMID: 31294885]

[82] Baláž P, Achimovičová M, Baláž M, et al. Hallmarks of mechanochemistry: from nanoparticles to technology. Chem Soc Rev 2013; 42(18): 7571-637.
[http://dx.doi.org/10.1039/c3cs35468g] [PMID: 23558752]

[83] Stolle A, Szuppa T, Leonhardt SES, Ondruschka B. Ball milling in organic synthesis: solutions and challenges. Chem Soc Rev 2011; 40(5): 2317-29.

[http://dx.doi.org/10.1039/c0cs00195c] [PMID: 21387034]

[84] Schreyer H, Eckert R, Immohr S, de Bellis J, Felderhoff M, Schüth F. Milling Down to Nanometers: A General Process for the Direct Dry Synthesis of Supported Metal Catalysts. Angew Chem Int Ed 2019; 58(33): 11262-5.
[http://dx.doi.org/10.1002/anie.201903545] [PMID: 31184405]

[85] Nicole L, Laberty-Robert C, Rozes L, Sanchez C. Hybrid materials science: a promised land for the integrative design of multifunctional materials. Nanoscale 2014; 6(12): 6267-92.
[http://dx.doi.org/10.1039/C4NR01788A] [PMID: 24866174]

[86] Sanchez C, Julián B, Belleville P, Popall M. Applications of hybrid organic–inorganic nanocomposites. J Mater Chem 2005; 15(35-36): 3559-92.
[http://dx.doi.org/10.1039/b509097k]

[87] Sanchez C, Belleville P, Popall M, Nicole L. Applications of advanced hybrid organic–inorganic nanomaterials: from laboratory to market. Chem Soc Rev 2011; 40(2): 696-753.
[http://dx.doi.org/10.1039/c0cs00136h] [PMID: 21229132]

[88] Abdollahian Y, Hauser JL, Rogow DL, Oliver AG, Oliver SRJ. Solvothermal synthesis and characterization of two inorganic–organic hybrid materials based on barium. Dalton Trans 2012; 41(40): 12630-4.
[http://dx.doi.org/10.1039/c2dt31518a] [PMID: 22968326]

[89] Melde BJ, Holland BT, Blanford CF, Stein A. Mesoporous Sieves with Unified Hybrid Inorganic/Organic Frameworks. Chem Mater 1999; 11(11): 3302-8.
[http://dx.doi.org/10.1021/cm9903935]

[90] Kickelbick G. Hybrid inorganic-organic mesoporous materials. Angew Chem Int Ed 2004; 43(24): 3102-4.
[http://dx.doi.org/10.1002/anie.200301751] [PMID: 15199556]

[91] König A, Malek A, Fehrenbacher U, Brunklaus G, Wilhelm M, Hirth T. Silane-functionalized Flame-retardant Aluminum Trihydroxide in Flexible Polyurethane Foam. J Cell Plast 2010; 46(5): 395-413.
[http://dx.doi.org/10.1177/0021955X10367703]

[92] Heintz AS, Fink MJ, Mitchell BS. Mechanochemical Synthesis of Blue Luminescent Alkyl/Alkenyl-Passivated Silicon Nanoparticles. Adv Mater 2007; 19(22): 3984-8.
[http://dx.doi.org/10.1002/adma.200602752]

[93] Heintz AS, Fink MJ, Mitchell BS. Silicon nanoparticles with chemically tailored surfaces. Appl Organomet Chem 2010; 24(3): 236-40.
[http://dx.doi.org/10.1002/aoc.1602]

[94] Gu D, Schmidt W, Pichler CM, *et al.* Surface-Casting Synthesis of Mesoporous Zirconia with a CMK-5-Like Structure and High Surface Area. Angew Chem Int Ed 2017; 56(37): 11222-5.
[http://dx.doi.org/10.1002/anie.201705042] [PMID: 28657163]

[95] Maciel GE. Solid-State NMR Spectroscopy of Inorganic Materials. ACS Symposium Series. Vol. 717: 326-56.
[http://dx.doi.org/10.1021/bk-1999-0717.ch012]

[96] Margolese D, Melero JA, Christiansen SC, Chmelka BF, Stucky GD. Direct Syntheses of Ordered SBA-15 Mesoporous Silica Containing Sulfonic Acid Groups. Chem Mater 2000; 12(8): 2448-59.
[http://dx.doi.org/10.1021/cm0010304]

[97] de Andrade GF, Soares DCF, Almeida RKS, Sousa EMB. Mesoporous Silica SBA-16 Functionalized with Alkoxysilane Groups: Preparation, Characterization, and Release Profile Study. J Nanomater 2012; 2012(1): 816496.
[http://dx.doi.org/10.1155/2012/816496]

[98] Vallet-Regí M, Balas F, Arcos D. Mesoporous materials for drug delivery. Angew Chem Int Ed 2007; 46(40): 7548-58.

[http://dx.doi.org/10.1002/anie.200604488] [PMID: 17854012]

[99] Song SW, Hidajat K, Kawi S. Functionalized SBA-15 materials as carriers for controlled drug delivery: influence of surface properties on matrix-drug interactions. Langmuir 2005; 21(21): 9568-75.
[http://dx.doi.org/10.1021/la051167e] [PMID: 16207037]

[100] Slowing II, Trewyn BG, Giri S, Lin VSY. Mesoporous Silica Nanoparticles for Drug Delivery and Biosensing Applications. Adv Funct Mater 2007; 17(8): 1225-36.
[http://dx.doi.org/10.1002/adfm.200601191]

[101] Chen Z, Hsu FC, Battigelli D, Chang HC. Capture and release of viruses using amino-functionalized silica particles. Anal Chim Acta 2006; 569(1-2): 76-82.
[http://dx.doi.org/10.1016/j.aca.2006.03.103]

[102] Dhara K, Sarkar K, Srimani D, Saha SK, Chattopadhyay P, Bhaumik A. A new functionalized mesoporous matrix supported Pd(ii)-Schiff base complex: an efficient catalyst for the Suzuki–Miyaura coupling reaction. Dalton Trans 2010; 39(28): 6395-402.
[http://dx.doi.org/10.1039/c003142a] [PMID: 20532296]

[103] Antochshuk V, Jaroniec M. Adsorption, Thermogravimetric, and NMR Studies of FSM-16 Material Functionalized with Alkylmonochlorosilanes. J Phys Chem B 1999; 103(30): 6252-61.
[http://dx.doi.org/10.1021/jp990314c]

[104] Karpov SI, Roessner F, Selemenev VF. Studies on functionalized mesoporous materials—Part I: characterization of silylized mesoporous material of type MCM-41. J Porous Mater 2014; 21(4): 449-57.
[http://dx.doi.org/10.1007/s10934-014-9791-x]

[105] Gupta S, Ramamurthy PC, Madras G. Synthesis and characterization of flexible epoxy nanocomposites reinforced with amine functionalized aluminananoparticles: a potential encapsulant for organic devices. Polym Chem 2011; 2(1): 221-8.
[http://dx.doi.org/10.1039/C0PY00270D]

[106] Meroni D, Lo Presti L, Di Liberto G, et al. A Close Look at the Structure of the TiO_2-APTES Interface in Hybrid Nanomaterials and Its Degradation Pathway: An Experimental and Theoretical Study. J Phys Chem C 2017; 121(1): 430-40.
[http://dx.doi.org/10.1021/acs.jpcc.6b10720] [PMID: 28191270]

[107] Woodley JM. New frontiers in biocatalysis for sustainable synthesis. Curr Opin Green Sustain Chem 2020; 21: 22-6.
[http://dx.doi.org/10.1016/j.cogsc.2019.08.006]

[108] Reetz MT. What are the Limitations of Enzymes in Synthetic Organic Chemistry? Chem Rec 2016; 16(6): 2449-59.
[http://dx.doi.org/10.1002/tcr.201600040] [PMID: 27301318]

[109] Hauer B. Embracing Nature's Catalysts: A Viewpoint on the Future of Biocatalysis. ACS Catal 2020; 10(15): 8418-27.
[http://dx.doi.org/10.1021/acscatal.0c01708]

[110] Bommarius AS, Paye MF. Stabilizing biocatalysts. Chem Soc Rev 2013; 42(15): 6534-65.
[http://dx.doi.org/10.1039/c3cs60137d] [PMID: 23807146]

[111] Mandpe P, Prabhakar B, Gupta H, Shende P. Glucose oxidase-based biosensor for glucose detection from biological fluids. Sens Rev 2020; 40(4): 497-511.
[http://dx.doi.org/10.1108/SR-01-2019-0017]

[112] Mano N, de Poulpiquet A. O_2 Reduction in Enzymatic Biofuel Cells. Chem Rev 2018; 118(5): 2392-468.
[http://dx.doi.org/10.1021/acs.chemrev.7b00220] [PMID: 28930449]

[113] Lojou E. Hydrogenases as catalysts for fuel cells: Strategies for efficient immobilization at electrode interfaces. Electrochim Acta 2011; 56(28): 10385-97.

[http://dx.doi.org/10.1016/j.electacta.2011.03.002]

[114] Lubitz W, Ogata H, Rüdiger O, Reijerse E. Hydrogenases. Chem Rev 2014; 114(8): 4081-148.
[http://dx.doi.org/10.1021/cr4005814] [PMID: 24655035]

[115] Alfano M, Cavazza C. Structure, function, and biosynthesis of nickel-dependent enzymes. Protein Sci 2020; 29(5): 1071-89.
[http://dx.doi.org/10.1002/pro.3836] [PMID: 32022353]

[116] Najafpour MM, Zaharieva I, Zand Z, et al. Water-oxidizing complex in Photosystem II: Its structure and relation to manganese-oxide based catalysts. Coord Chem Rev 2020; 409: 213183.
[http://dx.doi.org/10.1016/j.ccr.2020.213183]

[117] Morello G, Megarity CF, Armstrong FA. The power of electrified nanoconfinement for energising, controlling and observing long enzyme cascades. Nat Commun 2021; 12(1): 340.
[http://dx.doi.org/10.1038/s41467-020-20403-w] [PMID: 33436601]

[118] Stolarczyk K, Rogalski J, Bilewicz R. NAD(P)-dependent glucose dehydrogenase: Applications for biosensors, bioelectrodes, and biofuel cells. Bioelectrochemistry 2020; 135: 107574.
[http://dx.doi.org/10.1016/j.bioelechem.2020.107574] [PMID: 32498025]

[119] Alvarez-Malmagro J, García-Molina G, López De Lacey A. Electrochemical Biosensors Based on Membrane-Bound Enzymes in Biomimetic Configurations. Sensors (Basel) 2020; 20(12): 3393.
[http://dx.doi.org/10.3390/s20123393] [PMID: 32560121]

[120] Bollella P, Katz E. Bioelectrocatalysis at carbon nanotubes. In: Kumar CV, Pyle AM, Christianson DW, Eds. Methods in Enzymology: Nanoarmoring of Enzymes with Carbon Nanotubes and Magnetic Nanoparticles. Amsterdam, The Netherlands: Elsevier 2020; Vol. 630: pp. 215-47.
[http://dx.doi.org/10.1016/bs.mie.2019.10.012]

[121] Wu R, Ma C, Zhu Z. Enzymatic electrosynthesis as an emerging electrochemical synthesis platform. Curr Opin Electrochem 2020; 19: 1-7.
[http://dx.doi.org/10.1016/j.coelec.2019.08.004]

[122] Yuan M, Kummer MJ, Minteer SD. Strategies for Bioelectrochemical CO_2 Reduction. Chemistry 2019; 25(63): 14258-66.
[http://dx.doi.org/10.1002/chem.201902880] [PMID: 31386223]

[123] Mazurenko I, Wang X, de Poulpiquet A, Lojou E. H_2/O_2 enzymatic fuel cells: from proof-of-concept to powerful devices. Sustain Energy Fuels 2017; 1(7): 1475-501.
[http://dx.doi.org/10.1039/C7SE00180K]

[124] Xiao X, Xia H, Wu R, et al. Tackling the Challenges of Enzymatic (Bio)Fuel Cells. Chem Rev 2019; 119(16): 9509-58.
[http://dx.doi.org/10.1021/acs.chemrev.9b00115] [PMID: 31243999]

[125] Mazurenko I, de Poulpiquet A, Lojou E. Recent developments in high surface area bioelectrodes for enzymatic fuel cells. Curr Opin Electrochem 2017; 5(1): 74-84.
[http://dx.doi.org/10.1016/j.coelec.2017.07.001]

CHAPTER 9

Mechanochemical Green Grafting: Methods, Environmental Benefits, and Uses

Vijayshri Rokde[1], Kishor Danao[1], Deweshri Nandurkar[1], Shweta Saboo[2,*] and Ujwala Mahajan[3]

[1] *Department of Pharmaceutical Chemistry, Dadasaheb Balpande College of Pharmacy, Nagpur, Maharashtra 440037, India*

[2] *Department of Pharmacognosy, Government College of Pharmacy, Chhatrapati Sambhajinagar, Maharashtra 431005, India*

[3] *Department of Pharmacognosy, Dadasaheb Balpande College of Pharmacy, Nagpur, Maharashtra 440037, India*

Abstract: Mechanochemistry has garnered considerable attention as a potent, long-lasting, efficient, eco-friendly, and economical synthesis technique for creating new functional materials. This method is based on physicochemical reactions that are accelerated by mechanical force using milling and grinding. Mechanochemical synthesis is described as a chemical reaction that occurs through the absorption of mechanical energy. To facilitate chemical reactivity, reactions are carried out by grinding the reagents using ball-mill devices, such as vibrating, planetary, tumbler ball-mills, or single-screw devices, which employ mechanical forces. This technique is reported for the practical, solvent-free synthesis of superhydrophobic surfaces using a mechanochemical approach. One-step mechanochemical grafting was employed to generate thiol-functionalized montmorillonite, resulting in covalent organic frameworks (COFs) with excellent iodine capture properties as adsorbents. MOFs comprise the following widely used structures: ZIF-8, HKUST-1, MIL-101, UiO-66, and MOF-5. These approaches are used to prepare bio-inspired metal–organic frameworks, novel metallopharmaceuticals and metallodrugs, phenol hydrazone derivatives, cyclodextrin nanosponges (CD-NS) polymers, copper oxide nanoparticles, and silver nanoparticles for antibacterial activity; to perform mechanoactivation of silicon to synthesize alkoxysilanes; to synthesize heterocyclic derivatives using a ball mill; to produce pharmaceutical cocrystals; and to synthesize catalysts. The distinguishing features of mechanochemical processes over solution-based chemistry include more selective reactions, which allow for simpler work-up procedures.

Keywords: Ball mill, COF, Green grafting, Mechanochemistry, MOF, Nanoparticles.

---

[*] **Corresponding author Shweta Saboo:** Department of Pharmacognosy, Government College of Pharmacy, Chhatrapati Sambhajinagar, Maharashtra 431005, India; E-mail: Shweta.saboo1@gmail.com

INTRODUCTION

Mechanochemistry [1], which is well known in the fields of crystal engineering and polymorphism, is regaining popularity as an organic synthesis approach. While the technology is not new, it is developing beyond its original purpose of providing a solvent-free alternative. Despite numerous synthetic transformations having been recorded, those with reactivity differences from traditional solution-based reactions are likely the most fascinating. It is described as "a chemical reaction that is induced by the direct absorption of mechanical energy". Therefore, it complements the traditional techniques of activation: heat, irradiation, and electrochemistry. It is also connected to tribochemistry [2, 3], which involves chemical interactions on the surface/boundary of distinct environments. Despite its recent renaissance, mechanochemistry remains considerably less studied and understood than traditional methods of energy input [3 - 6]. For example, pericyclic reactions behave differently depending on whether they are triggered by light or heat. The response outcome is foreseeable using the Woodward-Hoffmann rules. Interestingly, when ultrasound is employed as the input energy for a pericyclic reaction, unexpected products have been detected; in fact, it has been proposed that Woodward-Hoffmann laws should not be used to describe mechanochemical pericyclic reactions [7].

Mechanochemistry, therefore, offers the chance to investigate a unique chemical space of reactions. Mechanochemistry can describe the impact and pulling forces. This approach focuses on the influence of milling. Wilhelm Ostwald is credited with coining the term "mechanochemistry" in 1907, along with electrochemistry, photochemistry, and thermochemistry [8-10]. Matthew Carey Lea first proposed the idea that mechanochemical reactions could differ fundamentally from thermochemical ones in 1892 [11-13]. Subsequently, the area of mechanochemistry has grown steadily, from Staudinger's pioneering work on polymer mechanochemistry in the 1930s to various investigations into supramolecular mechanochemistry in the 1980s and solid-solid organic synthesis techniques in the 1990s [14]. Presently, the IUPAC defines a mechanochemical reaction as a chemical reaction caused by the direct absorption of mechanical energy [15]. Despite the rapid development of shearing- and grinding-induced mechanochemistry and the frequent publishing of novel synthetic processes for diverse products, the mechanism at the microscopic and molecular levels remains unknown [16]. Reactive extruders and high-speed ball mills are sometimes referred to as "black boxes" because the chemical processes they carry out are protected by enclosures, most of which are composed of metal or ceramic. This leads to one of the key challenges of mechanochemistry: the lack of a good theory that completely describes the current reactions while being universally applicable to varied synthetic methods [17]. Mechanochemistry now has three theories: the hot-spot theory, the magma-plasma model, and the pseudo-fluid model, each of which may describe specific occurrences in distinct subdisciplines [18-20]. Mechanochemistry is a rapidly expanding field in chemistry. The efficient energy dispersion and mass movement caused by mechanical forces have permitted the removal of solvents, resulting in cleaner, quicker, and more straightforward chemical syntheses than conventional solvent-based processes. Mechanochemistry has established itself as a dependable synthesis technique, as evidenced by several reports. In particular, the use of mechanochemistry to synthesize molecules that are difficult or impossible to obtain using conventional methods has advanced many areas of chemistry, and mechanochemical activation is now recognized as a promising technique alongside thermal, photochemical, and electrochemical activation.

METHODS OF MECHANOCHEMICAL GREEN GRAFTING

Improving a surface polymer's wettability, biological compatibility, physical properties, etc., is the primary goal of surface modification. The different techniques of grafting include chemical grafting, free radical grafting, ionic grafting and photochemical grafting. By removing the need for reactive chemicals, enzyme application can provide a safer, more cost-effective, and environmentally friendly method of grafting procedures. Moreover, the specificity of enzymes may provide a possibility for precisely customizing desired macromolecular characteristics. Among them, it is predicated on the idea that the chemical/electrochemical grafting process may be started with the aid of an enzyme. By removing the need for reactive chemicals, enzyme application can provide a safer, more cost-effective, and environmentally friendly method of grafting procedures [21]. Moreover, the specificity of enzymes may provide a possibility for precisely customizing desired macromolecular characteristics. High expectations are placed on innovative materials by recent technology breakthroughs and the quest for new functionalities of designed solids, which are typically delivered by extremely complex procedures [22]. However, it is essential to create manufacturing methods that are straightforward, affordable, waste-free, high-yield, and scalable for both financial and environmental reasons [23]. Material synthesis has traditionally been dominated by wet chemistry procedures, which have been well-optimized throughout time to fulfill strict standards to limit waste creation. However, due to the inherent nature of solution processing, the production of solvent waste cannot be prevented. Furthermore, solution-based chemistries are linked with extensive treatment periods and post-synthesis drying procedures. Mechanochemistry has recently gained popularity as a viable method for carrying out chemical reactions and material synthesis under considerably gentler circumstances than solvothermal approaches. A noteworthy benefit of employing mechanochemistry is the ability to carry out the reactions without the need for solvents. Therefore, the main benefit of mechanochemical processes is their ability to prevent the waste of solvents. Scientists involved in the synthesis of organic materials and the creation of nanomaterials have previously taken notice of this. There are several benefits to using a mortar and pestle. In the first place, it gives researchers exact monitoring of the method of grinding, allowing them to adjust the particle size and distribution and form the MOF structure as given in Fig. (**1**) [24-26].

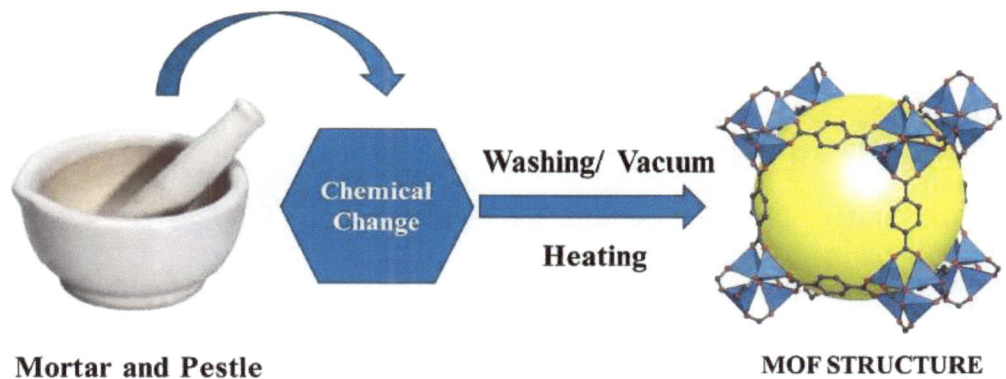

Fig. (1). Mechanochemical method.

Sometimes, certain conditions for a reaction are needed, or small-scale reactions are involved; using a mortar and pestle might be helpful. It is crucial to remember that the procedure is labor- and time-intensive due to its manually performed nature, which makes it less appropriate for large-scale use. So, automated ball mills are frequently chosen in these situations because they are more effective and can process bigger amounts. Ball mills are adaptable devices with a broad variety of milling functions.

Ball milling can significantly aid in the production of hybrid inorganic-organic materials (HIOMs). The latter is an important family of functional materials that combines the properties of inorganic solids and organic functional moieties. This provides a good opportunity to modify the features of the systems, and as a result, HIOMs have been investigated for a wide range of technical applications, including optics, microelectronics, health, energy, transportation, housing, and the environment. Typically, solvothermal methods, including direct sol-gel synthesis and post-synthetic functionalization, sometimes called grafting, are used to create HIOMs. The former method uses inorganic polymerization processes, such as the hydrolysis and condensation of metal alkoxides, and allows for the one-step creation of hybrid materials. For the prepared material's surface modification, the latter approach is recommended. While reports of reduced pore volume and low loading exist, post-synthetic grafting allows for the preservation of the original framework, for example, in ordered porous solids, where it is especially crucial to maintain the characteristics of the inorganic equivalent. Because of the moisture sensitivity of the organic silane compounds utilized, the grafting procedure is normally carried out in dry circumstances with organic solvents, and at elevated temperatures for lengthy treatment durations before solvent removal and drying. In this study, we present a broad and straightforward method for producing HIOMs at room temperature in a matter of minutes, based on a mechanochemical process. The method entails ball milling at room temperature both organosilicon compounds and commonly used inorganic oxides or mesoporous structures [27, 28]. The anchoring of the organic functional group and structural features of functionalized materials are characterized by a spectrum of methods, including attenuated total reflection Fourier transform infrared (ATR-FTIR), diffuse reflectance infrared Fourier transform spectroscopy (DRIFTS), thermogravimetric analysis coupled to mass spectrometry (TGA-MS), 29Si magic angle spinning nuclear magnetic resonance spectroscopy (MAS NMR), N_2 sorption, and X-ray diffraction (XRD) [29-31].

Fritsch GmbH patented the first planetary ball mill in 1961. Since then, the instrument has undergone several expansions, including the development of multi-purpose devices for parallel synthesis, and a plethora of new manufacturers have emerged on the market. Ball mills are currently widely used in laboratories because they are perfect for rapid sample preparation and comminution with low loss. A large variety of grinding media is now accessible, allowing for common uses in a number of industries, from chemical reactions and high-purity pharmaceutical operations to basic sample preparation. Ball mills are used for a variety of tasks, including the generation of emulsions and creams, mixing and homogenizing, manipulation of materials in an inert atmosphere, and chemical or physical modification of materials. Grinding machines with higher energy are now being developed for use in mechanochemical organic synthesis. The production potential remains in the range of tens to hundreds of grams, despite a daily increase in synthesis capacity [32-35].

Reactivity in chemicals is enabled via grinding compounds using ball-mill devices, such as vibrating ball mills (VBM), planetary ball mills (PBM), SPEX mills, tumbler ball mills, or single-screw devices (SSD), employing mechanical forces. A common and traditional instrument for hand grinding and milling operations is the mortar and pestle. Using a mortar and pestle to perform mechanochemical reactions requires the mechanical energy produced by grinding manually. After adding the mixture of solid reagents to the mortar, the pestle is then utilized to create the friction and pressure, which causes particles to come together and interact. Solid-state reactions are aided by this grinding process, which doesn't require any further boiling and solvents [36].

Mechanochemical processes have traditionally been carried out in a variety of mill types. In most vibratory mills, there is a jar that is agitated vigorously back and forth, impacting and colliding with the media of grinding and chemicals. The mechanical alloying of materials and the production of both organic and inorganic chemicals have long been frequent uses for this kind of mill. One or more grinding jars that spin both along their own axis, in addition to a central axis, make up the planetary ball mill motion. The reactants are effectively mixed and ground by this dual rotation. These devices can function on hundreds of grams scales, unlike vibratory mills, which can only execute responses at the multigram scale.

Comparable with a vibratory mill, the SPEX ball mill works by shaking or agitating both the vial and the grinding balls in a figure-of-eight motion. Minerals, polymers, metallic materials, and ceramics may all be ground quickly and effectively due to the mechanical action. This instrument is frequently used for material synthesis, sample pulverization, and mechanical alloying. Many sectors, including chemical manufacturing, food processing, pharmaceuticals, and polymer processing, utilize twin-screw extruders extensively. These are multipurpose instruments providing fine control over material mixing, compounding, and extrusion processes. Two intermeshing screws are kept in a barrel by twin-screw extruders. The screws are helical in position, although they revolve in the same direction. Numerous benefits come with twin-screw extruders, such as accurate monitoring of processing settings, high output, effective compounding, including mixing, and the capacity to handle a variety of materials. They are therefore extensively utilized in the production of plastic goods, processing of food, formulations for drugs, and other chemical processes that need exact control over mixing, compounding, and extrusion. Moreover, lastly, a recent experiment using a drill press and a counter-clockwise revolving drill bit showed that tight binding may lead to great mechanochemistry [37-39].

Currently, research is being done on a number of nanomaterial categories. MOFs comprise the following widely used structures: ZIF-8, HKUST-1, MIL-101, UiO-66, and MOF-5, as shown in Fig. (**2**). These approaches are used to prepare bio-inspired metal-organic frameworks [40].

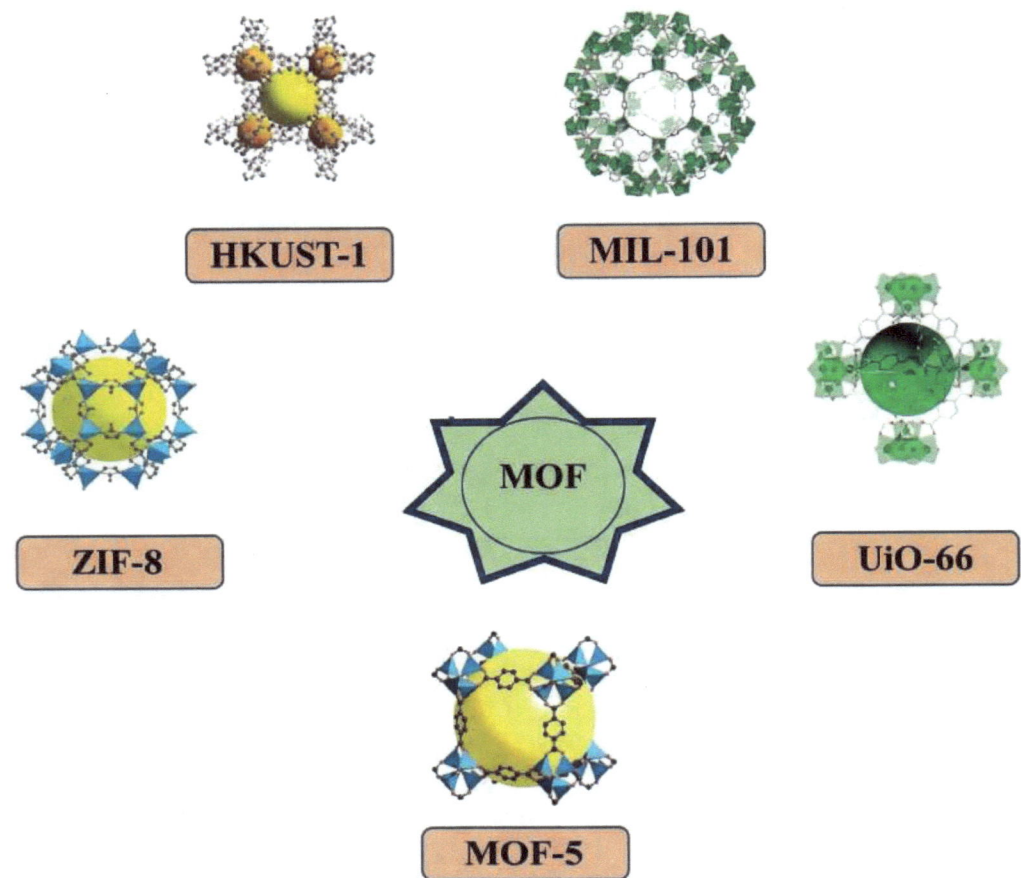

Fig. (2). Structures of different types of MOF.

By utilizing this technology, novel metallopharmaceuticals, metallodrugs, and phenol hydrazone derivative synthesis are employed, as well as cyclodextrin nanosponges (CD-NS) polymers, copper oxide nanoparticles, and silver nanoparticles, all of which are utilized for antibacterial activity. Mechano activation of silicon to synthesize alkoxysilanes prepared by authors with the removal of steps, namely silicon and catalyst preparation, this approach significantly streamlines the conventional multiple-step process and makes it compliant with green chemistry needs [41-42]. The oxide layer was eliminated by vibration milling, and the reactor working area is where the substantial silicon percentage (1000–2000 μm) is mechanoactivated. When the brass grinding bodies and reactor walls are worn down, a catalytic surface forms on silicon [43, 44]. Heterocyclic derivative synthesis [45-49] was performed by using ball mill [50,

51], and pharmaceutical cocrystals synthesis [52] by using solvent drop grinding technique (SDG) may also be used to produce all of the cocrystals that were grown from solution and identified using single crystal X-ray crystallography. For this reason, SDG seems to be an affordable, environmentally friendly, and dependable technique for both finding new cocrystals and preparing those that already exist., catalyst synthesis occurred by this method [53].

ROLE OF MECHANOCHEMICAL GREEN GRAFTING

Ball-milled Seashells as A Nano-Biocomposite

Ball-milled seashells, as a nano-biocomposite catalyst and natural source of $CaCO_3$ in its aragonite microcrystalline form with fixed CO_2, were optimized for the synthesis of isoamyl acetate (3-methylbutyl ethanoate) by response surface methodology with a five-level three-factor rotatable circumscribed central composite design [54, 55]. When acetic acid and isoamyl alcohol are used as starting materials for the solvent-free, environmentally safe [56] production of isoamyl acetate, seashell nano-biocomposite has shown itself to be a highly effective heterogeneous multifunctional catalyst [57]. Under the following ideal circumstances, the authors produced a high yield of 91% nanocomposite: a molar ratio of alcohol to acetic acid of 1:3.7, catalyst loading of 15.7 mg, reaction temperature of 98 °C, and reaction duration of 219 min. The protocol offered them several benefits, including the use of a cheap, readily available, and easily prepared nano-biocomposite material with appropriate thermal stability. Additionally, the process was simple and did not require the use of hazardous reagents, lower catalyst loading, or reaction temperature, or corrosive Bronsted acids or toxic azeotropic solvents, or water adsorbents. [58].

Fig. (3) represents the ball-milled seashells as a nano-biocomposites.

Fig. (3). Ball-milled seashells as a nano-biocomposite [58].

Mechanochemical Transformations of Biomass into Functional Materials

Providing renewable sources of chemicals and materials, biomass is a possible substitute for commodities obtained from petroleum and contributes significantly to our efforts to combat climate change. As biomass is chemically and structurally complex, converting it into value-added products requires creative thinking, including the combination of physical and chemical processes, as well as a thorough understanding of its composition at various scales. In this regard, mechanochemistry is increasingly being used in biomass valorization due to its potential as a productive, long-term, and ecologically benign method. The mechanochemical method has numerous benefits over the solution-based wet technique [59]. Mechanochemical processing eliminates the need for complicated post-treatment procedures, including solvent removal and product purification, while also reducing the requirement for reagent solubility and minimizing solvent usage as represented in Fig. (**4**). For instance, ball milling can be used to create Nb_2O_5 supported by polysaccharides in less than 30 minutes. [60, 61].

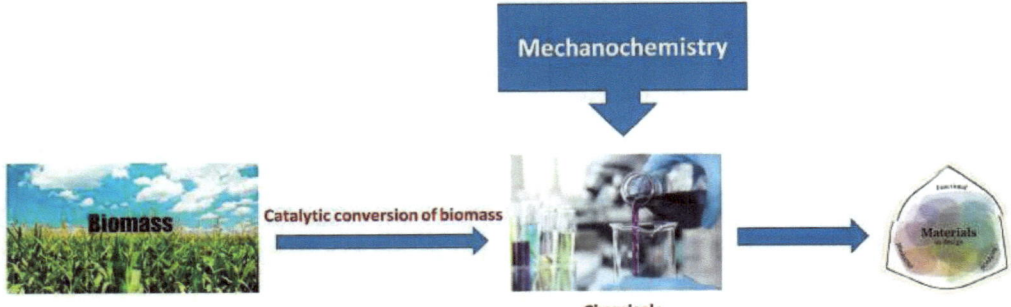

Fig. (4). Mechanochemical transformations of biomass into functional materials [62].

Solvent-free Synthesis of Amorphous Salts of Folic Acid

Mechanochemistry was used to create eight novel amorphous organic salts of folic acid (FA). FA can prevent cardiovascular and neurological diseases. By eliminating the use of hazardous solvents, mechanochemistry resolves significant problems with FA solubility. Low FA solubility prevents vitamins and medications from having therapeutic effects. Derivatization using intricate synthetic processes can increase the solubility of FA. Mechanochemistry was used to create an easy-to-follow, environmentally friendly chemistry synthesis procedure. Salts with good physicochemical properties were obtained by selectively combining the conformers of the amine derivative with FA. [63,64] The octanol-water partition coefficient values ($K_{octanol/water}$) of the new 1:1 and 1:2 amorphous FA salts are 10–10,000 times better in aqueous solution and 10–100 times better than those of FA alone. $K_{octanol/water}$ is considered a surrogate of cell permeability. The produced FA salts have no harmful effects on primary dermal fibroblasts in healthy humans. Because of their improved cell permeability and lower toxicity, the FA salts of choline hydroxide and its derivatives may be attractive candidates for future pharmacological and nutraceutical applications as represented in Fig. (**5**) [65- 67].

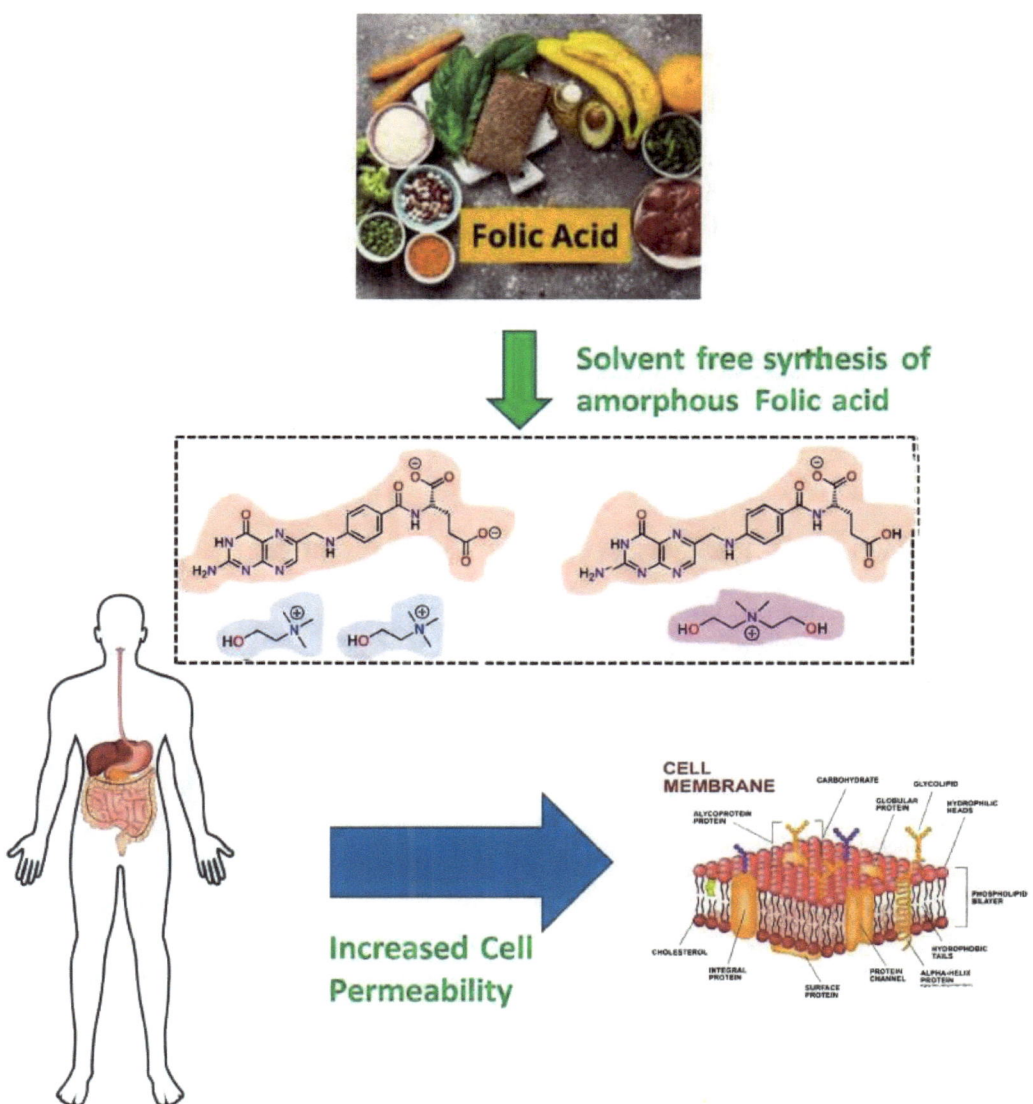

Fig. (5). Increased cell permeability by amorphous FA salts of choline hydroxide [67].

Mechanochemical Design of Nanomaterials

For the synthesis of a wide range of materials, including oxides, oxyfluorides, fluorides, MOs and MMOs, nanocomposites, and materials based on zeolites, mechanochemistry has proven to be an efficient method.

Mechanochemical treatment reduces particle size and introduces surface imperfections that influence catalysis, and in some cases, can lead to bond cleavage and bond formation. To circumvent diffusional problems, mechanochemistry can be used to introduce mesoporosity in microporous materials more successfully than with conventional chemical treatments. Mechanochemistry reduces the amount of solvent used and waste produced. It also demonstrates excellent efficiency in reactions involving gaseous reactants and may even lead to improved scalability [68, 69].

Furthermore, compared to materials generated using conventional techniques, mechanochemically synthesized materials have been shown to exhibit increased stabilities, which are frequently linked to enhanced catalytic activity. Milling often has a significant impact on the catalytic sites and surface structure, altering the final activities, which may be beneficial for the diagnosis of various diseases. [70-73].

ADVANTAGES OF ENZYMATIC GREEN GRAFTING

- The homogeneous mixing of the biomaterial in the membrane reactor is higher.
- The Packed bed reactor with grafted enzymes is useful in biocatalyst recycling.
- The Biopolymer, Poly(ethylene terephthalate) (PET) and enzymatic polymerization and kinetic resolution along with API synthesis is possible with mechanochemistry.
- The enzyme-based nanomaterial synthesis with a green chemistry approach.

FUTURE SCOPE

- The enzymatic green grafting method and mechanosynthesis can be used to make biofuel (such as biohalogen, biobutanol, bioethanol, biodiesel, and biogas). These could help us increase productivity, efficacy, and ecological sustainability.
- We may create green composites that have better elongation and flexural strain at breakage, as well as a plasticizing effect.
- Using mechanosynthesis and the enzymatic green grafting technique, we may be able to create free and immobilized biocatalysts for the removal of micropollutants from water and wastewater: Recent developments and obstacles.

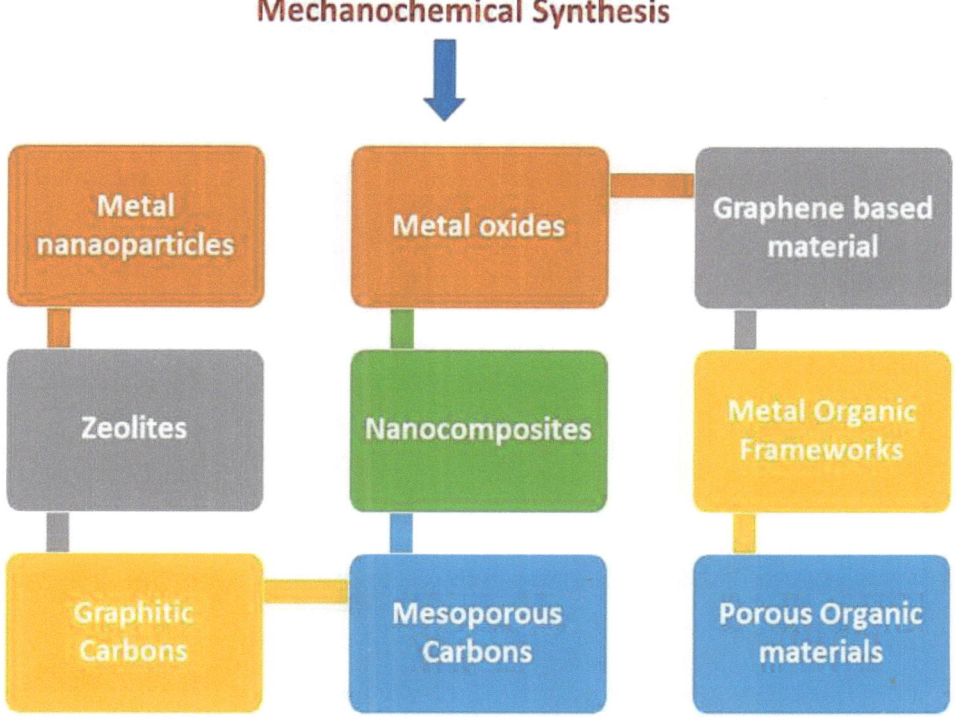

Fig. (6). Mechanochemical design of nanomaterials [74].

Fig. (**6**) represents the mechanochemical design of nanomaterials.

The Cosmo textile industry, which has been treated with cosmetic chemicals that provide effective stability, maintain dermal distribution, and prolong dermal cosmetic efficiency, may find enzyme grafting useful.

We think that putting environmentally viable technologies like mechanochemistry into practice would ultimately contribute to our species' survival and a more sustainable future.

CONCLUSION

Mechanochemistry has garnered significant interest as a powerful, durable, effective, affordable, and environmentally benign synthesis method for the production of novel functional materials. This process, which involves milling and grinding, is based on physicochemical processes that are accelerated by mechanical force. A chemical reaction created by absorbing mechanical energy is known as a mechanochemical reaction. The reagents are ground into powder using ball-mill devices, such as vibrating, planetary, tumbler ball-mills, or single-screw devices that use mechanical forces, to promote chemical reactivity. This method, which uses a one-step mechanochemical grafting process to create thiol-functionalized montmorillonite and covalent organic frameworks (COFs), adsorbents with superior iodine capture, is reported for the practical, solvent-free synthesis of superhydrophobic surfaces. MOFs are made up of the commonly used ZIF-8, HKUST-1, MIL-101, UiO-66, and MOF-5 structures. The aforementioned methods are employed in the synthesis of phenol hydrazone derivatives, novel metallopharmaceuticals, metallodrugs, copper oxide nanoparticles, silver nanoparticles for antibacterial activity, cyclodextrin nanosponges (CD-NS) polymers, mechanoactivation of silicon to synthesize alkoxysilanes, ball mill-based heterocyclic derivative synthesis, pharmaceutical cocrystals synthesis, and catalyst synthesis. The advantages of mechanochemical techniques over solution-based chemistry include easier work-up procedures due to their more selective reactions.

ABBREVIATIONS

IUPAC	International Union of Pure and Applied Chemistry
MOF	Metal–Organic Framework
HIOMs	Hybrid Inorganic–Organic Materials
FTIR	Fourier Transform Infrared Spectroscopy
DRIFTS	Diffuse Reflectance Infrared Fourier Transform Spectroscopy
TGA-MS	Thermogravimetric Analysis coupled with Mass Spectrometry
MAS NMR	Magic Angle Spinning Nuclear Magnetic Resonance Spectroscopy
XRD	X-ray Diffraction
VBM	Vibrating Ball Mill
PBM	Planetary Ball Mill
SSD	Single-Screw Device
ZIF-8	Zeolitic Imidazolate Framework-8
HKUST-1	Hong Kong University of Science and Technology-1
MIL-101	Matériaux de l'Institut Lavoisier-101
UiO-66	University of Oslo-66
MOF-5	Metal–Organic Framework-5
CD-NS	Cyclodextrin Nanosponges
SDG	Solvent Drop Grinding
FA	Folic Acid
PET	Polyethylene Terephthalate

ACKNOWLEDGEMENTS

My sincere gratitude goes out to our Principal, Mrs. (Dr.) Ujwala Mahajan, President, Mr. Manoj Balpande of Dadasaheb Balpande College of Pharmacy, Besa, Nagpur and all my co-authors for all of their support for the completion of this chapter.

REFERENCES

[1] James SL, Friščić T. Mechanochemistry. Chem Soc Rev 2013; 42(18): 7494-6.
[http://dx.doi.org/10.1039/c3cs90058d] [PMID: 23917489]

[2] Michalchuk AAL, Boldyreva EV, Belenguer AM, Emmerling F, Boldyrev VV. Tribochemistry, mechanical alloying, mechanochemistry: what is in a name? Front Chem 2021; 9: 685789.
[http://dx.doi.org/10.3389/fchem.2021.685789] [PMID: 34164379]

[3] Kajdas C. General approach to mechanochemistry and its relation to tribochemistry. Tribology in engineering. 2013 May 8: 209-40.
[http://dx.doi.org/10.5772/50507]

[4] Friščić T, Mottillo C, Titi HM. Mechanochemistry for Synthesis. Angew Chem Int Ed 2020; 59(3): 1018-29.
[http://dx.doi.org/10.1002/anie.201906755] [PMID: 31294885]

[5] Do JL, Friščić T. Mechanochemistry: a force of synthesis. ACS Cent Sci 2017; 3(1): 13-9.
[http://dx.doi.org/10.1021/acscentsci.6b00277] [PMID: 28149948]

[6] Gilman JJ. Mechanochemistry. science. 1996 Oct 4; 274(5284): 65.

[7] Fantozzi N, Volle JN, Porcheddu A, Virieux D, García F, Colacino E. Green metrics in mechanochemistry. Chem Soc Rev 2023; 52(19): 6680-714.
[http://dx.doi.org/10.1039/D2CS00997H] [PMID: 37691600]

[8] Jones W, Eddleston MD. Introductory Lecture: Mechanochemistry, a versatile synthesis strategy for new materials. Faraday Discuss 2014; 170: 9-34.
[http://dx.doi.org/10.1039/C4FD00162A] [PMID: 25266823]

[9] Šepelák V, Düvel A, Wilkening M, Becker KD, Heitjans P. Mechanochemical reactions and syntheses of oxides. Chem Soc Rev 2013; 42(18): 7507-20.
[http://dx.doi.org/10.1039/c2cs35462d] [PMID: 23364473]

[10] Krusenbaum A, Grätz S, Tigineh GT, Borchardt L, Kim JG. The mechanochemical synthesis of polymers. Chem Soc Rev 2022; 51(7): 2873-905.
[http://dx.doi.org/10.1039/D1CS01093J] [PMID: 35302564]

[11] Takacs LM. Carey Lea, the father of mechanochemistry. Bull Hist Chem 2003; 28(1): 27.

[12] Takacs L. Two important periods in the history of mechanochemistry. J Mater Sci 2018; 53(19): 13324-30.
[http://dx.doi.org/10.1007/s10853-018-2198-3]

[13] Takacs L. The historical development of mechanochemistry. Chem Soc Rev 2013; 42(18): 7649-59.
[http://dx.doi.org/10.1039/c2cs35442j] [PMID: 23344926]

[14] Lavalle P, Boulmedais F, Schaaf P, Jierry L. Soft-mechanochemistry: mechanochemistry inspired by nature. Langmuir 2016; 32(29): 7265-76.
[http://dx.doi.org/10.1021/acs.langmuir.6b01768] [PMID: 27396617]

[15] Baláž M, Baláž M. Mechanochemistry. Environmental Mechanochemistry: Recycling Waste into Materials using High-Energy Ball Milling. 2021: 1-52.

[16]   Claramunt RM, López C, Sanz D, Elguero J. Mechano heterocyclic chemistry: Grinding and ball mills. Adv Heterocycl Chem 2014; 112: 117-43.
[http://dx.doi.org/10.1016/B978-0-12-800171-4.00003-2]

[17]   Bonné R. Industrial production of porous materials. In: Handbook of Porous Materials: Synthesis, Properties, Modeling and Key Applications. Vol 4. Singapore: World Scientific; 2020. p. 177.

[18]   Pagola S. Outstanding advantages, current drawbacks, and significant recent developments in mechanochemistry: a perspective view. Crystals (Basel) 2023; 13(1): 124.
[http://dx.doi.org/10.3390/cryst13010124]

[19]   Barbero CA, Acevedo DF. Mechanochemical synthesis of polyanilines and their nanocomposites: A critical review. Polymers (Basel) 2022; 15(1): 133.
[http://dx.doi.org/10.3390/polym15010133] [PMID: 36616492]

[20]   Myna R, Hellmayr R, Georgiades M, et al. Can surface coating of circular saw blades potentially reduce dust formation? Materials (Basel) 2021; 14(18): 5123.
[http://dx.doi.org/10.3390/ma14185123] [PMID: 34576349]

[21]   Martinez V, Stolar T, Karadeniz B, Brekalo I, Užarević K. Advancing mechanochemical synthesis by combining milling with different energy sources. Nat Rev Chem 2022; 7(1): 51-65.
[http://dx.doi.org/10.1038/s41570-022-00442-1] [PMID: 37117822]

[22]   Alivisatos P, Barbara PF, Castleman AW, et al. From molecules to materials: Current trends and future directions. Adv Mater 1998; 10(16): 1297-336.
[http://dx.doi.org/10.1002/(SICI)1521-4095(199811)10:16<1297::AID-ADMA1297>3.0.CO;2-7]

[23]   Wei L, Abd Rahim S, Al Bakri Abdullah M, et al. Producing metal powder from machining chips using ball milling process: A review. Materials (Basel) 2023; 16(13): 4635.
[http://dx.doi.org/10.3390/ma16134635] [PMID: 37444950]

[24]   Cuccu F, De Luca L, Delogu F, et al. Mechanochemistry: new tools to navigate the uncharted territory of "impossible" reactions. ChemSusChem 2022; 15(17): e202200362.
[http://dx.doi.org/10.1002/cssc.202200362] [PMID: 35867602]

[25]   Amrute AP, Zibrowius B, Schüth F. Mechanochemical Grafting: A Solvent-less Highly Efficient Method for the Synthesis of Hybrid Inorganic–Organic Materials. Chem Mater 2020; 32(11): 4699-706.
[http://dx.doi.org/10.1021/acs.chemmater.0c01266]

[26]   Shergill RS, Farlow A, Perez F, Patel BA. 3D-printed electrochemical pestle and mortar for identification of falsified pharmaceutical tablets. Mikrochim Acta 2022; 189(3): 100.
[http://dx.doi.org/10.1007/s00604-022-05202-y] [PMID: 35152330]

[27]   Amrute AP, De Bellis J, Felderhoff M, Schüth F. Mechanochemical synthesis of catalytic materials. Chemistry 2021; 27(23): 6819-47.
[http://dx.doi.org/10.1002/chem.202004583] [PMID: 33427335]

[28]   Amrute AP, Zibrowius B, Schüth F. Mechanochemical Grafting: A Solvent-less Highly Efficient Method for the Synthesis of Hybrid Inorganic–Organic Materials. Chem Mater 2020; 32(11): 4699-706.
[http://dx.doi.org/10.1021/acs.chemmater.0c01266]

[29]   Lucia E. Hydrophobized Ti–SiO$_2$ catalysts for the liquid phase epoxidation of olefins [PhD dissertation]. Louvain-la-Neuve: Université catholique de Louvain; 2023.

[30]   Mourdikoudis S, Pallares RM, Thanh NTK. Characterization techniques for nanoparticles: comparison and complementarity upon studying nanoparticle properties. Nanoscale 2018; 10(27): 12871-934.
[http://dx.doi.org/10.1039/C8NR02278J] [PMID: 29926865]

[31]   Baláž M. Environmental mechanochemistry. Berlin/Heidelberg, Germany: Springer International Publishing 2021.

[http://dx.doi.org/10.1007/978-3-030-75224-8]

[32] El-Eskandarany MS. Mechanical Alloying: Nanotechnology, Materials Science and Powder Metallurgy. 3rd ed. Oxford: Elsevier; 2015.

[33] Tavares LM. A review of advanced ball mill modelling. Kona Powder Particle J 2017; 34(0): 106-24.
[http://dx.doi.org/10.14356/kona.2017015]

[34] Stolle A, Szuppa T, Leonhardt SES, Ondruschka B. Ball milling in organic synthesis: solutions and challenges. Chem Soc Rev 2011; 40(5): 2317-29.
[http://dx.doi.org/10.1039/c0cs00195c] [PMID: 21387034]

[35] Chattopadhyay PP, Manna I, Talapatra S, Pabi SK. A mathematical analysis of milling mechanics in a planetary ball mill. Mater Chem Phys 2001; 68(1-3): 85-94.
[http://dx.doi.org/10.1016/S0254-0584(00)00289-3]

[36] Leitão EPT. Comparison of traditional and mechanochemical production processes for nine active pharmaceutical ingredients (APIs). RSC Sustainability 2024; 2: 3655-3668.
[http://dx.doi.org/10.1039/D4SU00385C]

[37] Zolriasatein A, Shokuhfar A, Safari F, Abdi N. Comparative study of SPEX and planetary milling methods for the fabrication of complex metallic alloy nanoparticles. Micro & Nano Lett 2018; 13(4): 448-51.
[http://dx.doi.org/10.1049/mnl.2017.0608]

[38] Maurice D, Courtney TH. Milling dynamics: Part II. dynamics of a SPEX mill and a one-dimensional mill. Metall Mater Trans, A Phys Metall Mater Sci 1996; 27(7): 1973-9.
[http://dx.doi.org/10.1007/BF02651946]

[39] De Castro M, Cintho OM, Capocchi JD. Comparative Study of Mechanical Activation Improved by High Energy Ball Mills in Chromium Oxide Reduction. In: Materials Science Forum. Durnten-Zurich: Trans Tech Publications Ltd; 2014. p. 41-5.
[http://dx.doi.org/10.4028/www.scientific.net/MSF.802.41]

[40] Rokde V, Danao K, Dumore M, Mahajan U. Stability of metal organic frameworks. In: Metal Organic Frameworks. London: Elsevier; 2024. p. 125-35.
[http://dx.doi.org/10.1016/B978-0-443-15259-7.00023-1]

[41] Danao K, Rokde V, Nandurkar D, Fule R, Shivhare R, Mahajan U. Metal-organic frameworks for drug delivery: part B. In: Metal Organic Frameworks. London: Elsevier; 2024. p. 257-87.

[42] Danao K, Rokde V, Nagar A, Bendale AR, Karande N. Advances in food and nutrition research. In: Nutrigenomics and Nutraceuticals. Palmerton (PA): Apple Academic Press; 2024. p. 65-87.

[43] Teoh Y. Exploring new mechanochemical methodologies using SpeedMixer technology [master's thesis]. Montreal (QC): McGill University (Canada); 2023.

[44] Temnikov MN, Anisimov AA, Zhemchugov PV, et al. Mechanochemistry – a new powerful green approach to the direct synthesis of alkoxysilanes. Green Chem 2018; 20(9): 1962-9.
[http://dx.doi.org/10.1039/C7GC03862C]

[45] Rokde V, Danao K, Nimje J, et al. Design, Synthesis, Antimicrobial Evaluation of Novel 2-Oxo-4-Substituted Aryl-Azetidine Benzotriazole Derivatives. Chem Biodivers 2023; 20(7): e202300433.
[http://dx.doi.org/10.1002/cbdv.202300433] [PMID: 37306062]

[46] Nandurkar D, Menghani S, Danao K, et al. New Benzopyrrole Derivatives: Synthesis and Appraisal of Their Potential as Antimicrobial Agents. Chem Biodivers 2023; 20(7): e202300394.
[http://dx.doi.org/10.1002/cbdv.202300394] [PMID: 37300516]

[47] Rokde V, Sihare M. Synthesis Of A Novel Series Of Indole Clubbed Oxazepine Derivatives Using Conventional And Microwave Action. Journal for ReAttach Therapy and Developmental Diversities 2023; 6(10s 2): 2365-73.

[48] Nandurkar D, Danao K, Rokde V, Shivhare R, Mahajan U. Pyrazole Scaffold: Strategies toward the

Synthesis and Their Applications. In: Kumari P, Patel AB, Eds. Strategies for the Synthesis of Heterocycles and Their Applications. London: IntechOpen; 2022.

[49] Danao KR, Rokde VV, Nandurkar DM, Mahajan UN. Pyrazole Scaffold: Potential PTP1B Inhibitors for Diabetes Treatment. Curr Diabetes Rev 2025; 21(2): e130224226925.
[http://dx.doi.org/10.2174/0115733998280245240130075909] [PMID: 38351692]

[50] El-Sayed TH, Aboelnaga A, El-Atawy MA, Hagar M. Ball milling promoted N-heterocycles synthesis. Molecules 2018; 23(6): 1348.
[http://dx.doi.org/10.3390/molecules23061348] [PMID: 29867039]

[51] Kaur M, Banik BK, Kaur N. Green synthetic approach for biologically relevant heterocyclic compounds by using ball mill. InGreen Approaches in Medicinal Chemistry for Sustainable Drug Design 2024 Jan 1 (pp. 233-258). Elsevier.
[http://dx.doi.org/10.1016/B978-0-443-16164-3.00012-1]

[52] Weyna DR, Shattock T, Vishweshwar P, Zaworotko MJ. Synthesis and structural characterization of cocrystals and pharmaceutical cocrystals: mechanochemistry vs slow evaporation from solution. Cryst Growth Des 2009; 9(2): 1106-23.
[http://dx.doi.org/10.1021/cg800936d]

[53] De Jong KP, Geus JW. Carbon nanofibers: catalytic synthesis and applications. Catal Rev, Sci Eng 2000; 42(4): 481-510.
[http://dx.doi.org/10.1081/CR-100101954]

[54] Dekamin MG, Eslami M. Highly efficient organocatalytic synthesis of diverse and densely functionalized 2-amino-3-cyano-4H-pyrans under mechanochemical ball milling. Green Chem 2014; 16(12): 4914-21.
[http://dx.doi.org/10.1039/C4GC00411F]

[55] Gorrasi G, Sorrentino A. Mechanical milling as a technology to produce structural and functional bio-nanocomposites. Green Chem 2015; 17(5): 2610-25.
[http://dx.doi.org/10.1039/C5GC00029G]

[56] Ando H, Kurata A, Kishimoto N. Antimicrobial properties and mechanism of volatile isoamyl acetate, a main flavour component of Japanese sake (Ginjo-shu). J Appl Microbiol 2015; 118(4): 873-80.
[http://dx.doi.org/10.1111/jam.12764] [PMID: 25626919]

[57] Fattahi AH, Dekamin MG, Clark JH. Optimization of green and environmentally-benign synthesis of isoamyl acetate in the presence of ball-milled seashells by response surface methodology. Sci Rep 2023; 13(1): 2803.
[http://dx.doi.org/10.1038/s41598-023-29568-y] [PMID: 36797437]

[58] Wang S, Dai G, Yang H, Luo Z. Lignocellulosic biomass pyrolysis mechanism: A state-of-the-art review. Pror Energy Combust Sci 2017; 62: 33-86.
[http://dx.doi.org/10.1016/j.pecs.2017.05.004]

[59] Shirkavand E, Baroutian S, Gapes DJ, Young BR. Combination of fungal and physicochemical processes for lignocellulosic biomass pretreatment – A review. Renew Sustain Energy Rev 2016; 54: 217-34.
[http://dx.doi.org/10.1016/j.rser.2015.10.003]

[60] Liu WJ, Jiang H, Yu HQ. Development of biochar-based functional materials: toward a sustainable platform carbon material. Chem Rev 2015; 115(22): 12251-85.
[http://dx.doi.org/10.1021/acs.chemrev.5b00195] [PMID: 26495747]

[61] Shen F, Xiong X, Fu J, et al. Recent advances in mechanochemical production of chemicals and carbon materials from sustainable biomass resources. Renew Sustain Energy Rev 2020; 130: 109944.
[http://dx.doi.org/10.1016/j.rser.2020.109944]

[62] Srivastav S, Singh SK, Yadav AK, Srikrishna S. Folic acid supplementation ameliorates oxidative stress, metabolic functions and developmental anomalies in a novel fly model of Parkinson's disease.

[63] Pérez-Carreón K, Martínez LM, Videa M, Cruz-Angeles J, Gómez J, Ramírez E. Effect of Basic Amino Acids on Folic Acid Solubility. Pharmaceutics 2023; 15(11): 2544.
[http://dx.doi.org/10.3390/pharmaceutics15112544] [PMID: 38004524]

[64] Haghdoost-Yazdi H, Fraidouni N, Faraji A, Jahanihashemi H, Sarookhani M. High intake of folic acid or complex of B vitamins provides anti-Parkinsonism effect: No role for serum level of homocysteine. Behav Brain Res 2012; 233(2): 375-81.
[http://dx.doi.org/10.1016/j.bbr.2012.05.011] [PMID: 22610053]

[65] Wysoczanska K, Macedo EA, Sadowski G, Held C. Solubility enhancement of vitamins in water in the presence of covitamins: measurements and ePC-SAFT predictions. Ind Eng Chem Res 2019; 58(47): 21761-71.
[http://dx.doi.org/10.1021/acs.iecr.9b04302]

[66] Martins ICB, Forte A, Diogo HP, et al. A Solvent-free Strategy to Prepare Amorphous Salts of Folic Acid with Enhanced Solubility and Cell Permeability. Chem Methods 2022; 2(6): e202100104.
[http://dx.doi.org/10.1002/cmtd.202100104]

[67] Suut-Tuule E, Schults E, Jarg T, Adamson J, Kananovich D, Aarv R. Scalable Mechanochemical Synthesis of Biotin[6]uril. ChemSusChem. 2025 May; 18(9): e202402354. https://chemistry-europe.onlinelibrary.wiley.com/doi/10.1002/cssc.202402354

[68] Szczęśniak B, Borysiuk S, Choma J, Jaroniec M. Mechanochemical synthesis of highly porous materials. Mater Horiz 2020; 7(6): 1457-73.
[http://dx.doi.org/10.1039/D0MH00081G]

[69] Baláž P, Achimovičová M, Baláž M, et al. Hallmarks of mechanochemistry: from nanoparticles to technology. Chem Soc Rev 2013; 42(18): 7571-637.
[http://dx.doi.org/10.1039/c3cs35468g] [PMID: 23558752]

[70] Achar TK, Bose A, Mal P. Mechanochemical synthesis of small organic molecules. Beilstein J Org Chem 2017; 13(1): 1907-31.
[http://dx.doi.org/10.3762/bjoc.13.186] [PMID: 29062410]

[71] Brezny AC, Landis CR. Recent developments in the scope, practicality, and mechanistic understanding of enantioselective hydroformylation. Acc Chem Res 2018; 51(9): 2344-54.
[http://dx.doi.org/10.1021/acs.accounts.8b00335] [PMID: 30118203]

[72] Burange AS, Alothman ZA, Luque R. Mechanochemical design of nanomaterials for catalytic applications with a benign-by-design focus. Nanotechnol Rev 2023; 12(1): 20230172.
[http://dx.doi.org/10.1515/ntrev-2023-0172]

[73] Rokde V, Danao K, Bali N, Mahajan U. The Severity of COVID-19 in Diabetes Patients. Curr Diabetes Rev 2023; 19(5): e061022209633.
[http://dx.doi.org/10.2174/1573399819666221006103113] [PMID: 36201275]

[74] Amrute AP, De Bellis J, Felderhoff M, Schüth F. Mechanochemical Synthesis of Catalytic Materials. Chemistry 2021; 27(23): 6819-6847.
[http://dx.doi.org/10.1002/chem.202004583] [PMID: 33427335]

SUBJECT INDEX

A

Acoustic cavitations 107
Acrylic acid 28, 36, 43, 110, 182, 219
Adhesion improvement 180
Agro-based biofibers 142
Alginate hydrogel 12, 135
Amorphous salts 279, 280
Antibacterial performance 219
Antifouling coatings 211
Antimicrobial surfaces 12, 15, 188, 209, 228
Atom economy 4, 62, 75, 76
Addition reactions 76, 77, 78, 79
Atom transfer radical polymerization (ATRP) 2, 8, 25, 69, 70, 80, 130

B

Bacterial cellulose 142, 238
Ball mill devices 271, 275, 283
   planetary ball mill 275
   vibratory ball mill 275
Bio-based catalysts 2, 18, 251
Bio-based polymers 3, 5, 9, 85
Bio-sourced solvents 169
Biocompatibility 2, 4, 27, 45, 135, 189, 209, 227
Bioconjugation 240
Biodegradable polymers 1, 3, 5, 10, 14, 17, 81, 97, 126, 204
Biofuel production 259, 281
Biological grafting 40, 131
Biomass valorization 279
Biomedical devices 1, 3, 4, 11, 16, 154, 187, 197, 227
Biomaterial surface engineering 228
Bioresorbable implants 12, 147
Block copolymers 33, 65, 73, 188
Bromination 49, 95, 175

C

Carbon footprint 1, 9, 90, 150, 225
Carbon nanotubes 105, 191
Cascade-catalyzed RAFT 248
Catalysis 5, 6, 8, 102, 111, 137, 16, 250, 138, 220, 278, 111
Cationic living polymerization 33
Cell encapsulation 149
Cellulose modification 9, 43, 81, 133, 140, 142, 219, 240
Chain extenders 31
Chain transfer 30, 33, 81, 97, 176
Chemical grafting 27, 68, 128, 173, 202
Chitin nanocrystals 49
Chitosan grafting 9, 12, 15, 42, 82, 132, 135, 141, 219, 240
Click chemistry 50, 64, 94, 107, 174
   azide-alkyne cycloaddition (aac) 50, 94, 174
   diels-alder reactions 50, 253
   thiol-x reactions 50, 174
Cold atmospheric plasma (CAP) 228
Conducting polymers (CP) 46, 47, 60, 113
Conventional free radical polymerization (CFRP) 28, 29
Covalent organic frameworks (COFs) 271, 283

D

Deep eutectic solvent (DES) 88, 92, 177, 183
Dendron-polymer conjugates 50, 175
Density function theory (DFT) 141
Differential scanning calorimeter (DSC) 45, 135, 142
Direct dehydrogenation 104
Dispersive liquid-liquid microextraction (DLLME) 92, 93
Drug delivery systems (DDS) 1, 3, 4, 10, 14, 15, 40, 135, 187, 209, 257

E

Eco-efficient methods 124, 154, 155, 191
Electro-responsive hydrogels 200
Electron paramagnetic resonance (EPR) 196
Enantioselective organocatalysis 111
Enzyme immobilization 41, 103, 104, 253, 260
Enzyme-polymer hybrids 253, 262
Enzymatic green grafting 40, 41, 154, 238, 239, 250, 255
Environmental remediation 241, 259
Extrinsically conducting polymers (ECP) 47

F

Fatty acid methyl esters (FAMEs) 88
Flocculation properties 136
Flow chemistry 110, 112
Fourier transform infrared (FTIR) 182, 275
Free radical polymerization (FRP) 2, 22, 28, 62, 69, 130, 174
Functionalized scaffolds 14, 15, 16

G

Gamma radiation 39, 43, 81, 131, 188, 190, 191
G-value 194
Gelatinization 246
Glass transition temperature (Tg) 45
Grafting approaches 23, 52, 60
    grafting from 23, 25, 46, 64, 125, 127
    grafting through 23, 26, 52, 64, 125
    grafting to/onto 23, 24, 46, 52, 64, 125, 127, 253
Green electrospinning 124, 145, 150, 153, 155
Green solvents 1, 86, 89, 169, 171, 176, 210, 213, 216

H

Hemostatic dressing 145
Hierarchical nanozeolites 170
High-throughput screening (HTS) 16, 260
Hybrid inorganic-organic materials (HIOMs) 169, 274, 275
Hydrogel 3, 7, 28, 43, 130, 198
Hydroperoxides 40, 131, 194

I

Immobilization techniques 198, 261
In situ spectroscopy 6
Inherently safer chemistry 6, 108, 115, 178
Interpenetrating polymeric networks (IPN) 23, 85, 136
Intrinsically conducting polymers (ICP) 47
Ionic grafting 22, 31, 68, 128, 129, 132, 174
Ionic liquids 5, 7, 32, 86, 89, 111, 169, 177, 178, 182, 213

L

Laccase enzyme 238, 242
Life cycle sustainability 124, 155
Lignocellulosic biomass 137, 138
Lipase enzyme 8, 41, 238
Living polymerization 25, 27, 32, 34, 129, 130, 211

M

Macroinitiator 26, 30, 44
Mechanochemistry 271, 272, 279, 281
Medical implants 179, 187, 209, 227
Metal-free catalytic systems 103
Metal-organic frameworks 271, 276
Methane monooxygenase (MMO) 250
Microcrystalline cellulose (MCC) 143
Microwave-assisted grafting 5, 91
Miktoarm star copolymer 23
Molecularly imprinted polymer (MIP) 29

N

Nanocomposite films 142
Nanofiber air filters 148
Nanoparticle functionalization 4, 15, 147, 148, 149, 154, 252, 259, 282
Nitroxide-mediated polymerization (NMP) 22, 28, 30, 80
Non-thermal plasma 209, 210

O

Ordered porous solids 275
Organosilylation 170
Oxidative polymerization 49, 242, 246

P

PEGylation 11, 14
Photochemical grafting 22, 35, 128, 130, 173, 191, 211
Physisorption 26
Plasma-assisted green grafting (PAGG) 209, 210, 213, 215, 225, 227
Polydispersity index 33, 72, 75
Polymer functionalization 1, 6, 17, 22, 52, 113, 240, 343, 346
Polymer grafting approaches 23, 33, 44, 52, 60, 64, 113, 127
Process optimization 6, 16, 90, 154, 226, 242

R

Radiation-induced green grafting (RIGG) 39, 131, 187, 188, 192, 202, 204
Reactive extrusion polymerization (REX) 28, 30, 31, 69, 112
Regenerative medicine 13, 38, 51, 85, 213
Renewable feedstocks 5, 77, 81, 109, 178
Resorbable sutures 101, 102
Reversible addition-fragmentation chain transfer (RAFT) 22, 28, 29, 69, 71, 72, 130, 248, 249, 254

S

Selective grafting 44, 76, 216
Solid-state batteries 112
Solvent drop grinding (SDG) 278
Solvothermal synthesis 170
Stimuli-responsive materials 17
Supramolecular polymers 49
Surface activation 214, 217, 221

T

Targeted drug delivery 11, 180, 189, 240
Thermoresponsive gels 199
Tissue engineering scaffolds 17
Tribochemistry 272

U

Ultrasound-assisted microextraction 92, 93

UV radiation 22, 35, 36, 37, 46, 128, 133, 188, 191

W

Waste reduction 4, 6, 60, 67, 106, 215, 226
Water purification 22, 38, 148, 195, 217, 241
Wood preservative 238, 242, 243

X

X-ray diffraction (XRD) 45, 135, 139, 142, 154, 275
X-ray photoelectron spectroscopy (XPS) 139, 180

Z

Zwitterionic properties 109, 136

www.ingramcontent.com/pod-product-compliance
Lightning Source LLC
Chambersburg PA
CBHW041456280526
45792CB00004B/1027